Electrical Load Forecasting
Modeling and Model Construction

Soliman Abdel-hady Soliman, Ph.D.
Professor of Electrical Engineering
Misr University for Science and Technology
Egypt

Ahmad M. Al-Kandari, Ph.D.
Associate Professor of Electrical Engineering
College of Technological Studies
Kuwait

ELSEVIER

AMSTERDAM • BOSTON • HEIDELBERG • LONDON
NEW YORK • OXFORD • PARIS • SAN DIEGO
SAN FRANCISCO • SINGAPORE • SYDNEY • TOKYO
Butterworth–Heinemann is an imprint of Elsevier

Butterworth–Heinemann is an imprint of Elsevier
30 Corporate Drive, Suite 400, Burlington, MA 01803, USA
The Boulevard, Langford Lane, Kidlington, Oxford, OX5 1GB, UK

Library of Congress Cataloging-in-Publication Data
Soliman, S.A.
Electrical load forecasting : modeling and model construction / Soliman Abdel-hady Soliman (S.A. Soliman), Ahmad M. Al-Kandari.
p. cm.
Includes bibliographical references and index.
ISBN 978-0-12-810221-3 (alk. paper)
1. Electric power-plants–Load–Forecasting–Mathematics. 2. Electric power systems–Mathematical models. 3. Electric power consumption–Forecasting–Mathematics. I. Al-Kandari, Ahmad M. II. Title.
TK1005.S64 2010
333.793'213015195–dc22 2009048799

British Library Cataloguing-in-Publication Data
A catalogue record for this book is available from the British Library.

For information on all Butterworth–Heinemann publications visit our Web site at *www.elsevierdirect.com*

Typeset by: diacriTech, India

10 11 12 13 14 10 9 8 7 6 5 4 3 2 1

Electrical Load Forecasting

To my parents, I was in need of them during my operation
To my wife, Laila, with love and respect
To my kids, Rasha, Shady, Samia, Hadeer,
and Ahmad, I love you all
To everyone who has the same liver problem,
please do not lose hope in God
(S. A. Soliman)

To my parents, who raised me
To my wife, Noureyah, with great love and respect
To my sons, Eng.Bader and Eng.Khalied,
for their encouragement
To my beloved family and friends
(A. M. Al-Kandari)

Contents

Acknowledgments

In the market and the community of electric power system engineers, there is a shortage of books focusing on short-term electric load forecasting. Many papers have been published in the literature, but no book is available that contains all these publications. The idea of writing this book came to my mind two or three years ago, but the time was too limited to write such a big book. In the spring of 2009, I was diagnosed with liver cancer, and I have had to treat it locally through chemical therapy until a suitable donor is available and I can have a liver transplant. The president of Misr University for Science and Technology, Professor Mohammad Rafat, and my brothers, Professor Mostafa Kamel, vice president for academic affairs, and Professor Kamal Al-Bedawy, dean of engineering, asked me to stay at home to eliminate physical stress. As such, I had a lot of time to write such a book, especially because there are many publications in the area of short-term load forecasting. Indeed, my appreciation goes to them and chancellor of Misr University, Mr. Khalied Al-Tokhay.

My appreciation goes to my wife, Laila, who did not sleep, sitting beside me day and night while I underwent therapy. My appreciation also goes to my kids Rasha, Shady, Samia, Hadeer, and Ahmad, who raced to be the first donor for their dad.

My appreciation goes to my brothers-in-law, Eng. Ahmad Nabil Mousa, Professor Mahmoud Rashad, and Dr. Samy Mousa, who had a hard time because of my illness; he never left me alone even though he was out of the city of Cairo. Furthermore, my appreciation goes also to my sons-in-law, Ahmad Abdel-Azim and Mohammad Abdel-Azim.

My deep appreciation goes to Dr. Helal Al-Hamadi of Kuwait University, who was the coauthor with me for some materials we used in this book.

Many thanks also go to my friends and colleagues among the faculty of the engineering department at Misr University for Science and Technology. To them, I say, "You did something unbelievable." In addition, many thanks to my friends and colleagues among the faculty of the engineering department at Ain Shams University for their moral support. Special thanks go to my good friends Professor Mahmoud Abdel-Hamid and Professor Ibrahim Helal, who forgot the misunderstanding between us and came to visit me at home on the same day he heard that I was sick and took me to his friend, Professor Mohammad Alwahash, who is a liver transplant expert. Professor M. E. El-Hawary, of Dalhousie University, Nova Scotia Canada, my special friend, I miss you MO; I did everything that makes you happy in Egypt and Canada.

My deep appreciation goes to the team of liver transplantation and intensive care units at the liver and kidney hospital of Al-Madi Military Medical Complex; Professor Kareem Bodjema, the French excellent expert in liver transplantation; Professor

Magdy Amin, the man with whom I felt secure when he visited me in my room with his colleagues, who answered my calls any time during the day or night, and who supported me and my family morally; Professor Salah Aiaad, the man, in my first meeting with him, whom I felt I knew for a long time; Professor Ali Albadry; Professor Mahmoud Negm, who has a beautiful smile; Professor Ehab Sabry, the man who can easily read what's in my eyes; and Dr. Mohammad Hesaan, who reminds me of when I was in my forties—everything should go ideally for him.

Last, but not least, my deep appreciation and respect go to General Samir Hamam, the manager of Al-Madi Military Medical Complex, for helping to make everything go smoothly. To all, I say you did a good job in every position at the hospital. May God keep you all healthy and wealthy and remember these good things you did for me to the day after.

S.A. Soliman

It is a privilege to be a coauthor with as great a professor as Professor Soliman Abdel-hady Soliman. I learned a lot from him. I thank him for giving me the opportunity to coauthor this book, which will cover a needed area in load forecasting. I do thank Professor M.E. El-Hawary for teaching me and guiding me in the scope of the material of this book. Also, my appreciation goes to Professor Yacoub Al-Refae, general director of The Public Authority for Applied Education and Training in Kuwait, for his encouragements and notes.

A.M. Al-Kandari

The authors of this book would like to acknowledge the effort done by Ms. Sarah Binns for reviewing this book many times and we appreciate her time. To her we say, you did a good job for us, you were sincere and honest in every stage of this book.

Introduction

Economic development, throughout the world, depends directly on the availability of electric energy, especially because most industries depend almost entirely on its use. The availability of a source of continuous, cheap, and reliable energy is of foremost economic importance.

Electrical load forecasting is an important tool used to ensure that the energy supplied by utilities meets the load plus the energy lost in the system. To this end, a staff of trained personnel is needed to carry out this specialized function. Load forecasting is always defined as basically the science or art of predicting the future load on a given system, for a specified period of time ahead. These predictions may be just for a fraction of an hour ahead for operation purposes, or as much as 20 years into the future for planning purposes.

Load forecasting can be categorized into three subject areas—namely,

1. Long-range forecasting, which is used to predict loads as distant as 50 years ahead so that expansion planning can be facilitated.
2. Medium-range forecasting, which is used to predict weekly, monthly, and yearly peak loads up to 10 years ahead so that efficient operational planning can be carried out.
3. Short-range forecasting, which is used to predict loads up to a week ahead so that daily running and dispatching costs can be minimized.

In the preceding three categories, an accurate load model is required to mathematically represent the relationship between the load and influential variables such as time, weather, economic factors, etc. The precise relationship between the load and these variables is usually determined by their role in the load model. After the mathematical model is constructed, the model parameters are determined through the use of estimation techniques.

Extrapolating the mathematical relationship to the required lead time ahead and giving the corresponding values of influential variables to be available or predictable, forecasts can be made. Because factors such as weather and economic indices are increasingly difficult to predict accurately for longer lead times ahead, the greater the lead time, the less accurate the prediction is likely to be.

The final accuracy of any forecast thus depends on the load model employed, the accuracy of predicted variables, and the parameters assigned by the relevant estimation technique. Because different methods of estimation will result in different values of estimated parameters, it follows that the resulting forecasts will differ in prediction accuracy.

Over the past 50 years, the parameter estimation algorithms used in load forecasting have been limited to those based on the least error squares minimization criterion, even though estimation theory indicates that algorithms based on the least absolute value criteria are viable alternatives. Furthermore, the artificial neural network (ANN) had showed success in estimating the load for the next hour. However, the ANN used by a utility is not necessarily suitable for another utility and should be retrained to be suitable for that utility.

It is well known that the electric load is a dynamic one and does not have a precise value from one hour to another. In this book, fuzzy systems theory is implemented to estimate the load model parameters, which are assumed to be fuzzy parameters having a certain middle and spread. Different membership functions, for load parameters, are used—namely, triangular membership and trapezoidal membership functions. The problem of load forecasting in this book is restricted to short-term load forecasting and is formulated as a linear estimation problem in the parameters to be estimated. In this book, the parameters in the first part are assumed to be crisp parameters, whereas in the rest of the book these parameters are assumed to be fuzzy parameters. The objective is to minimize the spread of the available data points, taking into consideration the type of membership of the fuzzy parameters, subject to satisfying constraints on each measurement point, to ensure that the original membership is included in the estimated membership.

Outline of the Book

In this book, different techniques used in the past two decades are implemented to estimate the load model parameters, including fuzzy parameters with certain middle and certain spread. The book contains nine chapters:

Chapter 1, "Mathematical Background and State of the Art." This chapter introduces mathematical background to help the reader understand the problems formulated in this book. In this chapter, the reader will study matrices and their applications in estimation theory and see that the use of matrix notation simplifies complex mathematical expressions. The simplifying matrix notation may not reduce the amount of work required to solve mathematical equations, but it usually makes the equations much easier to handle and manipulate. This chapter explains the vectors and the formulation of quadratic forms, and, as we shall see, that most objective functions to be minimized (least errors square criteria) are quadratic in nature. This chapter also explains some optimization techniques and introduces the concept of a state space model, which is commonly used in dynamic state estimation. The reader will also review different techniques that, developed for the short term, give the state of the art of the various algorithms used during the past decades for short-term load forecasting. A brief discussion for each algorithm is presented in this chapter. Advantages and disadvantages of each algorithm are discussed. Reviewing the most recent publications in the area of short-term load forecasting indicates that most of the available algorithms treat the parameters of the proposed load model as crisp parameters, which is not the case in reality.

Chapter 2, "Static State Estimation." This chapter presents the theory involved in different approaches that use parameter estimation algorithms. In the first part of the chapter, the crisp parameter estimation algorithms are presented; they include the least error squares (LES) algorithm and the least absolute value (LAV) algorithm. The second part of the chapter presents an introduction to fuzzy set theory and systems, followed by a discussion of fuzzy linear regression algorithms. Different cases for the fuzzy parameters are discussed in this part. The first case is for the fuzzy linear regression of the linear models having fuzzy parameters with nonfuzzy outputs, the second case is for the linear regression of fuzzy parameters with fuzzy output, and the third case is for fuzzy parameters formulated with fuzzy output of left and right type (LR-type).

Chapter 3, "Load Modeling for Short-Term Forecasting." This chapter proposes different load models used in short-term load forecasting for 24 hours.

- Three models are proposed in this chapter—namely, models A, B, and C. Model A is a multiple linear regression model of the temperature deviation, base load, and either wind-chill factor for winter load or temperature humidity factor for summer load. The parameters of load A are assumed to be crisp parameters in this chapter. The term *crisp parameters* mean clearly defined parameter values without ambiguity.
- Load model B is a harmonic decomposition model that expresses the load at any instant, t, as a harmonic series. In this model, the weekly cycle is accounted for through use of a daily load model, the parameters of which are estimated seven times weekly. Again, the parameters of this model are assumed to be crisp.
- Load model C is a hybrid load model that expresses the load as the sum of a time-varying base load and a weather-dependent component. This model is developed with the aim of eliminating the disadvantages of the other two models by combining their modeling approaches. After finding the parameter values, one uses them to determine the electric load from which these parameter values are extracted, and this value is called the estimated load. Then the parameter values are used to predict the electric load for a randomly chosen day in the future, and it is called the predicted load for that chosen day.

Chapter 4, "Fuzzy Regression Systems and Fuzzy Linear Models." The objective of this chapter is to introduce principal concepts and mathematical notions of fuzzy set theory, a theory of classes of objects with non sharp boundaries.

- We first review fuzzy sets as a generalization of classical crisp sets by extending the range of the membership function (or characteristic function) from [0, 1] to all real numbers in the interval [0, 1].
- A number of notions of fuzzy sets, such as representation support, α-cuts, convexity, and fuzzy numbers, are then introduced. The resolution principle, which can be used to expand a fuzzy set in terms of its α-cuts, is discussed.
- This chapter introduces fuzzy mathematical programming and fuzzy multiple-objective decision making. We first introduce the required knowledge of fuzzy set theory and fuzzy mathematics in this chapter.
- Fuzzy linear regression also is introduced in this chapter; the first part is to estimate the fuzzy regression coefficients when the set of measurements available is crisp, whereas in the second part the fuzzy regression coefficients are estimated when the available set of measurements is a fuzzy set with a certain middle and spread.
- Some simple examples for fuzzy linear regression are introduced in this chapter.

- The models proposed in Chapter 3 for crisp parameters are used in this chapter. Fuzzy model A employs a multiple fuzzy linear regression model. The membership function for the model parameters is developed, where triangular membership functions are assumed for each parameter of the load model. Two constraints are imposed on each load measurement to ensure that the original membership is included in the estimated membership.
- Fuzzy model B, which is a harmonic model, also is proposed in this chapter. This model involves fuzzy parameters having a certain median and certain spread.
- Finally, a hybrid fuzzy model C, which is the combination of the multiple linear regression model A and harmonic model B, is presented in this chapter.

Chapter 5, "Dynamic State Estimation." The objective of this chapter is to study the dynamic state estimation problem and its applications to electric power system analysis, especially short-term load forecasting. Furthermore, the different approaches used to solve this dynamic estimation problem are also discussed in this chapter. After reading this chapter, the reader will be familiar with

The five fundamental components of an estimation problem:
- The variables to be estimated.
- The measurements or observations available.
- The mathematical model describing how the measurements are related to the variable of interest.
- The mathematical model of the uncertainties present.
- The performance evaluation criterion to judge which estimation algorithms are "best."

Formulation of the dynamic state estimation problem:
- Kalman filtering algorithm as a recursive filter used to solve a problem.
- Weighted least absolute value filter.
- Different problems that face Kalman filtering and weighted least absolute value filtering algorithms.

Chapter 6, "Load Forecasting Results Using Static State Estimation." The objective of this chapter is as follows:

In Chapter 3, the models are derived on the basis that the load powers are crisp in nature; the data available from a big company in Canada are used to forecast the load power in the crisp case.
- In this chapter, the results obtained for the crisp load power data for the different load models developed in Chapter 3 are shown.
- A comparison is performed between the two static LES and LAV estimation techniques.
- The parameters estimated are used to predict a load using both techniques, where we compare between them for summer and winter.

Chapter 7, "Load Forecasting Results Using Fuzzy Systems." Chapter 6 discusses the short-term load-forecasting problem, and the LES and LAV parameter estimation algorithms are used to estimate the load model parameters. The error in the estimates is calculated for both techniques. The three models, proposed earlier in Chapter 3, are used in that chapter to present the load in different days for different seasons. In this chapter, the fuzzy load models developed in Chapter 5 are tested. The fuzzy parameters of these models are estimated using the past history data for summer weekdays and weekend days as well as for winter weekdays and weekend days. Then these models are used to predict the fuzzy load power for 24 hours ahead, in both summer and winter seasons. The results are given in the form of tables and figures for the estimated and predicted loads.

Chapter 8, "Dynamic Electric Load Forecasting." The main objectives of this chapter are as follows:

- A one-year long-term electric power load-forecasting problem is introduced as a first step for short-term load forecasting.
- A dynamic algorithm, the Kalman filtering algorithm, is suitable to forecast daily load profiles with a lead-time from several weeks to a few years.
- The algorithm is based mainly on multiple simple linear regression models used to capture the shape of the load over a certain period of time (one year) in a two-dimensional layout (24 hours $\times$ 52 weeks).
- The regression models are recursively used to project the 2D load shape for the next period of time (next year). Load-demand annual growth is estimated and incorporated into the Kalman filtering algorithm to improve the load-forecast accuracy obtained so far from the regression models.

Chapter 9, "Electric Load Modeling for Long-Term Forecasting." The objectives of this chapter are as follows:

- This chapter provides a comparative study between two static estimation algorithms—namely, the least error squares (LES) and least absolute value (LAV) algorithms—for estimating the parameters of different load models for peak-load forecasting necessary for long-term power system planning.
- The proposed algorithms use the past history data for the load and the influence factors, such as gross domestic product (GDP), population, GDP per capita, system losses, load factor, etc.
- The problem turns out to be a linear estimation problem in the load parameters. Different models are developed and discussed in the text.

1 Mathematical Background and State of the Art

1.1 Objectives

The objectives of this chapter are

- Introducing a mathematical background to help the reader understand the problems formulated in this book.
- Studying matrices and their applications in estimation theory and showing that the use of matrix notation simplifies complex mathematical expressions. The simplifying matrix notation may not reduce the amount of work required to solve mathematical equations, but it usually makes the equations much easier to handle and manipulate.
- Explaining the vectors and the formulation of quadratic forms and, as we shall see, that most objective functions to be minimized (least error squares criteria) are quadratic in nature.
- Explaining some optimization techniques.
- Introducing the concept of a state space model, which is commonly used in dynamic state estimation.
- Reviewing the literature to introduce different techniques developed for short-term load forecasting.
- Explaining the merit of each technique used in the estimation of load forecasting and suitable places for implementation.
- In this chapter, we also try to compare different techniques used in electric load forecasting.

1.2 Matrices and Vectors

A matrix is an array of elements [1]. The elements of a matrix may be real or complex or functions of time. A matrix that has n rows and m columns is called an $n \times m$ (n by m) matrix. If $n = m$, the matrix is referred to as a square matrix. If A is an $n \times m$ matrix, then it can be written as

$$A = \begin{bmatrix} a_{11} & a_{12} & \cdots & a_{1m} \\ a_{21} & a_{22} & \cdots & a_{2m} \\ \vdots & \vdots & \cdots & \vdots \\ a_{n1} & a_{n2} & \cdots & a_{nm} \end{bmatrix} \tag{1.1}$$

In shorthand,

$$A = \left[a_{ij}\right]_{n\times m} \qquad \begin{matrix} i = 1,\ldots,n \\ j = 1,\ldots,m \end{matrix} \tag{1.2}$$

DOI: 10.1016/B978-0-12-381543-9.00001-4

Note that the determinant is also an array of elements with n rows and n columns (always square) and has a value. The matrix does not have a value but has a determinant.

Column Matrix: This type of matrix has only one column and more than one row; that is, an $m \times 1$ matrix, $m > 1$. Quite often, a column matrix is referred to as a column vector or simply an m-vector. For example, the column vector X is written as

$$X = \begin{bmatrix} x_1 \\ x_2 \\ \vdots \\ x_m \end{bmatrix} = \text{col.}(x_1, x_2, \ldots, x_n) \tag{1.3}$$

Row Matrix: This type of matrix has only one row and more than one column; that is, an $1 \times n$ matrix, $n > 1$. Quite often, we call it a row vector. For example, the row vector Y is given by

$$Y = [y_1, y_2, \ldots, y_n] = \text{row.}(y_1, y_2, \ldots, y_n) \tag{1.4}$$

Diagonal Matrix: This is a square matrix with all elements equal to zero except for the diagonal element; that is, $a_{ij} = 0$ for all $i \neq j$. For example,

$$A = \begin{bmatrix} a_{11} & 0 & 0 \\ 0 & a_{22} & 0 \\ 0 & 0 & a_{33} \end{bmatrix} \tag{1.5a}$$

or, in terms of a shortcut,

$$A = \text{diag.}(a_{11}, a_{22}, a_{33}) \tag{1.5b}$$

Symmetric Matrix: This type of matrix is a square matrix that satisfies the relation

$$a_{ij} = a_{ji} \quad \text{for all } i, j$$

The following example indicates this matrix:

$$A = \begin{bmatrix} 1 & 3 & 5 \\ 3 & 0 & 4 \\ 5 & 4 & 2 \end{bmatrix}$$

In terms of a shortcut:

$$A^T = A \quad \text{where } T = \text{transpose of}$$

Transpose of a Matrix: The transpose of a matrix is defined as a matrix obtained by interchanging the corresponding rows and columns in A. If A is an $n \times m$ matrix, which is represented by

$$A = [a_{i,j}]_{n \times m}$$

then the transpose of A, denoted by A^T, is given by

$$A^T = \left[a_{j,i}\right]_{m\times n} \tag{1.6}$$

Note that the order of A is $n \times m$, while the order of A^T is $m \times n$. For example, if

$$A = \begin{bmatrix} 4 & 3 & 2 & 1 \\ 5 & 6 & 0 & -2 \\ 6 & 1 & 4 & 3 \end{bmatrix}_{3\times4}$$

then

$$A = \begin{bmatrix} 4 & 5 & 6 \\ 3 & 6 & 1 \\ 2 & 0 & 4 \\ 1 & -2 & 3 \end{bmatrix}_{4\times3}$$

The following are some operations using the transpose of a matrix:

1. $\left[A^T\right]^T = A$ (1.7)

2. $[kA]^T = kA^T$, $k =$ scalar number (1.8)

3. $(A + B)^T = A^T + B^T$ (1.9)

4. $(AB)^T = B^T \times A^T$ (1.10)

1.3 Matrix Algebra

1.3.1 Addition of Matrices

If A is an $n \times m$ matrix, and B is also an $n \times m$ matrix, then the sum of the two matrices is given by

$$C = A + B \tag{1.11a}$$

where the elements of the matrix C are given by

$$c_{ij} = a_{ij} + b_{ij} \quad \text{for all } i,j \tag{1.11b}$$

For example, if

$$A = \begin{bmatrix} 2 & 3 & 0 \\ 4 & 5 & 1 \end{bmatrix}$$

and

$$B = \begin{bmatrix} -1 & 2 & 4 \\ 0 & -2 & 3 \end{bmatrix}$$

then

$$C = \begin{bmatrix} (2-1) & (3+2) & (0+4) \\ (4+0) & (5-2) & (1+3) \end{bmatrix}$$

$$C = \begin{bmatrix} 1 & 5 & 4 \\ 4 & 3 & 4 \end{bmatrix}$$

1.3.2 Matrix Subtraction (Difference)

The subtraction (difference) of matrices is similar to the addition of matrices if all the signs of the second matrix are changed from positive to negative and from negative to positive; that is,

$$C_{(n\times m)} = A_{(n\times m)} - B_{(n\times m)} \tag{1.12}$$

where

$$[c_{ij}]_{n\times m} = [a_{ij}]_{n\times m} + [-b_{ij}]_{n\times m} \tag{1.13a}$$

$$[c_{ij}]_{n\times m} = [a_{ij}]_{n\times m} - [b_{ij}]_{n\times m} \tag{1.13b}$$

or

$$c_{ij} = a_{ij} - b_{ij} \quad \text{for all } i \text{ and} j \tag{1.13c}$$

The following rules hold true for addition and subtraction:

1. $(A + B) + C = A + (B + C)$ (associate law) (1.14)

2. $A + B + C = B + C + A = C + A + B$ (commutative law) (1.15)

1.3.3 Matrix Multiplication

Let A be an $n \times m$ matrix and B be an $m \times p$ matrix. Then the product of A and B is defined as

$$C_{(n\times p)} = A_{(n\times m)} \times B_{(m\times p)} \tag{1.16}$$

Note that the number of columns in the first matrix, m, must be equal to the number of rows in the second matrix to carry out the multiplication.

The elements of the matrix C are given by

$$c_{ij} = \sum_{k=1}^{r} a_{ik} b_{kj} \qquad \begin{array}{l} i = 1, \ldots, n \\ j = 1, \ldots, p \end{array} \tag{1.17}$$

If, for example, the matrix A is given by

$$A = \begin{bmatrix} 2 & 3 \\ 1 & 4 \\ -1 & 3 \end{bmatrix}_{3\times 2}$$

and

$$B = \begin{bmatrix} 2 & -1 \\ -1 & 3 \end{bmatrix}_{(2\times2)}$$

then

$$C_{(3\times2)} = A_{(3\times2)} \times B_{(2\times2)}$$

$$C_{(3\times2)} = \begin{bmatrix} 2 & 3 \\ 1 & 4 \\ -1 & 3 \end{bmatrix}_{(3\times2)} \begin{bmatrix} 2 & -1 \\ -1 & 3 \end{bmatrix}_{(2\times2)}$$

$$= \begin{bmatrix} (2\times2)+(3)(-1) & (2\times-1+3\times3) \\ (1\times2-1\times4) & (1\times-1+4\times3) \\ (-1\times2-1\times3) & (-1\times-1+3\times3) \end{bmatrix}$$

$$C_{(3\times2)} = \begin{bmatrix} 1 & 7 \\ -2 & 11 \\ -5 & 10 \end{bmatrix}$$

If the matrix A is given by

$$A = \begin{bmatrix} 0 & 1 \\ -3 & -4 \end{bmatrix}_{(2\times2)}$$

and the vector matrix $X(t)$ is given by

$$X(t) = \begin{bmatrix} x_1(t) \\ x_2(t) \end{bmatrix}_{(2\times1)}$$

then

$$C = AX(t) = \begin{bmatrix} 0 & 1 \\ -3 & -4 \end{bmatrix}_{(2\times2)} \begin{bmatrix} x_1(t) \\ x_2(t) \end{bmatrix}_{(2\times1)}$$

$$= \begin{bmatrix} x_2(t) \\ -3x_1(t) - 4x_2(t) \end{bmatrix}_{(2\times1)}$$

It is possible in some cases to obtain the two products AB and BA. This could happen if A is an $r \times n$ matrix, and B is an $n \times r$ matrix. In this case, AB is an $r \times r$ matrix, whereas BA is an $n \times n$ matrix. Obviously, $AB \neq BA$, and we say that A and B do not commute, but if $AB = BA$, we say that A and B commute. If A and B are of the order $n \times n$, then AB, BA will be of the order $n \times n$. For example,

$$A = \begin{bmatrix} 1 & 3 & 2 \\ 5 & 6 & 7 \end{bmatrix}_{(2\times3)}$$

$$B = \begin{bmatrix} 2 & 3 \\ 1 & 5 \\ 4 & 2 \end{bmatrix}_{(3\times2)}$$

Then AB will be

$$AB_{2\times 2} = \begin{bmatrix} 1 & 3 & 2 \\ 5 & 6 & 7 \end{bmatrix} \begin{bmatrix} 2 & 3 \\ 1 & 5 \\ 4 & 2 \end{bmatrix} = \begin{bmatrix} 13 & 22 \\ 44 & 59 \end{bmatrix}$$

but BA will be

$$BA_{3\times 3} = \begin{bmatrix} 2 & 3 \\ 1 & 5 \\ 4 & 2 \end{bmatrix} \begin{bmatrix} 1 & 3 & 2 \\ 5 & 6 & 7 \end{bmatrix} = \begin{bmatrix} 17 & 24 & 25 \\ 26 & 33 & 37 \\ 14 & 24 & 22 \end{bmatrix}$$

Although the commutative law does not hold in general for matrix multiplication, the associative and distributive laws still apply. For the distributive law, we state that

$$A(B + C) = AB + AC \tag{1.18}$$

provided that the product is conformable.

For the associative law,

$$(AB)C = A(BC) \tag{1.19}$$

if the product is conformable.

1.3.4 *Inverse of a Matrix (Matrix Division)*

If A is a square matrix of which the determinant exists, and if B is another square matrix such that

$$AB = BA = I$$

then B is called the inverse of A, denoted by A^{-1}. Thus,

$$A \times A^{-1} = A^{-1}A = I$$

For matrices with low dimension, a straightforward procedure for matrix inversion is given by

$$A^{-1} = \frac{\text{adj}(A)}{|A|} \tag{1.20}$$

where adj(A) is the adjoint of A, and it is the transpose of the matrix of cofactors of A with elements

$$A_{ij} = (-1)^{i+j} M_{ij} \tag{1.21}$$

The minors M_{ij} are determinants of the $(n - 1) \times (n - 1)$ matrices obtained by deleting the ith row and jth column from A. The following example explains the steps involved.

Example 1.1

Find the inverse of A, where

$$A = \begin{bmatrix} 1 & -2 & 3 \\ 4 & 5 & 6 \\ 7 & 8 & 9 \end{bmatrix}$$

Calculate the determinant of A as

$$\begin{aligned} |A| &= 1(45-48) + 2(36-42) + 3(32-35) \\ &= -3-12-9 = -24 \end{aligned}$$

The matrix of cofactors is obtained as

$$\mathrm{Cof}(A) = \begin{bmatrix} -3 & 6 & -3 \\ 42 & -12 & -22 \\ -27 & 6 & 13 \end{bmatrix}$$

Thus, transposing Cof(A), we obtain adj(A) as

$$\mathrm{adj}(A) = \begin{bmatrix} -3 & 42 & -27 \\ 6 & -12 & 6 \\ -3 & -22 & 13 \end{bmatrix}$$

The inverse of A is obtained as

$$\begin{aligned} A^{-1} = \frac{\mathrm{adj}(A)}{|A|} &= \frac{-1}{24}\begin{bmatrix} -3 & 42 & -27 \\ 6 & -12 & -6 \\ -3 & -22 & 13 \end{bmatrix} \\ &= \begin{bmatrix} 0.125 & -1.75 & 1.125 \\ -0.25 & 0.50 & -0.25 \\ 0.125 & 0.91667 & -0.54166 \end{bmatrix} \end{aligned}$$

Some properties of the matrix inverse are

$$AA^{-1} = A^{-1}A = I \tag{1.22}$$

$$\left(A^{-1}\right)^{-1} = A \tag{1.23}$$

If A and B are square matrices and are nonsingular, then

$$(AB)^{-1} = B^{-1}A^{-1} \tag{1.24}$$

1.4 Rank of a Matrix

The rank of a matrix A is the maximum number of linearly independent columns of A, or it is the order of the largest nonsingular matrix contained in A. For example, the matrix

$$A = \begin{bmatrix} 0 & 1 \\ 0 & 0 \end{bmatrix} \text{ has a rank} = 1$$

and

$$A = \begin{bmatrix} 0 & 5 & 1 & 4 \\ 3 & 5 & 3 & 2 \end{bmatrix} \text{ has a rank} = 2$$

while

$$A = \begin{bmatrix} 3 & 9 & 2 \\ 1 & 3 & 0 \\ 2 & 6 & 1 \end{bmatrix} \text{ has a rank} = 2$$

but

$$B = \begin{bmatrix} 3 & 0 & 0 \\ 1 & 2 & 0 \\ 0 & 0 & 1 \end{bmatrix} \text{ has a rank} = 3$$

The following properties of the rank are useful in the determination of the rank of a matrix. Given an $n \times m$ matrix A,

1. Rank of A = Rank of A^T (1.25)

2. Rank of A = Rank of AA^T (1.26)

3. Rank of A = Rank of A^TA (1.27)

The rank of a matrix is of great importance in electric load forecasting using the static estimation algorithms.

1.5 Singular Matrix

If A is a square matrix, and if the determinant of A equals zero (i.e., $|A| = 0$), then the matrix A is called a singular matrix. On the other hand, if $|A|$ exists, the matrix A is called a nonsingular matrix. For example, the matrix A is

$$A = \begin{bmatrix} 2 & 1 \\ 4 & 2 \end{bmatrix} \text{ is a singular matrix because } |A| = 4 - 4 = 0.$$

If the matrix is a singular matrix, the inverse of this matrix does not exist. Even though the matrix A is singular, it has a rank. The preceding matrix A has a rank of one.

1.6 Characteristic Vectors of a Matrix

Given a matrix A with characteristic values or eigenvalues $\lambda_1, \ldots, \lambda_n$, the eigenvectors of the matrix satisfy the relations

$$AU_i = \lambda_i U_i \tag{1.28}$$

The U_is are called eigenvectors.

The matrix U of the eigenvectors is nonsingular if the eigenvectors are linearly independent:

$$U = [U_1 \quad U_2 \quad \ldots \quad U_n] \tag{1.29}$$

Let

$$U^{-1} = V = \begin{bmatrix} V_1^T \\ V_2^T \\ \vdots \\ V_n^T \end{bmatrix} \tag{1.30}$$

because

$$U^{-1}U = I$$

Thus, in expanded form, we have

$$\begin{bmatrix} V_1^T \\ V_2^T \\ \vdots \\ V_n^T \end{bmatrix} [U_1 \quad U_2 \quad \ldots \quad U_n] = \begin{bmatrix} V_1^T U_1 & V_1^T U_2 & \ldots & V_1^T U_n \\ V_2^T U_1 & V_2^T U_2 & \ldots & V_2^T U_n \\ \vdots & & & \\ V_n^T U_1 & V_n^T U_2 & \ldots & V_n^T U_n \end{bmatrix} \tag{1.31}$$

Therefore, we can conclude that component-wise

$$V_i^T U_i = 1; \quad j = 1, \ldots, n \tag{1.32}$$

$$V_i^T U_j = 0; \quad j = 1, \ldots, n \tag{1.33}$$

1.7 Diagonalization

Consider now the matrix product

$$\widetilde{A} = U^{-1}AU \tag{1.34}$$

Using equation (1.30), we have

$$\widetilde{A} = VAU \tag{1.35}$$

In terms of eigenvectors, we have

$$\widetilde{A} = V[AU_1, AU_2, \ldots, AU_n] \tag{1.36}$$

Using equation (1.28), we obtain

$$\widetilde{A} = V[\lambda U_1, \lambda U_2, \ldots, \lambda U_n] \tag{1.37}$$

Substituting for V in partitioned form, we get

$$\widetilde{A} = \begin{bmatrix} V_1^T \\ V_2^T \\ \vdots \\ V_n^T \end{bmatrix} [\lambda_1 U_1, \lambda_2 U_2, \ldots, \lambda_n U_n] \tag{1.38}$$

Performing the multiplication, we obtain

$$\widetilde{A} = \begin{bmatrix} \lambda_1 V_1^T U_1 & \lambda_2 V_1^T U_2 & \ldots & \lambda_n V_1^T U_n \\ \lambda_1 V_2^T U_1 & \lambda_2 V_2^T U_2 & \ldots & \lambda_n V_2^T U_n \\ \vdots & & & \\ \lambda_1 V_n^T U_1 & \lambda_2 V_n^T U_2 & \ldots & \lambda_n V_n^T U_n \end{bmatrix} \tag{1.39}$$

Substituting equations (1.32) and (1.33) into equation (1.39), we obtain

$$\widetilde{A} = \begin{bmatrix} \lambda_1 & 0 & 0 & 0 \\ 0 & \lambda_2 & 0 & 0 \\ \vdots & \vdots & \vdots & \vdots \\ 0 & 0 & 0 & \lambda_n \end{bmatrix} \tag{1.40}$$

Thus, the matrix $\widetilde{A}$ is a diagonal matrix of which the elements are the eigenvalues of A:

$$\widetilde{A} = U^{-1}AU \tag{1.41}$$

The expression in terms of the transformation T is

$$\widetilde{A} = TAT^{-1} \tag{1.42}$$

where

$$T^{-1} = U \tag{1.43}$$

The following example illustrates the preceding steps. Consider the matrix

$$A = \begin{bmatrix} -2 & 1 \\ 1 & -2 \end{bmatrix}$$

The eigenvalues are obtained as

$$\lambda I - A = \begin{bmatrix} \lambda + 2 & -1 \\ -1 & \lambda + 2 \end{bmatrix}$$

Thus, the characteristic polynomial is given by

$$\begin{aligned} P(\lambda) &= |\lambda I - A| = (\lambda + 2)^2 - 1 \\ &= \lambda^2 + 4\lambda + 3 \\ &= (\lambda + 1)(\lambda + 3) \end{aligned}$$

Thus, the eigenvalues are given by

$$\begin{aligned} \lambda_1 &= -1 \\ \lambda_2 &= -3 \end{aligned}$$

Next, we compute the eigenvectors as follows:

$$AU_1 = \lambda_1 U_1 \quad \text{for} \quad \lambda_1 = -1$$

Thus,

$$\begin{bmatrix} -2 & 1 \\ 1 & -2 \end{bmatrix} \begin{bmatrix} U_{11} \\ U_{21} \end{bmatrix} = \begin{bmatrix} -U_{11} \\ -U_{21} \end{bmatrix}$$

which gives

$$U_{11} = U_{21} \text{ (or assume that } U_{11} = 1, \text{ then } U_{21} = 1)$$

Therefore, the first eigenvector is given by

$$U_1 = \begin{bmatrix} 1 \\ 1 \end{bmatrix}$$

Following the same steps, we obtain the second eigenvector as

$$U_2 = \begin{bmatrix} -1 \\ 1 \end{bmatrix}$$

and hence the matrix U is given by

$$T^{-1} = [U_1 \quad U_2] = \begin{bmatrix} 1 & -1 \\ 1 & 1 \end{bmatrix}$$

Thus,

$$T = \begin{bmatrix} 0.5 & +0.5 \\ -0.5 & 0.5 \end{bmatrix}$$

Therefore, the transformed matrix A is given by

$$\widetilde{A} = TAT^{-1} = \begin{bmatrix} 0.5 & 0.5 \\ -0.5 & 0.5 \end{bmatrix} \begin{bmatrix} -2 & 1 \\ 1 & -2 \end{bmatrix} \begin{bmatrix} 1 & -1 \\ 1 & 1 \end{bmatrix} = \begin{bmatrix} -1 & 0 \\ -2 & -1 \end{bmatrix}$$

1.8 Partitioned Matrices

Partitioning is useful when applied to large matrices because manipulations can be carried out on the smaller blocks. More importantly, when one is multiplying partitioned matrices, the basic rule can be applied to the blocks as though they were single elements.

For example, the following 3×4 matrix is partitioned into four blocks:

$$A = \begin{bmatrix} 2 & 3 & \vdots & 1 & 5 \\ 0 & -2 & \vdots & 2 & 0 \\ \cdots & \cdots & \cdots & \cdots & \cdots \\ 1 & 0 & \vdots & -4 & 10 \end{bmatrix} = \begin{bmatrix} B & C \\ D & E \end{bmatrix}$$

where B, C, D, and E are the arrays indicated by dashed lines. The matrix entries of such a partitioned matrix are called submatrices. The main matrix is sometimes referred to as the supermatrix.

If A is square, and its only nonzero elements can be partitioned as principal submatrices, then it is called a block diagonal. A convenient notation that generalizes is to write A as

$$A = \text{diag.}(A_1, A_2, \ldots, A_n)$$

where the submatrices $A_1, A_2, \ldots, A_n$ are square matrices, not necessarily of equal dimension, which appear on the major diagonal. The inverse A^{-1} of $A = \text{diag.}(A_1, \ldots, A_n)$ is

$$A^{-1} = \text{diag.}\left(A_1^{-1}, A_2^{-1}, \ldots, A_n^{-1}\right)$$

Another advantage of partitioned matrices is that if the partitioned matrix A is multiplied by the matrix X, which is given by

$$X = \begin{bmatrix} X_1 \\ X_2 \end{bmatrix}$$

then

$$AX = \begin{bmatrix} B & C \\ D & E \end{bmatrix} \begin{bmatrix} X_1 \\ X_2 \end{bmatrix}$$

$$= \begin{bmatrix} BX_1 + CX_2 \\ DX_1 + EX_2 \end{bmatrix}$$

The only restriction is that the blocks must be conformable for multiplication, so that all the products BX_1, $CX_2, \ldots,$ etc., exist. This requires that in a product AX the number of columns in each block of A must equal the number of rows in the corresponding block of X.

1.9 Partitioned Matrix Inversion

It is difficult to obtain the inverse of matrices of high dimension by using the classical method. In this case, the partitioned form is useful. Suppose that F is a matrix in partitioned form as

$$F = \begin{bmatrix} A_{n\times n} & B_{n\times n} \\ C_{m\times m} & D_{m\times m} \end{bmatrix}$$

and

$$F^{-1} = \begin{bmatrix} W_{n\times n} & X_{n\times n} \\ Y_{m\times m} & Z_{m\times m} \end{bmatrix}$$

By definition of the matrix inverse,

$$FF^{-1} = I$$

so

$$\begin{bmatrix} A & B \\ C & D \end{bmatrix}\begin{bmatrix} W & X \\ Y & Z \end{bmatrix} = \begin{bmatrix} I_n & 0 \\ 0 & I_m \end{bmatrix}$$

and applying the rules of partitioned multiplication produces

$$AW + BY = I_n$$

$$AX + BZ = 0$$

$$CW + DY = 0$$

$$CX + DZ = I_m$$

By solving the preceding equations, we can obtain

$$W = A^{-1} - A^{-1}BY$$

$$Y = -(D - CA^{-1}B)^{-1}CA^{-1}$$

$$Z = (D - CA^{-1}B)^{-1}$$

$$X = -A^{-1}B(D - CA^{-1}B)^{-1}$$

provided that the matrix A is nonsingular.

Example 1.2

Consider the matrix F given by

$$F = \begin{bmatrix} 2 & 3 & 4 & -1 \\ 1 & 2 & 5 & -2 \\ 3 & 2 & 4 & 0 \\ 0 & 1 & -2 & 3 \end{bmatrix}$$

F can be partitioned as shown.

Thus, we can find

$$A = \begin{bmatrix} 2 & 3 \\ 1 & 2 \end{bmatrix}, \quad A^{-1} = \begin{bmatrix} 2 & -3 \\ -1 & 2 \end{bmatrix}$$

$$B = \begin{bmatrix} 4 & -1 \\ 5 & -2 \end{bmatrix}, \quad C = \begin{bmatrix} 3 & 2 \\ 0 & 1 \end{bmatrix}$$

$$\text{and} \quad D = \begin{bmatrix} 4 & 0 \\ -2 & 3 \end{bmatrix}$$

Then the matrices Y, W, Z, and X are given by

$$Y = -\left[\begin{bmatrix} 4 & 0 \\ -2 & 3 \end{bmatrix} - \begin{bmatrix} 3 & 2 \\ 0 & 1 \end{bmatrix}\begin{bmatrix} 2 & -3 \\ -1 & 2 \end{bmatrix}\begin{bmatrix} 4 & -1 \\ 5 & -2 \end{bmatrix}\right]^{-1} \times \begin{bmatrix} 3 & 2 \\ 0 & 1 \end{bmatrix}\begin{bmatrix} 2 & -3 \\ -1 & 2 \end{bmatrix}$$

$$= \frac{-1}{30}\begin{bmatrix} 18 & -18 \\ 19 & -14 \end{bmatrix}$$

$$W = \begin{bmatrix} 2 & -3 \\ -1 & 2 \end{bmatrix} + \frac{1}{30}\begin{bmatrix} 2 & -3 \\ -1 & 2 \end{bmatrix}\begin{bmatrix} 4 & -1 \\ 5 & -2 \end{bmatrix}\begin{bmatrix} 18 & -18 \\ 19 & -14 \end{bmatrix} = \frac{-1}{30}\begin{bmatrix} 10 & -20 \\ 21 & -6 \end{bmatrix}$$

$$Z = \left[\begin{bmatrix} 4 & 0 \\ -2 & 3 \end{bmatrix} - \begin{bmatrix} 3 & 2 \\ 0 & 1 \end{bmatrix}\begin{bmatrix} 2 & -3 \\ -1 & 2 \end{bmatrix}\begin{bmatrix} 4 & -1 \\ 5 & -2 \end{bmatrix}\right]^{-1}$$

$$= \begin{bmatrix} 13 & -6 \\ -8 & 6 \end{bmatrix}^{-1}$$

$$= \frac{1}{30}\begin{bmatrix} 6 & 6 \\ 8 & 13 \end{bmatrix}$$

$$X = \frac{-1}{30}\begin{bmatrix} 2 & -3 \\ -1 & 2 \end{bmatrix}\begin{bmatrix} 4 & -1 \\ 5 & -2 \end{bmatrix}\begin{bmatrix} 6 & 6 \\ 8 & 13 \end{bmatrix}$$

$$= \frac{1}{30}\begin{bmatrix} 10 & -10 \\ -12 & 3 \end{bmatrix}$$

Therefore,

$$F^{-1} = \begin{bmatrix} 0.333 & -0.6667 & 0.333 & -0.333 \\ 0.7 & -0.2 & -0.4 & 0.1 \\ -0.6 & 0.6 & 0.2 & 0.2 \\ -0.633 & 0.466 & 0.266 & 0.433 \end{bmatrix}$$

1.10 Quadratic Forms

An algebraic expression of the form

$$f_{(x,y)} = ax^2 + bxy + cy^2$$

is said to be a quadratic form. If we let

$$X = \begin{bmatrix} x \\ y \end{bmatrix}$$

then we obtain

$$f_{(x,y)} = [x \quad y] \begin{bmatrix} a & \frac{b}{2} \\ \frac{b}{2} & c \end{bmatrix} \begin{bmatrix} x \\ y \end{bmatrix}$$

or

$$f_{(X)} = X^T A X$$

The preceding equation, as mentioned previously, is in quadratic form. The matrix A in this form is a symmetrical matrix.

A more general form for the quadratic function can be written in a matrix form as

$$F_{(X)} = X^T A X + B^T X + C$$

where X is an $n \times 1$ vector, A is an $n \times n$ matrix, B is an $n \times 1$ vector, and C is 1×1 vector.

Example 1.3

Given the function

$$f_{(x,y)} = 2x^2 + 4xy - y^2 = 0$$

we need to write this function in a quadratic form.

We define the vector

$$X = \begin{bmatrix} x \\ y \end{bmatrix}$$

Then

$$f_{(x,y)} = \begin{bmatrix} x & y \end{bmatrix} \begin{bmatrix} 2 & 2 \\ 2 & -1 \end{bmatrix} \begin{bmatrix} x \\ y \end{bmatrix}$$

$$F_{(x)} = X^T A X$$

where

$$A = \begin{bmatrix} 2 & 2 \\ 2 & -2 \end{bmatrix}$$

Example 1.4

To obtain the quadratic form for the function

$$f_{(x_1,x_2)} = 3x_1^2 + 4x_1x_2 - 4x_2^2$$

we define the vector X as

$$X = \begin{bmatrix} x_1 & x_2 \end{bmatrix}^T$$

then

$$f_{(x_1,x_2)} = \begin{bmatrix} x_1 & x_2 \end{bmatrix}^T \begin{bmatrix} 3 & 2 \\ 2 & -4 \end{bmatrix} \begin{bmatrix} x_1 \\ x_2 \end{bmatrix}$$

then

$$f_{(X)} = X^T A X$$

where

$$A = \begin{bmatrix} 3 & 2 \\ 2 & -4 \end{bmatrix}$$

Let A be an $n \times n$ matrix and let X be an $n \times 1$ vector. Then, irrespective of whether A is symmetric,

$$X^T A X = \sum_{i=1}^{n}\sum_{j=1}^{n} x_i a_{ij} x_j$$

$$= \sum_{i=1}^{n}\sum_{j=1}^{n} x_i a_{ji} x_j$$

- An $n \times n$ matrix A is positive definite if, and only if,

  $$X^T AX > 0 \quad (\forall X \in R^n, X \neq 0)$$

- It is positive semidefinite if, and only if,

  $$X^T AX \geq 0 \quad (\forall X \in R^n)$$

- Similarly, A is negative definite if, and only if,

  $$X^T AX < 0 \quad (\forall X \in R^n, X \neq 0)$$

- A is negative semidefinite if, and only if,

  $$X^T AX \leq 0 \quad (\forall X \in R^n)$$

1.11 State Space Representation

The state space representation is an alternative method used to describe a system's dynamics, in which a compact standard notation is used for this description [3,6]. The concept of a system state x was defined by Kalman in 1963 [5] as:

> The state of a system is a mathematical structure consisting of a set of n time-dependent variables $x_1(t), x_2(t), \ldots, x_i(t), \ldots, x_n(t)$, referred to as the state variables. The system inputs $u_j(t)$ together with the initial conditions of the state variables $x_1(0), x_2(0), \ldots, x_i(0), \ldots, x_n(0)$ are sufficient to uniquely define the system's future response.

Let us explain this statement through an example, from a simple ac series R, L, C circuit.

The total voltage is given as

$$u(t) = u_R(t) + u_L(t) + u_c(t) \tag{1.44}$$

$$= Ri(t) + L\frac{di(t)}{dt} + \frac{1}{C}\int i(t)dt \tag{1.45}$$

Taking the Laplace transform of both sides, assuming zero initial conditions

$$V(s) = RI(s) + sLI(s) + \frac{1}{Cs}I(s)$$

$$\frac{V(s)}{I(s)} = \left(R + sL + \frac{1}{Cs}\right) \tag{1.46}$$

The input to the system is assumed to be the voltage $V(s)$, and the output is $I(s)$. Then

$$\frac{I(s)}{V(s)} = \frac{Cs}{s^2LC + RCs + 1} \tag{1.47}$$

The preceding equation is called the system transfer equation. If the voltage across the capacitor is taken as a state variable, then

$$\frac{dv_c(t)}{dt} = \frac{1}{C}i(t) \tag{1.48}$$

and

$$\frac{di(t)}{dt} = \frac{-1}{L}v_c(t) - \frac{R}{L}i(t) + \frac{1}{L}v(t) \tag{1.49}$$

Now, we define the following new variables as

$$x_1(t) = v_c(t) \tag{1.50}$$

$$x_2(t) = i(t) \tag{1.51}$$

Then equations (1.48) and (1.49) become

$$\dot{x}_1(t) = \frac{1}{C}x_2(t) \tag{1.52}$$

$$\dot{x}_2 = \frac{-1}{L}x_1(t) - \frac{R}{L}x_2(t) + \frac{1}{L}v(t) \tag{1.53}$$

$$\begin{bmatrix} \dot{x}_1(t) \\ \dot{x}_2(t) \end{bmatrix} = \begin{bmatrix} 0 & \frac{1}{C} \\ -\frac{1}{L} & -\frac{R}{L} \end{bmatrix} \begin{bmatrix} \dot{x}_1(t) \\ \dot{x}_2(t) \end{bmatrix} + \begin{bmatrix} 0 \\ \frac{1}{L} \end{bmatrix} v(t) \tag{1.54}$$

In compact form, equation (1.54) can be rewritten as

$$\dot{x}(t) = AX(t) + BU(t) \tag{1.55}$$

Equation (1.55) is the state space equation for the system. Let us observe here that there are two states, $x_1(t)$ and $x_2(t)$, for this system. The number of system states is equal to the order of the differential equation describing the system, which, at the same time, is equal to the number of initial conditions necessary to obtain a solution for the system.

The vector-matrix form can also be applied to nonlinear state equations. The general expression is

$$\frac{dx(t)}{dt} = f[x(t), u(t), t] \tag{1.56}$$

1.12 Difference Equations

In modern control systems, digital processors are used to perform the task of control, so it is important to establish equations that relate digital and discrete time signals. Whereas differential equations are used to represent systems with analog signals, difference equations are used for systems with discrete data. It is easier to use difference equations than differential equations on a digital computer, and these equations are generally easier to solve.

As an example of discrete approximation, we can use a forward difference process to approximate the derivative of a function at a given instant; that is,

$$\left.\frac{dy(t)}{dt}\right|_{t=kT} = \frac{y[(k+1)T] - y(kT)}{T} \tag{1.57}$$

where T is chosen to be some small value that will lead to a good approximation. When we use equation (1.57), a first-order differential equation in the form

$$\frac{dy(t)}{dt} + ay(t) = f(t) \tag{1.58}$$

can be transformed into

$$\frac{y[(k+1)T] - y(kT)}{T} + ay(kT) = f(kT) \tag{1.59}$$

$$y[(k+1)T] = [1 + aT]y(kT) + Tf(kT) \tag{1.60}$$

The preceding equation is called a first-order difference equation, and it gives the value of y one step ahead as a function of y and f one step back.

In general, a linear nth-order difference equation with constant coefficients can be written as

$$a_{n+1}y(k+n) + a_n y(k+n) + \ldots + a_2 y(k+1) + a_1 y(k) = f(k) \tag{1.61}$$

where $y(i)$, $i = k, k+1, \ldots, k+n$ denote the values of the discrete dependent variable y at the ith instant if the independent variable is discrete time. In general, the independent variable can represent any real physical quantity.

In a manner similar to analog systems, it is convenient to use a set of first-order difference equations (state equations) to represent a high-order difference equation. For the difference equation in equation (1.61), if we let

$$\begin{aligned} x_1(k) &= y(k) \\ x_2(k) &= x_1(k+1) = y(k+1) \\ &\vdots \\ x_{n-1}(k) &= x_{n-2}(k+1) = y(k+n-2) \end{aligned} \tag{1.62}$$

then the equation is written as

$$x_n(k+1) = -\frac{a_1}{a_{n+1}}x_1(k) - \frac{a_2}{a_{n+1}}x_2(k) - \cdots - \frac{a_n}{a_{n+1}}x_n(k) + \frac{1}{a_{n+1}}f(k) \quad (1.63)$$

The first $n - 1$ state equations are taken directly from equation (1.62), and the final one is given by equation (1.63). Writing these n first-order difference state equations in vector-matrix form, we have

$$x[(k+1)T] = Ax(kT) + Bu(kT) \quad (1.64)$$

where

$$x(kT) = \begin{bmatrix} x_1(kT) \\ x_2(kT) \\ \vdots \\ x_n(kT) \end{bmatrix} \quad (1.65)$$

is the $n \times 1$ state vector, and

$$A = \begin{bmatrix} 0 & 1 & 0 & 0 & \cdots & 0 \\ 0 & 0 & 1 & 0 & \cdots & 0 \\ 0 & 0 & 0 & 1 & \cdots & 0 \\ 0 & 0 & 0 & 0 & \cdots & 1 \\ -\frac{a_1}{a_{n+1}} & -\frac{a_2}{a_{n+1}} & -\frac{a_3}{a_{n+1}} & \cdot & \cdots & -\frac{a_n}{a_{n+1}} \end{bmatrix} \quad (1.66)$$

$$B = \begin{bmatrix} 0 \\ 0 \\ 0 \\ \vdots \\ 0 \\ \frac{1}{a_{n+1}} \end{bmatrix} \quad (1.67)$$

1.13 Some Optimization Techniques

In this section we discuss the general optimization problem without going into mathematical analysis details [2,4]. The first part of the section introduces unconstrained optimization that has many applications throughout this book, and the second part introduces the constrained optimization problem. Generally speaking, the optimization problem has the following form:

$$\text{Minimize} \quad f(x_1,\ldots,x_n) \tag{1.68}$$

$$\text{Subject to} \quad \phi_i(x_1,\ldots,x_n)=0, \quad (i=1,\ldots,\ell) \tag{1.69}$$

$$\psi_j(x_1,\ldots,x_n)\leq 0, \quad (j=1,\ldots,m) \tag{1.70}$$

Equation (1.69) represents ℓ, $\ell < n$ equality constraints, and equation (1.70) represents m inequality constraints. By using the vector notation, we may express the general constrained optimization problem as follows:

$$\text{Minimize} \quad f(X) \tag{1.71}$$

$$\text{Subject to} \quad \phi(X)=0 \tag{1.72}$$

$$\psi(X)\leq 0, \quad X\in R^n \tag{1.73}$$

The problem formulated in equations (1.71) to (1.73) is usually referred to as the general nonlinear programming problem. Any point X that satisfies these equations is called a feasible point.

1.13.1 Unconstrained Optimization

In the unconstrained optimization problem, we need to find the value of the vector $X=[x_1,\ldots,x_n]^T$ that minimizes the function

$$f(x_1,\ldots,x_n) \tag{1.74}$$

provided that the function f is continuous and has a first-order derivative.

To obtain the minimum and/or maximum of the function f, we set its first derivative, with respect to the x_is, to zero:

$$\frac{\partial f(x_1,\ldots,x_n)}{\partial x_1}=0 \tag{1.75}$$

$$\frac{\partial f(x_1,\ldots,x_n)}{\partial x_2}=0 \tag{1.76}$$

$$\vdots$$

$$\frac{\partial f(x_1,\ldots,x_n)}{\partial x_1}=0 \tag{1.77}$$

Equations (1.75) to (1.77) represent n equations in n unknowns. The solution of these equations produces candidate solution points. If the function f has second partial derivatives, then we calculate the Hessian matrix $H = \dfrac{\partial^2 f(x_1, \ldots, x_n)}{\partial x_i^2}$. If the matrix H is positive definite, then the function f is a minimum at the candidate points, but if the matrix H is negative definite, then f is a maximum at the candidate points. The following examples illustrate these steps.

Example 1.5

Minimize

$$f(x_1, x_2) = x_1^2 + x_1 x_2 + x_2^2 \quad (x \in R^2)$$

To obtain the candidate solution points, we have

$$\frac{\partial f(x_1, x_2)}{\partial x_1} = 2x_1 + x_2 = 0$$

and

$$\frac{\partial f(x_1, x_2)}{\partial x_2} = x_1 + 2x_2 = 0$$

Solving the preceding equations yields the candidate solution point as

$$[x_1^*, x_2^*]^T = [0, 0]^T$$

Next, we calculate the Hessian matrix using

$$H = \begin{bmatrix} \dfrac{\partial^2 f}{\partial x_1^2} & \dfrac{\partial^2 f}{\partial x_1 \partial x_2} \\ \dfrac{\partial^2 f}{\partial x_2 \partial x_1} & \dfrac{\partial^2 f}{\partial x_2^2} \end{bmatrix}$$

to obtain

$$H = \begin{bmatrix} 2 & 1 \\ 1 & 2 \end{bmatrix}$$

so

$$X^T H X = [x_1 \quad x_2] \begin{bmatrix} 2 & 1 \\ 1 & 2 \end{bmatrix} \begin{bmatrix} x_1 \\ x_2 \end{bmatrix}$$

$$= 2(x_1^2 + x_1 x_2 + x_2^2)$$

$$= 2\left\{ \left(x_1 + \frac{1}{2} x_2 \right)^2 + \frac{3}{4} x_2^2 \right\}$$

Therefore,

$$X^T HX > 0 \quad (\forall X \neq 0)$$

so H is positive definite, and the function f is a minimum at the candidate point.

Note that the positive definiteness of H can also be verified just by calculating the values of the different determinants, produced from H as

$$\Delta_1(H) = 2 = h_{11}$$
$$\Delta_2(H) = (4 - 1) = 3$$

Because all Δs are positive, then H is a positive definite matrix.

Example 1.6

Minimize

$$f(x_1, x_2) = 34x_1^2 - 24x_1x_2 + 41x_2^2$$

We set the first derivatives to zero to obtain

$$\frac{\partial f(x_1, x_2)}{\partial x_1} = 68x_1 - 24x_2 = 0$$

$$\frac{\partial f(x_1, x_2)}{\partial x_2} = -24x_1 + 82x_2 = 0$$

The solution to the preceding equation gives

$$\begin{bmatrix} x_1^* \\ x_2^* \end{bmatrix} = \begin{bmatrix} 0 \\ 0 \end{bmatrix}$$

Now we calculate the Hessian matrix as

$$H = \begin{bmatrix} 68 & -24 \\ -24 & 82 \end{bmatrix}$$

and check the definiteness for the Hessian matrix as

$$\Delta_1(H) = h_{11} = 68 > 0$$

$$\Delta_2(H) = 68 \times 82 - 24 \times 24 = 1328 > 0$$

Therefore, H is a positive definite matrix; or we calculate the quadratic form

$$\begin{aligned} X^T HX &= [x_1 \quad x_2]\begin{bmatrix} 68 & -24 \\ -24 & 82 \end{bmatrix}\begin{bmatrix} x_1 \\ x_2 \end{bmatrix} \\ &= 68x_1^2 - 48x_1x_2 + 82x_2^2 \\ &= 2(4x_1 - 3x_2)^2 + 32x_1^2 + 64x_2^2 \end{aligned}$$

so

$$X^T HX > 0 \quad (\forall X \neq 0)$$

Therefore, H is a positive definite and f is a minimum at the feasible points.

Example 1.7

Minimize

$$f(x_1, x_2) = x_1^3 - 2x_1^2x_2 + x_2^2$$

We have

$$\frac{\partial f(x_1, x_2)}{\partial x_1} = 3x_1^2 - 4x_1x_2 = 0$$

$$\frac{\partial f(x_1, x_2)}{\partial x_2} = -2x_1^2 + 2x_2 = 0$$

Solving the preceding two equations yields the critical points to be

$$x^* = \begin{bmatrix} x_1^* \\ x_2^* \end{bmatrix} = \begin{bmatrix} \frac{3}{4} \\ \frac{9}{16} \end{bmatrix}$$

The Hessian matrix is calculated as

$$H = \begin{bmatrix} (6x_1 - 4x_2) & -4x_1 \\ -4x_1 & 2 \end{bmatrix}$$

At the solution points, we calculate H as

$$H(x_1^*, x_2^*) = \begin{bmatrix} \frac{9}{4} & -3 \\ -3 & 2 \end{bmatrix}$$

$$\Delta_1(H) = \frac{9}{4} > 0$$

$$\Delta_2(H) = \frac{18}{4} - 9 = -\frac{18}{4} < 0$$

Therefore, $H(x_1^*, x_2^*)$ is a positive semidefinite, so nothing can be concluded about the nature of the solution point x^*. The solution point in this case is called a saddle point.

1.13.2 Constrained Optimization

The constrained optimization problem has the form:

Minimize

$$f(x_1, \ldots, x_n) \tag{1.78}$$

subject to satisfying

$$\phi_i(x_1, \ldots, x_n) = 0, \quad (i = 1, \ldots, \ell) \tag{1.79}$$

and

$$\psi_j(x_1, \ldots, x_n) \leq 0, \quad (j = 1, \ldots, m) \tag{1.80}$$

Let us consider, for instance, a case when an objective function is subject only to equality constraints. We form the augmented objective function by adjoining the equality constraints to the function via Lagrange's multipliers to obtain the alternative form:

Minimize

$$\tilde{f}(x_1, \ldots, x_n, \lambda_i) = f(x_1, \ldots, x_n) + \sum_{i=1}^{l} \lambda_i \phi_i(x_1, \ldots, x_n) \tag{1.81}$$

or, in vector form,

$$\tilde{f}(X, \lambda) = f(X) + \lambda^T \phi(x) \tag{1.82}$$

Putting the first derivative to zero, we obtain

$$\frac{\partial \tilde{f}(X, \lambda)}{\partial x_i} = \frac{\partial f(x)}{\partial x_i} + \sum_{j=1}^{\ell} \lambda j \frac{\partial \phi_j}{\partial x_i} = 0 \tag{1.83}$$

Equation (1.83) is a set of n equations in $(n + \ell)$ unknowns (x_i; $i = 1, \ldots, n$: λ_j; $j = 1, \ldots, \ell$). To obtain the solution, we must satisfy the equality constraints; that is,

$$\phi_i(x_1, \ldots, x_n) = 0 \quad i = 1, \ldots, \ell \tag{1.84}$$

Solving equations (1.83) and (1.84), we obtain x_i^* and λ_j^*. This scenario is illustrated in the following examples.

Example 1.8

Minimize

$$f(x_1, x_2) = x_1^2 + x_2^2$$

Subject to

$$\phi[x_1, x_2] = x_1 + 2x_2 + 1 = 0$$

For this problem $n = 2$, $\ell = 1$, $(n + \ell = 3)$. The augmented cost function is given by

$$\tilde{f}(x_1, x_2, \lambda) = x_1^2 + x_2^2 + \lambda(x_1 + 2x_2 + 1)$$

Putting the first derivatives to zero gives

$$\frac{\partial \tilde{f}}{\partial x_1} = 0 = 2x_1^* + \lambda$$

$$\frac{\partial \tilde{f}}{\partial x_2} = 0 = 2x_2^* + 2\lambda$$

and

$$\frac{\partial \tilde{f}}{\partial \lambda} = 0 = x_1^* + 2x_2^* + 1 \quad \text{(equality constraint)}$$

Solving the preceding three equations gives

$$x_1^* = -\frac{1}{5}, \quad x_2^* = -\frac{2}{5}, \quad \lambda = \frac{2}{5}$$

Example 1.9

Minimize

$$\tilde{f}(x_1, x_2, \lambda) = (10 + 5x_1 + 0.2x_1^2) + (20 + 3x_2 + 0.1x_2^2)$$

Subject to

$$x_1 + x_2 = 10$$

The augmented cost function is

$$\tilde{f}(x_1, x_2, \lambda) = (30 + 5x_1 + 0.2x_1^2 + 3x_2 + 0.1x_2^2) + \lambda(10 - x_1 - x_2)$$

Putting the first derivatives to zero, we obtain

$$\frac{\partial \tilde{f}}{\partial x_1} = 0 = 5 + 0.4x_1^* - \lambda$$

$$\frac{\partial \tilde{f}}{\partial x_2} = 0 = 3 + 0.2x_2^* - \lambda$$

$$\frac{\partial \tilde{f}}{\partial \lambda} = 0 = 10 - x_1^* - x_2^*$$

Solving the preceding three equations gives

$$x_1^* = 0, \quad x_2^* = 10, \quad \lambda = 5$$

and the minimum of the function is

$$f(0, 10) = 30 + 30 + 10 = 70$$

If there are inequality constraints, then the augmented function is obtained by adjoining these inequality constraints via Kuhn-Tucker multipliers to obtain

$$\tilde{f}(X, \lambda, \mu) = f(X) + \lambda^T \phi(X) + \mu^T \psi(X) \tag{1.85}$$

Putting the first derivative to zero, we obtain

$$\frac{\partial \tilde{f}}{\partial X} = 0 = \frac{\partial f(X^*)}{\partial X} + \lambda^T \frac{\partial \phi(X^*)}{\partial X} + \mu^T \frac{\partial \psi(X^*)}{\partial X} \tag{1.86}$$

and

$$\frac{\partial \tilde{f}}{\partial \lambda} = 0 = \phi(X^*) \tag{1.87}$$

with

$$\mu^T \psi(X^*) = 0 \tag{1.88}$$

If $\psi(X^*) > 0$, then $\mu = 0$.

Solving the preceding equations gives the candidate solution (X^*, λ, μ).

Example 1.10

Recall from the previous example that we have

Minimize

$$f(x_1, x_2) = 0.1x_2^2 + 0.2x_1^2 + 3x_2 + 5x_1 + 30$$

Subject to the following constraints

$$x_1 + x_2 = 10$$

with

$$x_1 \geq 0$$
$$0 \leq x_2 \leq 15$$

We form the augmented function as

$$\tilde{f}(x_1, x_2, \lambda, \mu_1, \mu_2, \mu_3) = f(x_1, x_2) + \lambda(10 - x_1 - x_2) + \mu_1 x_1 + \mu_2 x_2 + \mu_3(15 - x_2)$$

Putting the first derivatives to zero leads to

$$\frac{\partial \tilde{f}}{\partial x_1} = 0 = \frac{\partial f}{\partial x_1} - \lambda + \mu_1 + \mu_2$$

$$\frac{\partial \tilde{f}}{\partial x_2} = 0 = \frac{\partial f}{\partial x_2} - \lambda + \mu_2 - \mu_3$$

$$\frac{\partial \tilde{f}}{\partial \lambda} = 0 = 10 - x_1 - x_2$$

with

$$\mu_1 x_1 = 0$$
$$\mu_2 x_2 = 0$$
$$\mu_3(15 - x_2) = 0$$

Now we have six equations for six unknowns; however, solving these equations is very difficult. We assume that none of the variables violates its limits; thus, we obtain

$$\mu_1 = 0$$
$$\mu_2 = 0$$
$$\mu_3 = 0$$

and we must check the solution obtained for these conditions. The solution in this case is

$$x_1^* = 0, \quad x_2^* = 10, \quad \lambda = 5$$

Indeed, as we see, the variables do not violate their limits, and the optimal solution in this case is

$$x_1^* = 0, \quad x_2^* = 10, \quad \lambda^* = 5$$
$$\mu_1^* = 0, \quad \mu_2^* = 0, \quad \mu_3^* = 0$$

However, if we change the second inequality constraint to be

$$0 \le x_2 \le 8$$

then we can see that for the solution here, $x_2^* = 10$ violates the upper limit. In this case we put

$$x_2^* = 8; \text{with } \mu_3(8 - x_2^*) = 0$$

and recalculate x_1^* as

$$x_1^* = 10 - x_2^* = 10 - 8$$
$$x_1^* = 2, \ (x_1^* > 0)$$

Under this solution, $\mu_3 \neq 0$, but $\mu_1^* = 0$ and $\mu_2^* = 0$. To calculate λ^* and μ_3^*, we use the first two equations as

$$0 = 0.4x_1^* + 5 - \lambda^*$$

or

$$\lambda^* = 0.4(2) + 5$$
$$= 5.8$$

and

$$0 = 0.2x_2^* + 3 - \lambda^* - \mu_3^*$$

or

$$\mu_3^* = 1.6 + 3 - 5.8$$
$$= -1.2$$

1.14 State of the Art

Short-term load forecasting (STLF) is an integral part of power system operation, which is essential for securing an inexpensive supply of reliable electric energy. This type of forecasting is used to predict load demands up to a week ahead so that the day-to-day operation of a power system can be efficiently planned and so that the operating costs are minimized.

Short-term load forecasting can be performed in one of two modes—namely, online and offline forecasting. This categorization, as the names suggest, stems from the areas of application of the load predictors. Offline load forecasting is primarily implemented in the scheduling of the large generating units of which the startup times may vary from a few hours ahead to a few days ahead. The scheduling process is termed *unit commitment* and ensures that there is sufficient operating generation capacity to meet the variable load demand with specified reliability [7]. When load forecasting is poor, incorrect scheduling may occur, resulting in higher daily operational cost caused by use of higher-cost quick-start units in the event of underscheduling or, alternatively, resulting in the uneconomic operation of large generating units in the event of overscheduling [50].

Online operation of a power system, the economic load dispatching to various generating units, makes the generating mix dependent on calculations to minimize the cost function, which is based on the characteristics of the generating units. These calculations are based on values of load demand predicted a few hours in advance, and as such the optimum generating mix is dependent on the accuracy of the online forecasts.

It has been recognized for a long time that accurate short-term load predictors as well as a load model are basic necessities for the optimum economic operation of power systems. A prerequisite to the development of an accurate load-forecasting model is an understanding of the characteristics of the load to be modeled. This knowledge of load behavior is gained from experience with the load and thorough statistical analysis of past load data. Utilities with similar climatic and economic environments usually experience similar load behavior, and load models developed for one utility can usually be modified to suit another.

For short-term load forecasting, many factors should be included in the load prediction model. Reference [7] reviews the short-term load-demand modeling and forecasting for offline and online implementation. Included also in [7] is a review of most techniques used at that time; the merits and drawbacks of each approach are presented. Reference [8] presents an algorithm based on curve fitting of past load growth for forecasting distribution system loads. The proposed algorithm in this reference uses clustering of historical load at the small area level as the forecast algorithm. References [9,10] compare 14 methods of forecasting future distribution system loads in terms of forecasting accuracy, data needs, and resources. The tests of different forecast methods were carried out in as uniform a manner as possible. These references claim that the selection of a forecast method is based on a great deal more criteria than those discussed in the references. Data availability is usually an important factor; choice of a distribution load-forecasting method may also be constrained by many other factors, including available computer resources and the level of expertise of the users.

Reference [11] reviews some of the existing studies on 1- to 24-hour load-forecasting algorithms and presents an expert system-based algorithm as an alternative. This algorithm is developed on the logical and syntactical relationships between weather and load, and prevailing daily load shapes. It is found in this reference that the proposed algorithm is robust and accurate and has yielded results that are equally good, if not better, when compared to the regression-based forecasting techniques.

Reference [12] presents an adaptive nonlinear predictor with orthogonal escalator structure for short-term load forecasting. The proposed method in this reference uses a nonlinear time-varying functional relationship between load and temperature. Parameters in both linear and nonlinear parts of the predictor are updated systematically using a scalar orthogonalization procedure. Matrix operations are avoided in this reference, which results in a more robust and better numerical property algorithm. This reference claims that there is no need for seasonal offline model calibration or modification because the proposed adaptive algorithm has good model-tracking ability. Analysis and evaluation of five short-time load-forecasting techniques are performed in reference [13]. The five techniques are (1) multiple linear regression, (2) stochastic time series, (3) general exponential smoothing, (4) state space and Kalman filter, and (5) a knowledge-based approach.

The use of a statistical decision function is implemented in reference [14]. A hierarchical classification algorithm is applied to hourly temperature readings to divide the historical database into seasonal subsets. These subsets are identified statistically to fit a response function for each season. For a given day, an appropriate model is selected by performing discriminate analysis. It is found in this reference that the proposed algorithm is less sensitive to extreme values than other algorithms. Also, the parameters should be updated periodically using the most recent seasonal subsets.

Reference [15] presents a robust model for forecasting power system hourly load. The method exploits the convenience of the autocorrelation function, and the partial autocorrelation function of the resulting differences in previous load data identify a suboptimal model. The algorithm used in identifying the parameters of the proposed load is the iteratively reweighted least squares. Three-way decision variables in identifying an optimal model and the subsequent parameter estimates are used in this reference. These variables are (1) the weighting function, (2) the tuning constant, and (3) the sum of the squared residuals.

Reference [16] presents formulation and analysis of a short-term load-forecasting rule-based algorithm. Load parameters are classified into weather- and nonweather-related values. The rules are the product of identifiable statistical relationship and expert knowledge. The forecasting algorithm puts smaller weight on the temperature effect and depends on the natural diversity of the load with a reduced or enlarged base.

A knowledge-based expert system is proposed in reference [17] for short-term load forecasting of a power system. The expert system is developed using a five-year database. Eleven load shapes, each with different means of load calculations, are established in this reference. The effect of weather variables, such as temperature and humidity, on load forecasting is examined. The effect of thermal buildup is also studied. The proposed expert system is used to forecast the hourly loads of a power system over a whole year using the past five-year database. This reference claims that the developed expert system can serve as a valuable assistant to system operators in performing their daily load-forecasting duties.

Reference [18] describes a linear regression-based model for the calculation of short-term system load forecasts. The model in this reference has the merits of (1) innovative model building, including accurate holiday modeling by using binary variables; (2) temperature modeling by using heating and cooling degree functions;

(3) robust parameter estimation and parameter estimation by using weighted least squares linear regression techniques; (4) the use of reverse "errors-in-variables" techniques to mitigate the effects of potential errors in the explanatory variables on load forecasts; and (5) a distinction between time-independent daily peak-load forecasts and the maximum of the hourly load forecasts to prevent peak forecasts from being negatively biased. Taken together, the preceding merits result in accurate, robust, and adaptive responses to the changing conditions algorithm.

Reference [19] develops a composite load model for 1–24 hours' ahead prediction of hourly electric loads. The load model, in this reference, is composed of three components: the nominal load, type load, and residual load. The Kalman filter algorithm is used to estimate the parameters of the nominal load together with the exponentially weighted recursive least squares method. The type load component is extracted for weekend load prediction and updated by an exponential smoothing method. The autoregressive model predicts the residual load, and the parameters of the models are estimated using the recursive least squares method.

In the past two decades, artificial neural networks (ANNs) found wide applications in power system analysis and control. One of the successful applications is short-term load forecasting. References [20] and [21] present an approach using ANN for short-term load forecasting. In reference [20], a neural network based on self-organizing feature maps is used to identify those days with similar hourly load patterns. The load patterns of several days in the past are averaged to obtain the load pattern of the day under study. The averaging days are of the same type as the day under study. In reference [21], a feedforward multilayer neural network is designed to predict daily peak and valley loads. Once the peak load and valley load and the hourly load patterns are available, the desired hourly loads can be readily computed. The authors of these two references point out that the self-organizing feature map is capable of identifying a new type of load pattern before the operators can recognize the new day type.

An adaptive load-forecasting algorithm for a one-hour-ahead time period is developed in reference [22]. The major enhancement of this algorithm is the ability to forecast total hourly system load as far ahead as five days. An important benefit of the adaptive algorithm is the ability to predict load shapes in addition to daily peak loads. System operators are able to utilize the predicted load shapes of several-hour-old one-day-ahead or five-day-ahead forecasts, even when the individual hourly errors are rather large.

Reference [23] presents an ANN method to forecast the short-term load for a large power system. The load is assumed to have two distinct patterns: weekday and weekend. A nonlinear load model is proposed, together with several structures of ANN. This reference claims that the neural network, when grouped into different load patterns, gives a good load forecast. It is found that the backpropagation algorithm is robust in estimating the weights in nonlinear equations.

A multilayer neural network with an adaptive learning algorithm is proposed in reference [24] for short-term load forecasting. Effects of learning rate, momentum, and other factors on the efficiency and accuracy of the backpropagation-momentum learning method are studied in this reference. The proposed adaptive learning algorithm converges much faster than the learning rate, and the initial value of the momentum will not affect the conventional backpropagation-momentum learning method and the convergence property of the adaptive learning algorithm.

Reference [25] presents an improved neural network approach to produce short-term electric load forecasts. In this approach, a minimum distance measurement is used to identify the appropriate historical patterns of load and temperature readings to estimate the network weights. When one uses this strategy, the problems of holidays and drastic changes in weather patterns are overcome. This algorithm also includes a combination of linear and nonlinear terms that map past load and temperature inputs into the load forecast output. This reference demonstrates that even a simple three-layer network produces results that are quite favorable compared to those typically seen in the literature, with smaller absolute errors. Reference [26] reviews short-term load-forecasting techniques to find a standard for comparison. Size of error can be used as a measure for comparison standard.

Reference [27] presents a nonfully connected ANN model for short-term forecasting. The model used in this reference consists of one main ANN and three supporting ANNs. The main ANN is used to provide the models' basic forecast reference. Three supporting ANNs are added to increase the learning capacity of the proposed model. These supporting ANNs enable the model to better extract the relationships among different input categories and achieve improved accuracy. In addition, three feedforward connections are established in the main ANN. These feedforward connections provide the most recent load and temperature references and greatly improve learning efficiency. It is found that the model, compared with a fully connected ANN, requires less training time and has better performance.

Reference [28] presents an expert system using fuzzy set theory for short-term load forecasting. The uncertainties in weather variables and statistical models are taken into account by using fuzzy set theory. Also incorporated into the system are the operator's heuristic rules. Two approaches based on the minimum-maximum algorithm and the equal-area criterion algorithm are proposed to determine the most desirable change in peak load from separate sources of fuzzy information.

The ANN model in reference [29] is claimed to be a useful tool for short-term load forecasting. Radically different from statistical methods, these models have shown promising results in load forecasting. Reference [29] concludes that, on the basis of the results obtained, there is no firm criterion to select a suitable network structure for a set of hourly load and temperature data. Models are not unique, and systems with different load characteristics require different structures. However, once a model is identified for a given system, the model need not be modified frequently. Neural network models are sensitive to bad data, so intelligent data filtering techniques need to be designed to be able to maintain acceptable accuracy in the ANN models-based load forecasts.

Reference [30] presents a generalized short-term load-forecasting algorithm. This algorithm combines features from a knowledge base and statistical techniques. The technique is based on a generalized model for the weather-load relationship, which makes it site-independent. However, adding site-dependent characteristics easily customizes it. Such characteristics are formulated in the form of selection and adjustment rules. Once added, these rules are expected to improve the performance of the algorithm for a specific site. The technique in this reference has been proven to be fairly robust, is inherently updateable, and allows operator intervention if necessary. It does not require more than three years' worth of past data.

Based on the attractive features of both distributed artificial intelligence and existing load-forecasting techniques, a distributed problem-solving system for short-term load forecasting is presented in reference [31]. Such a distributed paradigm is a multi-agent system, each processing agent of which can compute autonomously and cooperate with other agents to reason an accurate and satisfactory solution for load forecasting. The designed load-forecasting system solves problems using three basic modules: a backboard module, knowledge sources, and a control mechanism. In this reference, the existing techniques are embedded in the domain knowledge source.

Reference [32] presents an algorithm using an unsupervised/supervised learning concept and the historical relationship between the load and temperature for a given season, day type, and hour of the day to forecast hourly electric load with a lead time of 24 hours. An additional approach using functional link net, temperature variables, average load, and the last one-hour load of the previous day is introduced and compared with the ANN model with one hidden-layer load forecast. Examination of load shapes indicates that the five working days, Saturdays, Sundays, and holidays should be separately treated.

References [33] and [34] present the applications of ANN to short-term load forecasting. Reference [33] investigates the effectiveness of ANN in short-term load forecasting. It has been shown that the application of a combined solution using artificial neural networks and expert systems yields a good short-term load forecast that neither system alone can provide.

Reference [34] applies another type of neural network, called the radial basis function network (RBFN), to the STLF. The results obtained using both a radial basis function network and backpropagation network (BPN) indicate that the RBFN model performs better than the BPN model. It is claimed that the RBFN model can also compute reliability measures, which is an added advantage of the RBFN model. These measures provide confidence intervals for the forecasts and an extrapolation index to determine when the model is extrapolating beyond its original training data.

Reference [35] presents an adaptive neural network–based short-term load-forecasting system. The system accounts for seasonal and daily characteristics, as well as abnormal conditions such as cold fronts, heat waves, holidays, and other conditions. The algorithm in this reference is capable of forecasting load with a lead time of one hour to seven days. The adaptive mechanism is used to train the neural networks when online.

Reference [36] presents an adaptive autoregressive moving average (ARMA) model for STLF of a power system. In this reference, the Box-Jenkins transfer function is considered as one of the better, more accurate methods, but it has limited accuracy without adapting the forecasting errors available to update the forecast. The adaptive approach first derives the error-learning coefficients by virtue of minimum mean square error (MMSE) theory and then updates the forecasts based on the one-step-ahead forecast errors and the coefficients. The proposed algorithm in this reference can deal with any unusual system condition. It is shown that the proposed adaptive ARMA is more accurate than the conventional Box-Jenkins approach.

Reference [37] presents a survey for applying fuzzy systems in power systems. It discusses five forecasting methods. These methods are also presented in reference [24]. Reference [38] presents a highly adaptable and robust short-term load-forecasting algorithm. Adaptive general exponential smoothing augmented with power spectrum analysis is used to account for the changing base load component. The algorithm includes an adaptive autoregressive modeling technique enhanced with partial autocorrelation analysis to model the random component of the load. The load consists of a base load, weather-sensitive load, and random load components. The Akaike information criterion (AIC) is employed to generate model parsimony. The weighted recursive least squares estimation algorithm with variable forgetting factors is applied to estimate model parameters. A nonlinear weather-sensitive model is used to present the influence of weather changes on energy consumption. This reference claims that the approach has the capacity to better track load-changing patterns, and the human intervention of this technique is a minimum, which enhances the suitability of the approach for online applications.

Reference [39] presents a hybrid model for short-term load forecasting that integrates artificial neural networks with fuzzy expert systems. The load is obtained in two steps. In the first step, the ANNs are trained with load patterns corresponding to the desired forecasted hour, and the trained ANNs obtain the provisional forecasted load. In the second step, the fuzzy expert systems modify the provisional forecasted load considering the possibility of load variation due to changes in temperature and the nature of the day if it is a holiday.

References [40,41] present a fuzzy system for STLF. The fuzzy system has the net structure and training procedures of a neural network and is called fuzzy neural network (FNN). An FNN initially creates a rule base from historical load data. The parameters of the rule base are then turned through a training process so that the output of the FNN matches the available historical load data adequately. Once trained, the FNN can be used to forecast future load.

Reference [42] proposes an optimal fuzzy inference method for short-term load forecasting. This reference constructs an optimal structure of the simplified fuzzy inference that minimizes model errors and the number of membership functions to grasp the nonlinear behavior of power system short-term loads. Simulated annealing and the steepest descent method identify the model parameters in this reference.

Reference [43] proposes an evolutionary programming (EP) approach to identify the parameters of an autoregressive moving average with exogenous (ARMAX) variable model for one-day- to one-week-ahead hourly load-demand forecasts. The surface of the forecasting error function possesses multiple local minimum points. Solutions of the traditional gradient search–based identification technique, therefore, may stall at the local optimal points, which results in an inadequate model. By simulating the natural evolutionary process, the EP algorithm offers the capability of converging toward the global extreme of a complex error surface. The results obtained using this approach indicate that this algorithm provides a method to simultaneously estimate the appropriate order and parameter values of the ARMAX model for diverse types of load data.

Reference [44] presents the application of ANN to determine the short-term load forecasting while paying attention to accurate modeling of holidays. A single neural network with 24 outputs is used for the short-term forecasting for all day types. Reference [45] compares three techniques: fuzzy logic (FL), neural network (NN), and autoregressive (AR) for very short-term load forecasting. The authors find a simple satisfying dynamic forecaster to predict the very short-term load trends online. FL and NN are good candidates for short-term load forecasting. A neural network technique for electric load forecasting based on weather compensation is presented in references [46] and [47]. The method is a nonlinear generalization of the Box-Jenkins approach for nonstationary time-series prediction. A nonlinear autoregressive integrated (NARI) model is identified to be the most appropriate model to include the weather compensation in short-term electric load forecasting. A weather compensation neural network based on a NARI model is implemented for one-day-ahead electric load forecasting. This weather compensation neural network can accurately predict the change of electric load consumption for the coming day. Based on the results obtained, the authors claim that this methodology is capable of providing a more accurate load forecast.

Previous experience with basic ANN architectures has shown that, even though these architectures provide results comparable with those obtained by human operators for most normal days, they show some deficiencies in the accuracy when applied to "anomalous" load conditions occurring during holidays and long weekends [48]. Reference [48] proposes a specific procedure based on a combined unsupervised/supervised approach. In the unsupervised stage, a preventive classification of historical load data by means of a Kohonen self-organizing map is provided, whereas in the supervised stage, the proper forecasting activity is obtained by using a multilayer perception with a backpropagation learning algorithm.

Reference [49] proposes a method for short-term load forecasting that would help demand-side management. The proposed method is based on the Kalman filtering algorithm with the incorporation of a "fading memory." The load is forecasted in two stages. In the first stage, the mean is first predicted, whereas in the second stage, a correction is incorporated in real time using error feedback from the previous hours. The authors claim that the proposed algorithm is suitable for developing countries where the total load is not large, especially at substation levels, and the data available are grossly inadequate. In this reference, the fading-memory Kalman filter assigns variable weight to past data. This causes reduction of the dependence on data far back into the past and also improves the accuracy of prediction to a certain extent. Also, it was suggested that the space for storage and the time taken for computation are both significantly low and make this method highly suitable for use in small computers.

Reference [50] compares two linear static parameter estimation techniques as they apply to the 24-hour offline forecasting problem. Three 24-hour load models are used. The least error squares and the least absolute value–based linear programming algorithm are the two parameter estimation approaches used to estimate the parameters of the three models. The three load models are (1) a multiple linear regression model, (2) harmonic decomposition model, and (3) hybrid multiple linear regression/harmonic decomposition model. It is concluded that if the data source is free of errors, both techniques produce the same degree of accuracy for the three models. However, if the data source is contaminated with gross errors, then the use of the least absolute value (LAV) criterion will result in greater prediction accuracy.

A method of forecasting the hourly load demand on power systems and use of threshold autoregressive models with a stratification rule is presented in [51]. When one uses the threshold model algorithm, fewer parameters are required to capture the random component in load dynamics. Based on the results obtained, the authors conclude that (a) the optimum stratification rule attempts to remove any judgmental input and to render the threshold process entirely mechanistic; and (b) the simplicity of the proposed threshold autoregressive model varies under different perspectives, such as the piecewise linear algorithms and the threshold procedures of the stratification to effectively handle nonstationary in the load variations. Therefore, the simplicity consists of finding architectures that are autoregressive to model the nonlinearity of the series and economical in terms of parameters. Finally, (c) the high level of achievement is due primarily to a more accurate AR modeling in a threshold model, resulting on the threshold AR model's ability to respond rapidly to sudden changes.

Reference [52] develops a forecasting model for one day ahead. This model identifies a "normal" or weather-insensitive load component, and a weather-sensitive load component–linear regression analysis of past load and weather is used to identify the normal load model. The weather-sensitive component of the load is estimated using the parameters of the regression analysis. In this reference, an automated load-forecasting system is presented that includes adaptability to changing operational conditions, computational economy, and robustness. Also presented in this reference are the monthly error statistics of forecast load for only one day ahead for recorded weather conditions.

Reference [53] presents a functional-link network-based short-term electric load-forecasting system for real-time implementation. The load and weather parameters are modeled as nonlinear ARMA processes, and parameters of this model are obtained using the functional approximation capabilities of an auto-enhanced functional link network. The adaptive mechanism with a nonlinear learning rule is used to train the link network online. The results obtained in this reference indicate that the functional link net-based load-forecasting system produces robust and more accurate load forecasts in comparison to a simple adaptive neural network or statistical-based approach.

Reference [54] describes a load-forecasting system called Artificial Neural Network Short-Term Load Forecasting (ANNSTLF). This system is suggested for use now by many utilities across North America. The effects of temperature and relative humidity on the load are considered. ANNSTLF also contains forecasts that can generate the hourly temperature and relative humidity forecasts needed by the system. ANNSTLF is based on a multiple ANN strategy that captures various trends in the data. The building block of the forecasters is a multilayer neural network trained with the error backpropagation learning rule. To adjust the ANN weights during online forecasting, it employs an adaptive scheme. The forecasting models are site-independent, and only the number of hidden-layer nodes of ANNs needs to be adjusted for a new database.

Reference [55] presents a "quasi-optimal" neural network to solve the short-term load-forecasting problem. Rules for building a "quasi-optimal" neural network to solve the short-term load forecasting are derived. It is demonstrated that the "quasi-optimal" neural network is superior to an automated Box-Jenkins seasonal ARIMA model in solving the STLF problem. Most significantly, the authors demonstrate how orthogonal fractional factorial designs can be used to understand how technical issues that arise in creating a neural network affect singularly and in pairs the performance of the network in solving the STLF problem.

An algorithm using a cascaded learning algorithm together with the historical load and weather data is presented in [56] to forecast half-hour load for the next 24 hours. This cascaded neural network algorithm (CANN) includes peak, minimum, and daily energy prediction as additional input data for the final forecast stage. These additional input data are predicted using the first ANN's model.

The use of a weighted least squares procedure when training a neural network to solve the short-term load-forecasting problem is presented in [57]. It is shown that a neural network that implements the weighted least squares procedure outperforms a neural network that implements the least squares procedure during the on-peak period for the two performance criteria specified—mean absolute error and cost—and during the entire period for the cost criterion. This reference shows the potential benefit of using a cost-based weighted least squares training approach.

Reference [58] postulates that the load can be modeled as the output of some dynamic system influenced by a number of weather, time, and other environmental variables. Recurrent neural networks, being members of a class of connectionist models exhibiting inherent dynamic behavior, can thus be used to construct empirical models for this dynamic system. This reference claims that due to the nonlinear dynamic nature of these models, the behavior of the load prediction system can be captured in a compact and robust representation.

Reference [59] presents a self-organizing model of fuzzy autoregressive moving average with exogenous (FARMAX) variables for one-day-ahead hourly load forecasting of power systems. A comparison between the existing and ARMAX model values shows reduction in error for forecasting results.

An efficient modeling technique based on fuzzy curve notation is presented in reference [60] to generate fuzzy models for short-term load forecasting. The steps in this approach are (a) prediction of the load curve extremals (peak and valley loads) using separate fuzzy models, (b) formulation of the representative day load curve to the forecasted peak values to obtain the predicted day load curves, and (c) transformation of the representative day load curve to fit the forecasted peak and valley loads to obtain the final next days' load curve forecast.

Reference [61] presents an approach to short-time load forecasting by the application of nonparametric regression. The method is derived from a load model in the form of a probability density function of load and load-affecting factors. A load forecast is a conditional expectation of load given the time, weather conditions, and other explanatory variables. This forecast can be calculated directly from historical data as a load average of past observed loads with the size of the local neighborhood and the specific weights on the load defined by a multivariate product kernel. The procedure requires a few parameters that can be easily calculated from historical data by applying the cross-validation technique.

Reference [62] describes a method for input variable selection for artificial neural network–based short-term load forecasting. The method is based on the phase-space embedding of a load-time series. The accuracy of the method is enhanced by the addition of temperature and cycle variables. This reference compares it favorably to the ones reported in the literature, indicating that a more parsimonious set of input variables can be used in STLF without sacrificing the accuracy of the forecast. This allows more compact ANNs, smaller training sets, and easier training.

Reference [63] studies a short-term electric load-forecasting technique using a multilayered feedforward ANN and a fuzzy set–based classification algorithm. The hourly data are subdivided into various classes of weather conditions using the fuzzy set representation of weather variables, and then the ANNs are trained and used to perform the load forecasting accurately up to 120 hours ahead.

Reference [64] presents an architecture that is substantially changed from the earlier neural network techniques. It includes only two ANN forecasters: one predicts the base load, and the other forecasts the change in load. The final forecast is computed by adaptive combinations of these two forecasts. The effects of humidity and wind speed are considered through a linear transformation of temperature. This algorithm significantly improves the accuracy of the holiday forecasts.

Reference [65] presents a method that is suitable for power system operational planning studies. Bayesian estimation is used to predict multiple-step-ahead peak forecasts using peak and average temperature forecasts as explanatory variables. Furthermore, the authors claim that better results can be obtained, with more attention paid to model identification.

Reference [66] describes the application of ANN in forecasting short-term load using a multilayer perception. ANN combines both time-series and regression approaches to predict load demand. A functional relationship between weather variable and electrical load is not needed because ANN can generate the functional relationship in learning and training the data.

A fuzzy modeling method is developed in reference [67] for short-term load forecasting. In this method, identification of the premise part and consequent part is separately accomplished via the orthogonal least squares (OLS) technique. The OLS is first employed to partition the input space and determine the number of fuzzy rules and the premise parameters. In the sequel, a second orthogonal estimator determines the input terms that should be included in the consequent part of each fuzzy rule and calculate its parameters. Different models are developed for each day type in every season.

Reference [68] presents a self-supervised adaptive neural network to perform STLF for a large power system covering a wide service area with several heavy load centers. The self-supervised network is used to extract the correlation features from temperature and load data. The authors' design provides a good adaptability using a rapid, online training mode that is crucial in applications, where the source statistics are nonstationary or where the forecaster is used with different power systems.

The behavior of electric power systems and networks varies considerably due to their characteristics. There does not appear to be one forecasting method that fits all power systems. In fact, the electric load on each system may be forecasted using different techniques to suit different situations.

Long-term electric peak-load forecasting is an important issue in effective and efficient planning. Over- or underestimation can greatly affect the revenue of the electric utility industry. Overestimation of the future load may lead to spending more money in building new power stations to supply this load. Moreover, underestimation of load may cause trouble in supplying this load from the available electric supplies and produce a shortage in the spinning reserve of the system that may lead to an insecure and unreliable system. Therefore, an accurate method is needed to forecast loads, as is an accurate model that takes into account the factors that affect the growth of the load over a number of years. Furthermore, an accurate algorithm is needed to estimate the parameters of such models. The growth in electricity consumption in many developing countries has outstripped existing projections, and accordingly, the uncertainties of forecasting have increased. Variables such as economic growth, population, and efficiency standards, coupled with other factors inherent in the mathematical development of forecasting models, make accurate projection difficult. Unfortunately, an accurate forecast depends on the judgment of the forecaster, and it is impossible to rely strictly on analytical procedures to obtain an accurate forecast.

The objective of the long-term load-forecasting task is to provide energy and peak-load predictions that meet planning requirements in a consistent and credible manner. Numerous techniques for short-term load forecasting (hour-by-hour forecasting) are available in the literature; they include the autoregressive moving average, Kalman filtering algorithm, artificial neural network, expert system, fuzzy systems, etc. A few of them have been applied to long-term annual load forecasting. These range from the simplest approach, such as use of the most recent observation as forecast, to highly complex approaches such as an econometric system of simultaneous equations.

The methods used for forecasting electrical peak load and energy for long-term planning fall within two main categories [69–84]—namely, the econometric method and extrapolation method. The least absolute value estimation algorithm is implemented to estimate the parameters of the annual peak load model. The model used is a function of the time only (one year is equivalent to one time step). Different orders for the peak-load models are developed. However, all of them are linear in the parameters to be estimated.

References

[1] M.E. El-Hawary, Control System Engineering, Reston Publishing Company, Inc., Reston, VA, 1984.

[2] R. Horst, P.M. Pardalos (Eds.), Handbook of Global Optimization, Kluwer Academic Publishers, Netherlands, 1995.

[3] B.C. Kuo, Automatic Control Systems, fourth ed., Prentice-Hall, Inc., Englewood Cliffs, NJ, 1982.

[4] G.L. Nemhauser, A.H.G. Rinnooy Kan, M.J. Todd (Eds.), Optimization, Elsevier Science Publishers, Netherlands, 1989.
[5] M.A. Wolfe, Numerical Methods, for Unconstrained Optimization: An Introduction, Van Nostrand Reinhold Company, New York, 1978.
[6] D.G. Zill, M.R. Cullen, Advanced Engineering Mathematics, PWS Publishing Company, Boston, 1992.
[7] M.A. Abu-El-Magd, N.K. Sinha, Short term load demand modeling and forecasting: a review, IEEE Trans. Syst. Man Cybern. 12 (3) (1982) 370–382.
[8] M.L. Willis, A.E. Schauer, J.E.D. Northcote, T.D. Vismor, Forecasting distribution system loads using curve shape clustering, IEEE Trans. Power Apparatus Syst. 102 (4) (1983) 893–901.
[9] M. Lee Willis, R.W. Powell, D.L. Wall, Load transfer coupling regression curve fitting for distribution load forecasting, IEEE Trans. Power Apparatus Syst. 103 (5) (1984) 1070–1076.
[10] M. Lee Willis, J.E.D Northcote-Green, Comparison tests of fourteen distribution load forecasting methods, IEEE Trans. Power Apparatus Syst. 103 (6) (1984) 1190–1197.
[11] S. Rahman, R. Bhatnager, An expert system based algorithm for short term load forecast, IEEE Trans. Power Syst. 3 (2) (1988) 392–399.
[12] Q.C. Lu, W.M. Grady, M.M. Crawford, G.M. Anderson, An adaptive nonlinear predictor with orthogonal escalator structure for short term load forecasting, IEEE Trans. Power Syst. 4 (1) (1989) 158–164.
[13] I. Moghram, S. Rahman, Analysis and evaluation of five short term load forecasting techniques, IEEE Trans. Power Syst. 4 (4) (1989) 1484–1491.
[14] N.F. Hubele, C.S. Cheng, Identification of seasonal short term load forecasting models using statistical decision functions, IEEE Trans. Power Syst. 5 (1) (1990) 40–45.
[15] M.E. El-Hawary, G.A.N. Mbamalu, Short term power system load forecasting using the iteratively re-weighted least squares algorithm, Electr. Power Syst. Res. 19 (1990) 11–22.
[16] S. Rahman, Formulation and analysis of a rule-based short-term load forecasting algorithm, IEE Proc. Gener. Transm. Distrib. 78 (5) (1990) 805–816.
[17] K.-l. Ho, Y.-Y. Hsu, C.-C. Lising, T.-S. Lai, Short term load forecasting of Taiwan power system using a knowledge based expert system, IEEE Trans. Power Syst. 5 (4) (1990) 1214–1221.
[18] A.D. Papalexopoulos, T.C. Hesterberg, A regression-based approach to short-term load forecasting, IEEE Trans. Power Syst. 5 (4) (1990) 1535–1550.
[19] J.H. Park, Y.M. Park, K.Y. Lea, Composite modeling for adaptive short term load forecasting, Power IEEE Trans. Syst. 6 (2) (1991) 450–457.
[20] Y.-Y. Msu, C.-C. Yang, Design of artificial neural networks for short term load forecasting. Part I: self-organizing feature maps for day type identification peak load and valley load forecasting, IEEE Proc. Gener. Trans. Distrib. 138 (5) (1991) 407–413.
[21] Y.-Y. Msu, C.-C. Yang, Design of artificial neural networks for short term load forecasting. Part II: multilayer feed forward networks for peak load and valley load forecasting, IEE Proc. 138 (5) (1991) 414–418.
[22] W.M. Grady, L.A. Groce, T.M. Huebner, Q.C. Lu, M.M. Crawford, Enhancement implementation and performance of an adaptive short term load forecasting algorithm, IEEE Trans. Power Syst. 6 (4) (1991) 1404–1410.
[23] K.Y. Lee, Y.T. Cha, J.H. Park, Short term load forecasting using an artificial neural network, IEEE Trans. Power Syst. 7 (1) (1992) 124–132.
[24] K.-L. Ho, Y.-Y. Msu, C.-C. Yang, Short term load forecasting using a multilayer neural network with an adaptive learning algorithm, IEEE Trans. Power Syst. 7 (1) (1992) 141–149.

[25] T.M. Peng, N.F. Huebele, G.G. Karady, Advancement in the application of neural networks for short-term load forecasting, IEEE Trans. Power Syst. 7 (1) (1992) 250–257.
[26] S. Rahman, I. Drezga, Identification of a standard for comparing short-term load forecasting techniques, Electr. Power Syst. Res. 25 (1992) 149–158.
[27] S.T. Chen, D.C. Yu, A.R. Moghaddamjo, Weather sensitive short-term load forecasting using non-fully connected artificial neural network, IEEE Trans. Power Syst. 7 (3) (1992) 1098–1105.
[28] Y.-Y. Hsu, K.L. Ho, Fuzzy expert systems: an application to short-term load forecasting, IEEE Proc. C 139 (6) (1992) 471–477.
[29] C.N. Lu, H.T. Wu, S. Vemuri, Neural network based short-term load forecasting, IEEE Trans. Power Syst. 8 (1) (1993) 336–342.
[30] S. Rahman, O. Hazim, A generalized knowledge based short-term load forecasting technique, IEEE Trans. Power Syst. 8 (2) (1993) 508–514.
[31] J.-L. Chen, R. Tsai, S.-S. Liang, A distributed problem solving system for short-term load forecasting, Electr. Power Syst. Res. 26 (1993) 219–224.
[32] M. Djukanovic, B. Babic, D.J. Sobajic, Y.-H. Pao, Unsupervised/supervised learning concept for 24–hour load forecasting, IEE Proc. Gener. Transm. Distrib. C 140 (4) (1993) 311–318.
[33] Azzam-ul-Asar, J.R. McDonald, A specification of neural network applications in the load forecasting problem, IEEE Trans. Control Syst. Technol. 2 (2) (1994) 135–141.
[34] D.K. Ranaweara, N.F. Hubele, A.D. Pagalexopoules, Application of radical basis function neural network model for short-term load forecasting, IEE Proc. Gener. Transm. Distrib. 142 (1) (1995) 45–50.
[35] O. Mohamed, D. Park, R. Merchant, T. Dinh, C. Tong, A. Azeem, et al., Practical experiences with an adaptive neural network short-term load forecasting system, IEEE Trans. Power Syst. 10 (1) (1995) 254–265.
[36] J.-F. Chen, W.-M. Wang, C.-M. Huang, Analysis of an adaptive time-series autoregressive moving-average (ARMA) model for short-term load forecasting, Electr. Power Syst. Res. 34 (1995) 187–196.
[37] D. Srinivasan, A.C. Liew, C.S. Chang, Applications of fuzzy systems in power systems, Electr. Power Syst. Res. J. 35 (1995) 39–43.
[38] A.A. El-Keib, X. Ma, H. Ma, Advancement of statistical based modeling techniques for short term load forecasting, Electr. Power Syst. Res. J. 35 (1995) 51–58.
[39] K.-H. Kim, J.-K. Park, K.-J. Hwang, S.-H. Kim, Implementation of hybrid short-term load forecasting system using artificial neural networks and fuzzy expert systems, IEEE Trans. Power Syst. 10 (3) (1995) 1534–1539.
[40] A.G. Bakirtzis, J.B. Theocharis, S.J. Kiartzis, K.J. Satsios, Short term load forecasting using fuzzy neural networks, IEEE Trans. Power Syst. 10 (3) (1995) 1518–1524.
[41] J.A. Momoh, K. Tomsovic, Overview and literature survey of fuzzy set theory in power systems, IEEE Trans. Power Syst. 10 (3) (1995) 1676–1690.
[42] H. Mori, H. Kobayashi, Optimal fuzzy inference for short term load forecasting, IEEE Trans. Power Syst. 11 (1) (1996) 390–396.
[43] H.-T. Yang, C.-M. Haung, C.-L. Haung, Identifications of ARMAX model for short term load forecasting: an evolutionary programming approach, IEEE Trans. Power Syst. 11 (1) (1996) 403–408.
[44] A.G. Bakirtzis, V. Petrldls, S.J. Klartzls, M.C. Alexladls, A.H. Malssis, A neural network short term load forecasting model for the Greek power system, IEEE Trans. Power Syst. 11 (2) (1996) 858–863.

[45] K. Liu, S. Subbaratan, R.R. Shoults, M.T. Manry, C. Kwan, F.L. Lewis, et al., Comparison of very short-term load forecasting techniques, IEEE Trans. Power Syst. 11 (2) (1996) 877–882.
[46] T.W.S. Chow, C.T. Leung, Nonlinear autoregressive integrated neural network model for short term load forecasting, IEE Proc. Gener. Transm. Distrib. 143 (5) (1996) 500–506.
[47] T.W.S. Chow, C.T. Leung, Neural network based short-term load forecasting using weather compensation, IEEE Trans. Power Syst. 11 (4) (1996) 1736–1742.
[48] R. Lamedica, A. Prudenzi, M. Sforna, M. Caciotta, V. Orsolini Cencellli, A neural network based technique for short-term forecasting of anomalous load periods, IEEE Trans. Power Syst. 11 (4) (1996) 1749–1756.
[49] S. Sargunaraj, D.P. Sen Gupta, S. Devi, Short-term load forecasting for demand side management, IEE Proceedings. Gener. Transm. Distrib. 144 (1) 68–74, January 1997.
[50] S.A. Soliman, S. Persaud, K. EL-Nagar, M.E. EL-Hawary, Application of least absolute value parameter estimation based on linear programming to short-term load forecasting, Electr. Power Energy Syst. 19 (3) (1997) 209–216.
[51] S.R. Huang, Short-term load forecasting using threshold autoregressive models, Online, IEE Proc. Gener. Transm. Distrib. 144 (477) (1997) 1144.
[52] O. Hyde, P.F. Hodnett, An adaptive automated procedure for short-term electricity load forecasting, IEEE Trans. Power Syst. 12 (1) (1997) 84–94.
[53] P.K. Dash, H.P. Satpathy, A.C. Liew, S. Rahman, A real-time short-term load forecasting system using functional link network, IEEE Trans. Power Syst. 12 (1997) 675–680.
[54] A. Khotanzad, R. Afkhami-Rohani, T.-L. Lu, A. Abaye, M. Davis, D.J. Maratukulam, ANNSTLF—a neural- network basic electric load forecasting system, IEEE Trans. Neural Netw. 8 (4) (1997) 835–846.
[55] M. Hisham Choueiki, C.A. Mount-Campbell, S.C. Ahalt, Building a quasi optimal neural network to solve short-term load forecasting problem, IEEE Trans. Power Syst. 12 (4) (1997) 1432–1439.
[56] A.S. Al-Fuhaid, M.A. EL-Sayed, M.S. Mahmoud, Cascaded artificial neural networks for short-term load forecasting, IEEE Trans. Power Syst. 12 (4) (1997) 1524–1529.
[57] M. Hisham Choueiki, C.A. Mount-Campbell, S.C. Ahalt, Implementing a weighted least squares procedure in training a neural network to solve the short-term load forecasting problem, IEEE Trans. Power Syst. 12 (4) (1997) 1689–1694.
[58] J. Vermaak, E.C. Botha, Recurrent neural networks for short-term load forecasting, IEEE Transactions on Power Systems, 13, (1) (February 1998) 126–132.
[59] Hong-Tzer Yang, Chao-Ming Haung, A new short-term load forecasting approach using self-organizing fuzzy ARMAX models, IEEE Trans. Power Syst. 13 (1) (1998) 217–225.
[60] S.E. Papadakis, J.B. Theocharis, S.J. Kiartzis, A.G. Bakertzis, A novel approach to short-term load forecasting using fuzzy neural networks, IEEE Trans. Power Syst. 13 (2) (1998) 480–492.
[61] W. Charytoniuk, M.S. Chen, P. Van Olinda, Nonparametric regression based short-term load forecasting, IEEE Trans. Power Syst. 13 (3) (1998) 725–730.
[62] I. Drezga, S. Rahman, Input variable selection for ANN-based short-term load forecasting, IEEE Trans. Power Syst. 13 (4) (1998) 1238–1244.
[63] M. Daneshdoost, M. Lotfalian, G. Bumroonggit, J.P. Ngoy, Neural network with fuzzy set-based classification for short-term load forecasting, IEEE Trans. Power Syst. 13 (4) (1998) 1386–1391.
[64] A. Khotanzed, R. Afkhami-Rohani, D. Maratukulam, ANNSTLF—artificial neural network short-term load forecasting—generation three, IEEE Trans. Power Syst. 13 (4) (1998) 1413–1422.

[65] A.P. Douglas, A.M. Breipohl, F.N. Lee, R. Adapa, The impacts of temperature forecast uncertainty on bayesian load forecasting, IEEE Trans. Power Syst. 13 (4) (1998) 1507–1513.
[66] R. Aggarwal, Y. Song, Artificial neural networks in power systems—Part 3: examples of applications in power systems, Tutorial: ANNs in power systems, Power Eng. J. (1998) 279–287.
[67] P.A. Mastorocostas, J.B. Theocharis, A.G. Bakirtzis, Fuzzy modeling for short-term load forecasting using the orthogonal least squares method, IEEE Trans. Power Syst. 14 (1) (1999) 29–36.
[68] H. Yoo, R.L. Pimmel, Short-term load forecasting using a self-supervised adaptive neural network, IEEE Trans. Power Syst. 14 (2) (1999) 779–784.
[69] H.L. Willis, L.A. Finley, M.J. Buri, Forecasting electric demand of distribution system in rural and sparsely populated regions, IEEE Trans. Power Syst. 10 (4) (1995) 2008–2013.
[70] P.H. Henault, R.B. Eastvedt, J. Peschon, L.P. Hajdu, Power system long term planning in the presence of uncertainty, IEEE Trans. Power Apparatus Syst. PAS-89 (1970) 156–164.
[71] G.S. Christensen, A. Rouhi, S.A. Soliman, A new technique for unconstrained and constrained LAV parameter estimation, Can. J. Elect. Comp. Eng. 14 (1) (1989) 24–30.
[72] Ministry of Electricity and Energy, Egyptian electricity authority, load and energy forecast for the period 1996/1997 to 2009/2010, Report, February 1998.
[73] H.K. Temraz, K.M. El-Nagar, M.M.A. Salama, Applications of non-iterative least absolute value estimation for forecasting annual peak electric power demand, Can. J. Elect. Comp. Eng. 23 (4) (1998) 141–146.
[74] D. Srinivasan, T.S. Swee, C.S. Cheng, E.K. Chan, Parallel neural network-fuzzy expert system strategy for short-term load forecasting: system implementation and performance evaluation, IEEE Trans. Power Syst. 14 (3) (1999) 1100–1106.
[75] I. Drezga, S. Rahman, Short-term load forecasting with local ANN predictors, IEEE Trans. Power Syst. 14 (3) (1999) 844–850.
[76] A.A. Ding, Neural-network prediction with noisy predictors, IEEE Trans. Neural Netw. 10 (5) (1999) 1196–1203.
[77] H.C. Wu, C. Lu, Automatic fuzzy model identification for short term load forecast, IEE Proc. Gener. Transm. Distrib. 146 (5) (1999) 477–482.
[78] W. Charytoniuk, M.S. Chen, Very short-term load forecasting using artificial neural networks, IEEE Trans. Power Syst. 15 (1) (2000) 263–268.
[79] K.H. Kim, H.S. Youn, Y.C. Kang, Short-term forecasting for special days in anomalous load conditions using neural networks and fuzzy inference method, IEEE Trans. Power Syst. 15 (2) (2000) 559–565.
[80] P.A. da Silva, L.S. Moulin, Confidence intervals for neural network based short-term load forecasting, IEEE Trans. Power Syst. 15 (4) (2000) 1191–1196.
[81] S.A. Villalba, C.A. Bel, Hybrid demand model for load estimation and short-term load forecasting in distribution electric systems, IEEE Trans. Power Syst. 15 (2) (2000) 764–769.
[82] R.H. Liang, C.C. Cheng, Combined regression-fuzzy approach for short-term load forecasting, IEE Proc. Gener. Transm. Distrib. 147 (4) (2000) 261–266.
[83] H.S. Hippert, C.E. Pedreira, R.C. Souza, Neural networks for short-term load forecasting: a review and evaluation, IEEE Trans. Power Syst. 16 (1) (2001) 44–55.
[84] M. Huang, H.T. Yang, Evolving wavelet-based networks for short-term load forecasting, IEE Proc. Gener. Transm. Distrib. 148 (3) (2001) 222–228.

2 Static State Estimation

2.1 Objectives

The main objectives of this chapter are

- Introducing the static state estimation problem and different techniques used for its solution.
- Becoming familiar with formulation of linear static estimation problems:
  - Least error squares (LES) algorithm as a technique to solve this linear estimation problem.
  - Linear constrained LES problem and its solution.
  - Recursive linear constrained LES problem and its solution.
  - Nonlinear LES problem and its solution.
  - Least absolute value (LAV) algorithm to solve the linear static estimation problem.
  - Linear constrained LAV problem and its solution.
  - Nonlinear LAV problem and its solution.

State estimation is the process of assigning a value to unknown system state variables and filtering out erroneous measurements before they enter into the computing process. A commonly used and familiar criterion in state estimation is that of minimizing the sum of the squares of the difference between the estimated and true (measured) value of a function. The concept of least error squares estimation has been known and used since the middle of the eighteenth century. Another valuable technique of state estimation is based on minimizing the absolute value of the difference between the measured and calculated quantities, and we call it the least absolute value. A few authors have developed a least absolute value estimator based on linear programming. These estimators, initially, were not widely accepted because they require excessive memory space, which was not available at that time, and they require computing time. The main advantage of the LAV algorithm, as we will see in this chapter, is the ability of the algorithm to reject the bad data points in the estimation process; that is, it is insensitive to outliers.

2.2 The Static Estimation Problem Formulation

The static estimation problem can be simply stated as: Given the system measurements described by the linear equation

$$\underline{Z} = H\underline{\theta} + \underline{v} \tag{2.1}$$

where Z is an $m \times 1$ vector of system measurements (known), $\underline{\theta}$ is an $n \times 1$ vector of parameters to be estimated (unknown), H is an $m \times n$ matrix describing the

DOI: 10.1016/B978-0-12-381543-9.00002-6

mathematical relationship between the measurements and the system parameters vector (known), and $\underline{v}$ is an $m \times 1$ vector of measurement errors (unknown) to be minimized. The task is to estimate the parameter vector $\underline{\theta}$, which minimizes the error vector $\underline{v}$.

The best parameter estimate $\hat{\theta}$ must be chosen to minimize a given objective or cost function. A general form of the cost function is

$$J_p(\underline{\theta}) = \left\{ \sum_{i=1}^{m} |Z_i - H_i\theta|^p \right\}^{\frac{1}{p}} \tag{2.2}$$

or

$$J_p(\underline{\theta}) = \left\{ \sum_{i=1}^{m} |r_i(\underline{\theta})|^p \right\}^{\frac{1}{p}} \tag{2.3}$$

where

$J_p(\underline{\theta})$ = the cost function to be minimized;
p = some number ≥ 1, which defines the nature of the cost function;
Z_i = the ith measurement;
H_i = the row of H corresponding to the ith measurement;
r_i = the residual of the ith measurement; that is, $r_i = Z_i - H_i\underline{\theta}$.

For $p = 2$, the cost function is the sum of the squares of the residuals. For $p = 1$, the cost function is the sum of the absolute values of the residuals.

If the number of measurements (m) equals the number of unknown parameters (n), then a solution for $\underline{\theta}$ can be obtained as

$$\underline{\hat{\theta}} = [H]^{-1}\underline{Z} \tag{2.4}$$

For this problem, the estimated parameters vector exactly fits the measurements set, that is,

$$\underline{Z} - H\underline{\hat{\theta}} = \underline{v} = 0 \tag{2.5}$$

2.2.1 Linear Least Error Squares Estimation

In most cases, the number of measurements (m) exceeds the number of system parameters (n); that is, $m > n$. Thus, the measurement error can be filtered out in the estimation process, and good-quality estimates are obtained. In the LES estimation, the objective is to minimize the sum of the squares of the errors or residuals.

For $p = 2$, equation (2.2) can be rewritten in vector form as

$$J_2(\underline{\theta}) = \frac{1}{2}\{(\underline{Z} - H\underline{\theta})^T(\underline{Z} - H\underline{\theta})\}^{\frac{1}{2}} \tag{2.6}$$

or

$$J_2'(\underline{\theta}) = \frac{1}{2}\{(\underline{Z} - H\underline{\theta})^T(\underline{Z} - H\underline{\theta})\} \tag{2.7}$$

It should be noted that minimizing the sum of the squares is equivalent to minimizing the square root of the sum of the squares [1,2].

Setting the first derivatives of equation (2.7), $dJ_2'(\theta)/d\theta$, to zero yields

$$-H^T\underline{Z} - H^T\underline{Z} + 2H^TH\hat{\underline{\theta}} = 0 \tag{2.8}$$

This gives

$$\hat{\underline{\theta}} = \left[H^TH\right]^{-1}H^T\underline{Z} \tag{2.9}$$

or

$$\hat{\underline{\theta}} = [H]^+\underline{Z} \tag{2.10}$$

where $[H]^+ = [H^T H]^{-1} H^T$ is called the left pseudo-inverse of H, and $\hat{\underline{\theta}}$ is the optimal or best LES estimate of $\underline{\theta}$ [1–3]. It should be noted that the second-order partial derivative is given by

$$\frac{\partial^2 J'(\underline{\theta})}{\partial \underline{\theta}^2} = H^TH \tag{2.11}$$

This matrix is positive definite as long as H is of full-column rank, the rank of $H = n$. Therefore, the value of $\hat{\underline{\theta}}$ given by equation (2.10) is unique and minimizes $J_2(\underline{\theta})$.

An LES estimator finds the mean value of a set of measurements [1]. The mean value is generally accepted to be the best estimate when the set of measurements has a Gaussian error distribution. However, for other error distributions, the LES will not produce the best estimate [2]. The LES estimate is also adversely affected by the presence of bad data, and most LS estimators use some sort of bad data suppression.

2.2.2 Weighted Linear Least Error Squares (WLES) Estimation

In the LES, explained in Section 2.3, all measurements are treated equally. In this case, less accurate measurements will affect the outcome as much as the more accurate measurements. As a result, the final set of data obtained from the least error squares estimation process will contain large errors due to the influence of bad measurements that contain gross errors. When one introduces a weighting matrix to distinguish the more accurate measurements from the less accurate measurements, the process can then force the results to coincide with the more accurate measurements. A sensible way of choosing the weights is to make them inversely proportional to the variance of the measurements. This approach means that a larger weighting is placed on measurements with smaller variance and a smaller weighting on measurements with larger variance.

The cost function to be minimized, in this case, is given as

$$J_2(\underline{\theta}) = \sum_{i=1}^{m} \frac{(Z_i - H_i\underline{\theta})^2}{\sigma_i^2} \tag{2.12}$$

or

$$J_2(\theta) = \sum_{i=1}^{m} \omega_i (Z_i - H_i\underline{\theta})^2 \tag{2.13}$$

where

σ_i is the standard deviation of the ith measuring device;
σ_i^2 is the variance of the ith measurement;
ω_i is the weight assigned to the ith measurement.

In vector form, equation (2.13) can be rewritten as

$$J_2(\underline{\theta}) = (\underline{Z} - H\underline{\theta})^T W (\underline{Z} - H\underline{\theta}) \tag{2.14}$$

It can be shown that the optimal weighted least error squares estimate is given by

$$\hat{\underline{\theta}} = \left[H^T W H\right]^{-1} H^T W \underline{Z} \tag{2.15}$$

The following examples, from the area of curve fitting, illustrate the linear LES.

Example 2.1

Given a set of observations $\underline{Z}$ (measurements $\underline{Z}$) as

$$\underline{Z}^T = [3 \quad 7 \quad 5 \quad 6]$$

estimate the parameters a_1 and a_2, if the measurement equation is given by

$$Z_i = a_1 x_i + a_2, \quad i = 1, \ldots, 4$$

where $\underline{x}$ is given by

$$\underline{x}^T = [1 \quad 2 \quad 3 \quad 4]$$

The measurement equation can be written as

$$\underline{Z} = H\underline{\theta}$$

where

$$H = \begin{bmatrix} 1 & 1 \\ 2 & 1 \\ 3 & 1 \\ 4 & 1 \end{bmatrix}, \quad \underline{\theta} = \begin{bmatrix} a_1 \\ a_2 \end{bmatrix}$$

Substituting for H and $\underline{Z}$ into equation (2.9), we obtain the LES solution for $\underline{\theta}$ as

$$\hat{\underline{\theta}} = \begin{bmatrix} 0.7 \\ 3.5 \end{bmatrix} = \begin{bmatrix} \hat{a}_1 \\ \hat{a}_2 \end{bmatrix}$$

The residual vector generated from this solution is

$$\underline{v} = \underline{Z} - H\hat{\underline{\theta}} = \begin{bmatrix} -1.2 \\ 2.1 \\ -0.6 \\ -0.3 \end{bmatrix}$$

Example 2.2

Estimate the parameters a_1 and a_2, given by the measurement equation as

$$Z_i = a_1 x_i + a_2$$

Assume that the following observations are provided:

$$(x_i, Z_i) = \{(0, 1.52), (1, 1.025), (2, 0.475), (3, 0.01), (4, -0.075), (5, -1.005)\}$$

The measurement equation can be written as

$$\underline{Z} = H\underline{\theta}$$

where

$$\underline{Z} = \begin{bmatrix} 1.520 \\ 1.025 \\ 0.475 \\ 0.010 \\ -0.475 \\ -1.005 \end{bmatrix}, \quad H = \begin{bmatrix} 0 & 1 \\ 1 & 1 \\ 2 & 1 \\ 3 & 1 \\ 4 & 1 \\ 5 & 1 \end{bmatrix}, \text{ and } \underline{\theta} = \begin{bmatrix} a_1 \\ a_2 \end{bmatrix}$$

Substituting in equation (2.9) or (2.10), we obtain the LES solution as

$$\hat{\underline{\theta}} = \begin{bmatrix} -0.5026 \\ 1.5148 \end{bmatrix} = \begin{bmatrix} \hat{a}_1 \\ \hat{a}_2 \end{bmatrix}$$

The LES residuals vector generated from this solution is

$$\underline{v} = \underline{Z} - H\underline{\hat{\theta}} = \begin{bmatrix} 0.00524 \\ 0.01281 \\ -0.03462 \\ 0.00295 \\ 0.02052 \\ -0.00690 \end{bmatrix}$$

Several examples from the area of curve fitting can be found in reference [3].

2.2.3 Constrained Least Error Squares (CLES) Estimation

In this section, we discuss the solution of the constrained linear least error squares problem, which requires finding the state vector $\underline{\hat{\theta}}$ that minimizes [1,3]

$$J_2(\theta) = \frac{1}{2}(\underline{Z} - H\underline{\theta})^T(\underline{Z} - H\underline{\theta}) \tag{2.16}$$

subject to satisfying linear constraints given by

$$C\underline{\theta} = \underline{d} \tag{2.17}$$

where

C is an $l \times n$ matrix that represents the relation between $\underline{\theta}$ and $\underline{d}$;
d is an $l \times 1$ vector that represents the constraints measurement.

We can form an augmented cost function by adjoining the equality constraints of equation (2.17) to the cost function in equation (2.16) via Lagrange's multiplier vector λ to obtain

$$J_2(\underline{\theta}) = \frac{1}{2}(\underline{Z} - H\underline{\theta})^T(\underline{Z} - H\underline{\theta}) + \underline{\lambda}^T(C\underline{\theta} - \underline{d}) \tag{2.18}$$

The cost function of equation (2.18) is a minimum when

$$\partial J_2(\underline{\theta})/\partial\underline{\theta} = 0 = \frac{1}{2}\left[-2H^T\underline{Z} + 2H^TH\underline{\theta}\right] + C^T\lambda$$

which gives

$$\underline{\hat{\theta}} = \left[H^TH\right]^{-1}\left[H^T\underline{Z} - C^T\underline{\lambda}\right] \tag{2.19}$$

The Lagrange's multiplier vector $\underline{\lambda}$ is to be determined such that the equality constraints of equation (2.17) are satisfied. Premultiplying equation (2.19) by C, we obtain

$$\underline{\lambda} = \left[C\left[H^T H\right]^{-1} C^T\right]^{-1} \left[C\left[H^T H\right]^{-1} H^T \underline{Z} - \underline{d}\right] \tag{2.20}$$

Thus, the state $\hat{\underline{\theta}}$ can be obtained by substituting equation (2.20) into (2.19) to obtain

$$\hat{\underline{\theta}} = \left[H^T H\right]^{-1} \left[H^T \underline{Z} - C^T \left[C\left[H^T H\right]^{-1} C^T\right]^{-1} \left[C\left[H^T H\right]^{-1} H^T \underline{Z} - \underline{d}\right]\right] \tag{2.21}$$

One should notice that if $C = 0$, there are no constraints; then $\hat{\underline{\theta}}$ turns out to be the optimal estimate for the unconstrained linear least error squares estimates given by equation (2.9). The following example illustrates the application of constrained LES estimation.

Example 2.3

Design an estimator of which the equation is given by

$$Z(x) = a_1 + a_2 x$$

to fit

$$\{(1, 2), (2, 2), (3, 3), (4, 4), (5, 3)\}$$

and satisfy the following constraint:

$$Z(6) = 5$$

Solution

The estimator equation can be written as

$$\underline{Z} = H\underline{\theta}$$

where

$$\underline{Z} = \begin{bmatrix} 2 \\ 2 \\ 3 \\ 4 \\ 3 \end{bmatrix}, \quad H = \begin{bmatrix} 1 & 1 \\ 1 & 2 \\ 1 & 3 \\ 1 & 4 \\ 1 & 5 \end{bmatrix}, \quad \underline{\theta} = \begin{bmatrix} a_1 \\ a_2 \end{bmatrix}$$

The equality constraints can be written as

$$C\underline{\theta} = \underline{d}$$

where

$$C = [1 \quad 6], \quad \underline{d} = [5]$$

The estimator solution is

$$\hat{\underline{\theta}} = \begin{bmatrix} 0.964 \\ 0.673 \end{bmatrix} = \begin{bmatrix} \hat{a}_1 \\ \hat{a}_2 \end{bmatrix}$$

The residuals vector from this solution is

$$\hat{\underline{v}} = \begin{bmatrix} 0.363 \\ -0.310 \\ 0.017 \\ 0.344 \\ -1.329 \end{bmatrix}$$

2.2.4 Recursive Least Error Squares (RLES) Estimation

The least error squares estimators described earlier are "batch processing" algorithms, in that all measurements are processed together to provide the estimate of a constant vector. If a new measurement appears, one way to solve the problem is to append the new data to $\underline{Z}$ and repeat the entire process. As a more efficient alternative, the prior estimate can be used as the starting point for a sequential estimation algorithm that assigns proper relative weighting to the old and new data.

Given m_1 measurements, $\underline{Z}_1$; the corresponding model and weight matrices, H_1 and W_1; and the resulting estimate $\hat{\underline{\theta}}_1$,

$$\underline{Z}_1 = H_1\underline{\theta}_1 + \underline{v}_1 \tag{2.22}$$

$$\hat{\underline{\theta}}_1 = \left[H_1^T W_1 H\right]^{-1} H_1^T W_1 \underline{Z}_1 \tag{2.23}$$

The new measurements vector $\underline{Z}_2$ with dimension m_2 is

$$\underline{Z}_2 = H_2\hat{\underline{\theta}}_2 + \underline{v}_2 \tag{2.24}$$

W_2 is $m_2 \times m_2$ containing the weights assigned for the new measurements. The overall $(m_1 + m_2)$ measurement vector is given by

$$\underline{Z} = \begin{bmatrix} \underline{Z}_1 \\ \underline{Z}_2 \end{bmatrix} \tag{2.25}$$

The cost function for the whole set can be partitioned as

$$J(\underline{Z}_1, \underline{Z}_2) = \frac{1}{2}\left[\left(\underline{Z}_1 - H_1\underline{\hat{\theta}}_1\right)^T \left(\underline{Z}_2 - H_2\underline{\hat{\theta}}_2\right)^T\right]\begin{bmatrix} W_1 & 0 \\ 0 & W_2 \end{bmatrix}\begin{bmatrix} \underline{Z}_1 - H_1\underline{\hat{\theta}}_1 \\ \underline{Z}_2 - H_2\underline{\hat{\theta}}_2 \end{bmatrix} \tag{2.26}$$

In equation (2.26), $\underline{\hat{\theta}}_2$ is the state estimate obtained by using all the measurements. Taking the derivatives of $J(\underline{Z}_1, \underline{Z}_2)$ and setting the values equal to zero provides the least squares estimate $\underline{\hat{\theta}}_2$,

$$\underline{\hat{\theta}}_2 = \underline{\hat{\theta}}_1 + \underline{K}_2\left(\underline{Z}_2 - H_2\underline{\hat{\theta}}_1\right) \tag{2.27}$$

where K_2 is the recursive weighted least squares estimator gain matrix given by

$$\underline{K}_2 = \underline{P}_1 H_2^T \left(H_2 \underline{P}_1 H_2^T + \underline{W}_2^{-1}\right)^{-1} \tag{2.28}$$

and

$$P_1^{-1} = H_1^T \underline{W}_1 H_1 \tag{2.29}$$

Equation (2.27) looks like a recursive filter, and measurements taken over a period of time could be used to update the old estimate as they appear. Redefining k as a time index and letting the observation vector at time k have r components, the recursive mean value estimator is

$$\underline{\hat{\theta}}_k = \underline{\hat{\theta}}_{k-1} + \underline{K}_k\left(\underline{Z}_k - H_k\underline{\hat{\theta}}_{k-1}\right) \tag{2.30}$$

with

$$\underline{K}_k = \underline{P}_{k-1} H_k^T \left(H_k \underline{P}_{k-1} H_k^T + \underline{W}_k^{-1}\right)^{-1} \tag{2.31}$$

and

$$\underline{P}_k = \left(\underline{P}_{k-1}^{-1} + H_k^T \underline{W}_k H_k\right)^{-1} \tag{2.32}$$

Note that $\underline{K}_k$ is an $(n \times r)$ gain matrix, and $\underline{P}_k$ is an $(n \times n)$ matrix that represents the estimation error at the kth sampling instant.

2.2.5 Nonlinear Least Error Squares (NLLES) Estimation

In the preceding sections, we discussed the linear least error squares estimation problem, where there is a direct linear relationship between the measurement value and the estimation parameters so that the solution is obtained directly without any iteration [6]. If the relationship between the measurements and the measurement parameters is nonlinear, we need to linearize the cost function by using first-order Taylor series expansion and solving the nonlinear parameter estimation problem using the linear least error squares algorithm explained earlier.

The nonlinear least squares problem is to estimate the parameter vector $\underline{\theta}$, which minimizes

$$J_2(\underline{\theta}) = \sum_{i=1}^{m} \frac{(Z_i - f_i(\underline{\theta}))^2}{\sigma_i^2} \tag{2.33}$$

The gradient of $J_2(\underline{\theta})$ is given by

$$\nabla J_2(\underline{\theta}) = \begin{bmatrix} \frac{\partial J_2(\underline{\theta})}{\partial \theta_1} \\ \frac{\partial J_2(\underline{\theta})}{\partial \theta_2} \\ \vdots \\ \frac{\partial J_2(\underline{\theta})}{\partial \theta_m} \end{bmatrix} \tag{2.34}$$

This can be rewritten as

$$\frac{1}{2}\nabla J_2(\underline{\theta}) = \begin{bmatrix} \frac{\partial f_1}{\partial \theta_1} & \frac{\partial f_1}{\partial \theta_2} & \cdots \\ \frac{\partial f_2}{\partial \theta_1} & \frac{\partial f_2}{\partial \theta_2} & \cdots \\ \frac{\partial f_m}{\partial \theta_1} & \frac{\partial f_m}{\partial \theta_2} & \cdots \end{bmatrix} \begin{bmatrix} \frac{1}{\sigma_1^2} & & \\ & \frac{1}{\sigma_2^2} & \\ & & \frac{1}{\sigma_m^2} \end{bmatrix} \begin{bmatrix} Z_1 - f_1(\theta) \\ Z_1 - f_1(\theta) \\ Z_1 - f_1(\theta) \end{bmatrix} \tag{2.35}$$

Equation (2.35) can be rewritten in compact form as

$$\nabla J_2(\underline{\theta}) = -2H^T W \Delta \underline{Z} \tag{2.36}$$

where we define the $m \times n$ matrix H as

$$H = \begin{bmatrix} \frac{\partial f_1}{\partial \theta_1} & \frac{\partial f_1}{\partial \theta_2} & \cdots & \frac{\partial f_1}{\partial \theta_n} \\ \frac{\partial f_2}{\partial \theta_1} & \frac{\partial f_2}{\partial \theta_2} & \cdots & \frac{\partial f_2}{\partial \theta_n} \\ \cdots & & & \\ \frac{\partial f_m}{\partial \theta_1} & \frac{\partial f_m}{\partial \theta_2} & \cdots & \frac{\partial f_m}{\partial \theta_n} \end{bmatrix} \tag{2.37}$$

$=$ Jacobian of $f(\underline{\theta})$.

$$W = m \times m \text{ weighting matrix} = \begin{bmatrix} \frac{1}{\sigma_1^2} & & \\ & \frac{1}{\sigma_2^2} & \\ & & \frac{1}{\sigma_m^2} \end{bmatrix}$$, and $\Delta \underline{Z} = m \times 1$ is a matrix of the difference between the measured values and the estimated values.

To make $\nabla J(\underline{\theta})$ equal to zero, we implement the Newton-Raphson method. Thus, we obtain

$$\Delta \underline{\theta} = \left[\frac{\partial \nabla J(\underline{\theta})}{\partial \underline{\theta}} \right]^{-1} [-\nabla J_2(\underline{\theta})] \tag{2.38}$$

The Jacobian matrix of $\nabla J_2(\theta)$ is calculated by treating $[H]$ as a constant matrix. Thus,

$$\frac{\partial \nabla J_2(\underline{\theta})}{\partial \underline{\theta}} = -2H^T W[-H]$$

Then

$$\Delta \underline{\hat{\theta}} = \left[H^T W H\right]^{-1} \left[H^T W \Delta \underline{Z}\right] \tag{2.39}$$

The procedures to solve the nonlinear state estimation problem can be stated as follows:

Step 1 Assume an initial guess for the parameter vector $\underline{\theta}$.
Step 2 Compute the measurement vector $\Delta \underline{Z}$ using the initial guess.
Step 3 Calculate the matrix H at the guess values.
Step 4 Solve for $\Delta \underline{\theta}$ using equation (2.39).
Step 5 If $\Delta \underline{\theta}$ satisfies certain specified terminating criterion, stop the iterations; otherwise, go to step 6.
Step 6 Update the parameter vector $\underline{\theta}$ as $\underline{\theta}_n = \underline{\theta}_0 + \Delta \underline{\theta}$ and go to step 2.

Example 2.4

Find the parameter vector $\dot{\underline{\theta}} = [\theta_1 \ \ \theta_2]^T$ that minimizes

$$F(\underline{\theta}) = \sum_{i=1}^{3} (Z_i - f_i(\underline{\theta}))^2$$

where

$$f_1(\theta_1) = \theta_1^2 + \theta_2 - 10$$
$$f_2(\theta_2) = \theta_1 + \theta_2^2 - 7$$
$$f_3(\theta_3) = \theta_1^2 + \theta_2^3 - 1$$

Assume an initial guess $\underline{\theta}_0 = [1 \quad 1]^T$. The procedure is as follows:

1. Calculate the initial measurement vector $\Delta\underline{Z}$ as

$$\Delta\underline{Z} = \begin{bmatrix} \theta_1^2 + \theta_2 - 10 \\ \theta_1 + \theta_2^2 - 7 \\ \theta_1^2 - \theta_2^3 - 1 \end{bmatrix}_{\underline{\theta}=\underline{\theta}_0} = \begin{bmatrix} -8 \\ -5 \\ -1 \end{bmatrix}$$

2. The matrix H is computed as

$$H = \begin{bmatrix} 2\theta_1 & 1 \\ 1 & 2\theta_2 \\ 2\theta_1 & -3\theta_2^2 \end{bmatrix}_{\underline{\theta}=\underline{\theta}_0} = \begin{bmatrix} 2 & 1 \\ 1 & 2 \\ 2 & -3 \end{bmatrix}$$

3. Solve for $\Delta\hat{\underline{\theta}}$ as

$$\Delta\hat{\underline{\theta}} = \left[H^T H\right]^{-1} H^T \Delta\underline{Z}$$

$$\Delta\hat{\underline{\theta}} = \begin{bmatrix} -2.885 \\ -1.4836 \end{bmatrix}$$

4. Update $\hat{\underline{\theta}}$ to be

$$\hat{\underline{\theta}}_n = \begin{bmatrix} 1 \\ 1 \end{bmatrix} + \begin{bmatrix} -2.885 \\ -1.4836 \end{bmatrix} = \begin{bmatrix} -1.885 \\ -0.4836 \end{bmatrix}$$

Iteration #2

5. Recalculate $\Delta\underline{Z}$ as

$$\Delta\underline{Z} = \begin{bmatrix} -6.931 \\ -8.6511 \\ 2.44 \end{bmatrix}$$

and H as

$$H = \begin{bmatrix} -3.77 & 1 \\ 1 & -0.967 \\ -3.77 & -0.702 \end{bmatrix}$$

6. The new value of $\Delta\hat{\underline{\theta}}$ is recalculated as

$$\Delta\hat{\underline{\theta}} = \begin{bmatrix} 0.28931 \\ 0.13712 \end{bmatrix}$$

and the iterations continue. After three iterations with convergence criteria for 1 and 2 selected as

$$\Delta\theta_1 = 0.0001, \text{ and } \Delta\theta_2 = 0.003$$

the least error squares solution is

$$\hat{\theta}_1 = 2.8682,\ \hat{\theta}_2 = 1.9419$$

and the residual vector $\underline{\hat{v}}$ is

$$\underline{\hat{v}} = \begin{bmatrix} -0.169 \\ -0.356 \\ -0.096 \end{bmatrix}$$

Example 2.5

Minimize

$$\sum_{i=1}^{3} [f_i(\underline{\theta})]^2$$

where

$$f_1(\theta_1) = \theta_1^2 + \theta_2^2 + \theta_1\theta_2$$
$$f_2(\theta_2) = \sin\theta_1$$
$$f_3(\theta_3) = \cos\theta_2$$

Assume an initial guess as $\underline{\theta}_0 = [3 \quad 1]^T$.

Repeating the same steps explained in the preceding example, we obtain an approximate LES solution after four iterations given by

$$\hat{\theta}_1 = -0.2974, \quad \hat{\theta}_2 = -0.8159$$

while the magnitude of the residual vector is

$$\underline{v}^T = [0.2918 \quad 0.2917 \quad 0.707]$$

2.3 Properties of Least Error Squares Estimation

Least error squares estimation results are easy to compute and possess a number of interesting properties. The least squares are the best estimates (maximum likelihood) when the measurement errors follow a Gaussian or normal distribution and the weighting matrix is equal to the inverse of the error covariance matrix. The least error squares estimates can be easily calculated.

Where the measurement error distribution does not follow a Gaussian distribution and the number of measurements greatly exceeds the number of unknown parameters, the method of least error squares yields very good estimates. However, there are many estimation problems for which the error distribution is not a Gaussian and the number of measurements does not greatly exceed the number of unknown parameters. In such cases, the least error squares estimations are adversely affected by bad data. This problem has been recognized and addressed by several researchers who have proposed different ways of refining the least error squares method to make estimation less affected by the presence of bad data. In the next section, we discuss an alternative technique to the LES estimation. This technique is based on least absolute value approximation.

2.4 Least Absolute Value Static State Estimation

The least absolute value estimation is based on minimizing the sum of the absolute value of the residuals [4,8]. The basic difference in calculating the actual approximation using the least absolute value formulation and least error squares formulation is that with the former, a best approximation is determined by interpolating a minimum subset of the available measurements, whereas with the latter, the best approximation is derived from the mean of the available measurements.

The idea of least absolute value approximation is not new. However, it has only been recently that algorithms to calculate least absolute value approximation have become available. The main purpose of this section is to give some insight into least absolute value approximation theory. The first part of this section gives a historical perspective on the development of least absolute value approximation. Then we discuss some of the algorithms that are available to obtain LAV state estimation. Then we introduce an algorithm based on LAV to obtain the best state estimate. Finally, we compare the least error squares method and the LAV technique through some illustrative examples.

2.4.1 Historical Perspective

The development of least absolute value method, as well as the least error squares, can be tracked back to the middle of the eighteenth century. The development of the methods was a result of trying to find the best method to summarize the information obtained from a number of measurements. The pioneers in regression analysis—Bascovish (1757), Laplace (1781), Gauss (1809), and Glaisher (1872)—proposed criteria for determining the best-fitting straight line through three or more points.

In 1781, Laplace presented a procedure for finding the best set of measurements based on minimizing the sum of the absolute deviations—namely, the least absolute value.

In 1809, Gauss demonstrated that the method of least squares is a consequence of the Gaussian Law of Error (normal distribution). Glaisher (1872) later showed that for a Laplacian (double exponential) distribution, the least absolute value estimator gives the most probably true value. In the early 1800s, regression analysis work focused on the conditions under which least squares regression and least absolute value regression give the best estimates. Laplace in 1812 showed that, for a large sample size, the method of least squares is superior. Houber in 1830 evaluated the least error squares and the least absolute value estimators and showed that for a Gaussian distribution, the least squares estimator gives the best results. Meanwhile, he noted that one advantage of the least absolute value estimator is that unbiased estimates can be obtained for any symmetrical distribution. However, Laurent in 1875 questioned the exactness of the Gaussian distribution, and on the basis of actual studies of measurements, he concluded that the method of least error squares should not be used when one has only a small number of observations. Jefferys, in 1939, showed that there is equivalence between the following three statements:

1. The Gaussian distribution is correct.
2. The mean value is the best value.
3. The method of the LES gives the best estimates.

He also pointed out that, for other symmetrical distributions, the method of least error squares should not be used, and stated that there is much to be said for the use of least absolute value when the distribution law is unknown, because it is less affected by large residuals than the least error squares.

The debates took place before the advent of computers and fast efficient methods of calculating the least error squares and least absolute value estimates; therefore, the discussions were primarily restricted to relatively small-size problems. Larger problems were primarily solved using analytical methods. In general, only least error squares regression was used because there were no efficient techniques for obtaining least absolute value estimates. In the early 1950s, emphasis was placed on developing efficient computational procedures for solving large problems. However, by this time the method of least squares was well established as the method for doing regression analysis.

The popularity of least error squares continued to grow even though it was known that it does not lead to the best available estimates of unknown parameters when the law of error (distribution) is other than Gaussian. But if the number of independent observations available is much larger than the number of parameters to be estimated, the method of least squares can usually be counted on to yield nearly best estimates.

In summary, the results of research to date indicate that the least absolute value of error technique gives better approximation when the errors in the measurements set have unknown distribution and also when the sample size is small.

2.4.2 Least Absolute Value of Error Estimation

The cost function in the case of LAV is given by "for $p = 1$" in equation (2.2):

$$J_1(\underline{\theta}) = \sum_{i=1}^{m} |Z_i - H_i\underline{\theta}| \tag{2.40}$$

As mentioned earlier, the minimum of $J_1\left(\hat{\underline{\theta}}\right)$ corresponds to the best LAV estimate $\hat{\underline{\theta}}$ of the system parameters.

Important characteristics of the LAV solution are given by the following two theorems.

Theorem 1

If the column rank of the $m \times n$ matrix H is k, $k \leq \underline{n}$ (for maximal rank $k = \underline{n}$), then there exists a vector $\hat{\underline{\theta}}$ corresponding to a best approximation that interpolates at least k points of the measurement set.

This theorem simply states that, if we have m measurements Z_i, $i = 1, 2, \ldots, m$ and n unknowns, then the optimal hyperplane based on LAV will pass through at least n points of the measurements set. This is in contrast to the least squares approximation, which does not necessarily pass through any of the measurement points of the set Z.

Theorem 2

If N_1 is the number of measurement points above the optimal hyperplane under the LAV plane, and N_2 is the number of points below the hyperplane, provided that $n+1$ points do not lie on a hyperplane in n-dimension, then

$$|N_1 - N_2| \leq n$$

These two theorems state the interpolation property of the LAV solution. Because the LAV solution interpolates data points, it will reject bad data points, provided that none of the bad data points are among the points interpolated. Thus, the problem reduces to selecting the n (n = the number of the parameter variables to be estimated) data points that, when used to find $\hat{\underline{\theta}}$, minimize the LAV cost function. The popular method of finding $\hat{\underline{\theta}}$ has been through linear programming. The formulation of linear programming problems can be carried out as explained Section 2.4.3.

2.4.3 *Least Absolute Value Based on Linear Programming*

In this section, we explain efficient algorithms used to solve the LAV estimation problem. Minimize the cost function of

$$J_1(\underline{\theta}) = \sum_{i=1}^{m} |v_i| \tag{2.41}$$

subject to satisfying

$$v_i \geq 0 \tag{2.42}$$

$$v_i + \left| Z_i - \sum_{j=1}^{n} H_{ij}\theta_i \right| \geq 0, \quad i = 1, \ldots, m \tag{2.43}$$

Equation (2.43) can be rewritten as

$$v_i \geq Z_i - \sum_{j=1}^{n} H_{ij}\theta_i, \quad i = 1, \ldots, m \tag{2.44}$$

and

$$v_i \geq \sum_{j=1}^{n} H_{ij}\theta_i - Z_i, \quad i = 1, \ldots, m \tag{2.45}$$

Thus, the linear programming problem is to minimize equation (2.41) subject to satisfying the constraints given by equations (2.44) and (2.45). We can see from equations (2.44) and (2.45), if one of the two constraints is negative, the other will be positive, and v_i must be positive to meet linear programming requirements. The following example demonstrates the formulation of a simple LAV estimation problem as a linear programming problem.

Example 2.6

In this example, we fit the given set of measurements {(0, 2), (0.1, −2), (2, −1), (2.2, 1)} with a straight line in the form of $y = ax + b$.

Solution

Let

$$\underline{Z} = \begin{bmatrix} 2 \\ -2 \\ -1 \\ 1 \end{bmatrix}, \quad H = \begin{bmatrix} 0.0 & 1 \\ 0.1 & 1 \\ 2.0 & 1 \\ 2.2 & 1 \end{bmatrix}, \quad \text{and} \quad \underline{\theta} = \begin{bmatrix} a \\ b \end{bmatrix}$$

The cost function is given by

Minimize

$$J = \sum_{i=1}^{4} \left| Z_i - \sum_{j=1}^{2} H_{ij}\theta_i \right|$$

Define v_i as

$$v_i = Z_i - \sum_{j=1}^{2} H_{ij}\theta_i, \quad i = 1, \ldots, 4$$

Then

Minimize

$$J = \sum_{i=1}^{4} |v_i|$$

subject to

$$v_1 + b \geq 2$$
$$v_2 + 0.1a + b \geq -2$$
$$v_3 + 2a + b \geq -1$$
$$v_4 + 2.2a + b \geq 1$$

and

$$v_1 - b \geq -2$$
$$v_2 - 0.1a - b \geq 2$$
$$v_3 - 2a - b \geq 1$$
$$v_4 - 2.2a - b \geq -1$$

In solving the least absolute value estimation problem, feasible values for a, b, and v_1 to v_4 will be found first; then these values will be iterated to decrease the value of the cost function. When the cost function reaches its minimum value, subject to satisfying the constraints, the estimation procedure is completed, and the values of a and b represent the LAV estimates. The values of v_1 to v_4 represent the absolute value of the residuals. The main steps involved in this algorithm are as follows:

- Select n points from the set of measurements.
- Evaluate the cost function.
- Select new points that decrease the cost function.

When the cost function becomes a minimum, the LAV solution has been reached. It has been shown that the size of the matrix to be stored and manipulated is $[2(m+n) \times n]$ [2].

The main disadvantages of the linear programming approach are

- It is an iterative technique, which requires considerable compute time.
- It needs a great deal of memory to store and manipulate a matrix of size $2(m+n) \times n$.
- The solution obtained may not be unique.

More recent algorithms attempt to overcome these difficulties.

2.4.4 Schlossmacher Iterative Algorithm

In 1973, Schlossmacher [4] presented an iterative technique that uses successive weighted least squares estimation to find the least absolute value estimates. The following steps are involved in this algorithm:

1. Obtain a weighted least error squares solution with all the weighting factors set to one.
2. Use the generated weighted least error squares solution to calculate the residuals $[v_1, i = 1, \ldots, m]$.

3. Set $W_i = \frac{1}{|v_i|}$, $i = 1, 2, \ldots, m$. If any $v_i \cong 0$, set $W_i = 0$.

4. Repeat steps 2 and 3 until the changes between successive iteration approach zero.

Although Schlossmacher's technique gives approximate estimates, it is an iterative technique and has been criticized as being computationally inefficient.

2.4.5 Sposito and Hand Algorithm

In 1976–1977, Sposito and Hand [8] suggested that least squares estimates can be used as starting points for a least absolute value estimator based on linear programming. Their research indicates that starting a linear programming problem at the least error squares estimate saves much iteration. They found that the total computing time is equal to the time to calculate the least error squares estimate plus the time needed to calculate the least absolute estimation by linear programming. This computing time is less than the time needed to calculate the least absolute value estimation based on linear programming from a flat start.

The main drawback of their technique is that it still requires a linear programming algorithm to calculate the LAV estimates.

Ten years later, Soliman and Christensen proposed a technique for solving the least absolute value estimation problem [1,2]. They implemented their technique in many areas in power systems analysis, including power systems state estimation, frequency relaying in power systems, electric short-term load forecasting, etc. The algorithm, as we shall see shortly, uses the interpolation characteristics of the LAV estimator given by the two theorems explained earlier.

2.4.6 Soliman and Christensen Algorithm

Given the measurement equation described in equation (2.1), the following are the main steps involved in the Soliman and Christensen LAV algorithm, unconstrained problem.

For a least absolute value estimator based on linear programming:

Step 1 Calculate the LES solution, as defined earlier, using the equation

$$\underline{\hat{\theta}} = \left[H^T H\right]^{-1} H^T \underline{Z}$$

Step 2 Calculate the LES residuals generated from this solution as

$$v_i = Z_i - H_i \underline{\hat{\theta}}$$

Step 3 Calculate the standard deviation of the calculated residuals as

$$\sigma^2 = \frac{1}{m-n+1} \sum_{i=1}^{m} (v_i - v_{av})^2 = \text{variance}$$

$$\sigma = \sqrt{\text{variance}} = \left[\frac{1}{m-n+1} \sum_{i=1}^{m} (v_i - v_{av})^2\right]^{\frac{1}{2}}$$

Step 4 Reject the outliers with residuals greater than the standard deviation σ, providing that the system is observable.

Step 5 Recalculate the new LES estimates using the remaining measurements and calculate the new corresponding residuals for these measurements.

Step 6 Select the n measurements that correspond to the smallest least error squares residual and form the corresponding $\hat{\underline{Z}}$ and $\hat{H}$.

Step 7 Solve for the least absolute value estimate, $\underline{\theta}^*$, using

$$\underline{\theta}^* = \left[\hat{H}\right]^{-1}\hat{\underline{Z}}$$

The following examples illustrate the steps involved.

Example 2.7

Fit the data point {(1, 3), (3, 6) (4, 6) (6, 12) (9, 11)} with a straight line in the form of $y=a_1x+a_2$ (the data represents exactly the line $y=x+2$, except for points 2 and 4):

$$\underline{Z} = \begin{bmatrix} y_1 \\ y_2 \\ y_3 \\ y_4 \\ y_5 \end{bmatrix} = \begin{bmatrix} 3 \\ 6 \\ 6 \\ 12 \\ 11 \end{bmatrix}, \quad H = \begin{bmatrix} 1 & 1 \\ 3 & 1 \\ 4 & 1 \\ 6 & 1 \\ 9 & 1 \end{bmatrix}, \quad \underline{\theta} = \begin{bmatrix} a_1 \\ a_2 \end{bmatrix}$$

Step 1 Calculate the least error squares estimates $\underline{\theta}^*$ by using

$$\underline{\theta}^* = \left[H^T H\right]^{-1} H^T \underline{Z} = \begin{bmatrix} a_1 \\ a_2 \end{bmatrix} = \begin{bmatrix} 1.10753 \\ 2.50538 \end{bmatrix}$$

Step 2 Calculate the least error squares residuals generated from this solution

$$\underline{v} = \underline{Z} - H\underline{\theta}^*$$

$$\underline{v} = \begin{bmatrix} -0.6129 \\ 0.1720 \\ -0.9355 \\ 2.8495 \\ -1.4731 \end{bmatrix}$$

Step 3 Calculate the standard deviation of residuals by using

$$\sigma = \sqrt{\frac{1}{m-n+1}\sum_{i=1}^{n}\left(v_i - v_{av}\right)^2}$$

where $v_{av} = \sum_{i=1}^{m} \frac{r_i}{m}$ to obtain

$$\sigma = 1.70073$$

Step 4 Reject the measurements of which the residuals are greater than the standard deviation, σ. From the values of v, the residuals greater than the standard deviation belong to point 4. If this measurement is rejected, then the new measurement vector $\underline{Z}$ and H matrix will be

$$\underline{Z} = \begin{bmatrix} 3 \\ 6 \\ 6 \\ 11 \end{bmatrix}, \quad H = \begin{bmatrix} 1 & 1 \\ 3 & 1 \\ 4 & 1 \\ 9 & 1 \end{bmatrix}$$

Step 5 Recalculate the least squares estimates θ^*:

$$\underline{\theta}^*_{\text{new}} = \left[H^T H\right]^{-1} H^T \underline{Z} = \begin{bmatrix} 0.96403 \\ 2.40288 \end{bmatrix}$$

Step 6 Recalculate the least error squares residuals generated from this solution:

$v_{\text{new}} = Z - H\theta^*_{\text{new}}$, which gives

$$v_{\text{new}} = \begin{bmatrix} -0.36691 \\ 0.70504 \\ -0.25899 \\ -0.07913 \end{bmatrix}$$

Step 7 Because the rank of the matrix H is 2, the estimator fits at least two points and these two points have the smallest LES residuals. These smallest residuals are v_4 and v_3. Form $\hat{\underline{Z}}$ and $\hat{H}$ as

$$\hat{\underline{Z}} = \begin{bmatrix} 11 \\ 6 \end{bmatrix}, \quad \hat{H} = \begin{bmatrix} 9 & 1 \\ 4 & 1 \end{bmatrix}$$

Step 8 Solve for the least absolute value estimates by using

$$\hat{\underline{\theta}} = \hat{H}^{-1} \hat{\underline{Z}}$$

which gives

$$\hat{\underline{\theta}} = \begin{bmatrix} 9 & 1 \\ 4 & 1 \end{bmatrix}^{-1} \begin{bmatrix} 11 \\ 6 \end{bmatrix} \hat{\underline{\theta}} = \begin{bmatrix} 1 \\ 2 \end{bmatrix}$$

Step 9 Calculate the least absolute value residuals by using

$$\hat{v} = \underline{Z} - H\hat{\underline{\theta}}$$

$$\hat{\underline{v}} = \begin{bmatrix} 0.0 \\ 1.0 \\ 0.0 \\ 4.0 \\ 0.0 \end{bmatrix}, \text{with } \|v\| = 5.0$$

From this example, we note that the first least squares solution is more affected by the presence of bad data (points 2 and 4), but the second solution is less affected by bad data and gives a more acceptable solution than the first solution. Rejecting bad data by using the standard deviation gives an effective method for rejecting the bad data and determining the number of bad data that must be rejected to obtain a suitable solution. All residuals of least absolute value are zero except the measurement number 2, and if we calculate the residual for point 4, which was rejected before the second solution of least squares, it will be equal to 4 (i.e., the least absolute value technique fits all but the second and fourth measurements). Thus, the two bad data points that do not fall on the line $y=x+2$ are rejected by the new algorithm. In contrast, the two solutions of the least squares estimates do not interpolate any of the data points and are affected by bad data.

Example 2.8

Fit the set of measurements {(0, 1), (1, 3), (2, −5), (3, −5), (4, −3), (5, 1), (6, 1), (7, 15)} with a polynomial of the form $y=a_1x^2+a_2x+a_3$ (the data represent the polynomial $y=x^2-5x+1$ except for points 2 and 7):

$$\underline{Z}=\begin{bmatrix}1\\3\\-5\\-5\\-3\\1\\1\\15\end{bmatrix},\quad H=\begin{bmatrix}0&0&1\\1&1&1\\4&2&1\\9&3&1\\16&4&1\\25&5&1\\36&6&1\\49&7&1\end{bmatrix},\quad \underline{\theta}=\begin{bmatrix}a_1\\a_2\\a_3\end{bmatrix}$$

Step 1 Calculate the least error squares estimates $\underline{\theta}^*$ by using

$$\underline{\theta}^*=\left[H^TH\right]^{-1}H^T\underline{Z}=\begin{bmatrix}1.00595\\-5.69643\\3.45830\end{bmatrix}$$

Step 2 Calculate the least error squares residuals using

$$\underline{v}=Z-H\underline{\theta}^*$$

$$v=\begin{bmatrix}-2.4583\\4.2321\\-1.0893\\-0.4226\\0.2321\\0.875\\-3.494\\2.125\end{bmatrix}$$

Step 3 Calculate the standard deviation σ using

$$\sigma = \sqrt{\frac{1}{m-n+1}\sum_{i=1}^{8}(v_i - v_{av})^2}$$

$$\sigma = 2.6727$$

Step 4 Reject the measurements having residuals greater than the standard deviation σ. Measurement numbers 2 and 7 have residuals greater than the standard deviation, and they will be rejected to obtain

$$\underline{Z} = \begin{bmatrix} 1 \\ -5 \\ -5 \\ -3 \\ 1 \\ 15 \end{bmatrix}, \quad H = \begin{bmatrix} 0 & 0 & 1 \\ 4 & 2 & 1 \\ 9 & 3 & 1 \\ 16 & 4 & 1 \\ 25 & 5 & 1 \\ 49 & 7 & 1 \end{bmatrix}$$

Step 5 Recalculate the least error squares estimates $\underline{\theta}^*$:

$$\underline{\theta}^*_{\text{new}} = \left[H^T H\right]^{-1} H^T \underline{Z} = \begin{bmatrix} 1 \\ -5 \\ 3 \end{bmatrix}$$

Step 6 Recalculate the least error squares residuals:

$$\underline{v}^T = [0.0 \quad 0.0 \quad 0.0 \quad 0.0 \quad 0.0 \quad 0.0]$$

Step 7 Select any three measurements to solve for least absolute value. Because all measurements have zero residuals,

$$\underline{\hat{\theta}} = \begin{bmatrix} 0 & 0 & 1 \\ 4 & 2 & 1 \\ 9 & 3 & 1 \end{bmatrix}^{-1} \begin{bmatrix} 1 \\ -5 \\ -5 \end{bmatrix} = \begin{bmatrix} 1 \\ -5 \\ 1 \end{bmatrix}$$

with $\hat{v}^T = [0.0, 6.0, 0.0, 0.0, -6.0, 0.0]$,

$$\|\underline{\hat{v}}\| = 12$$

This example demonstrates the rejection of bad data by using the standard deviation method. Where all bad data (points 2 and 7) are rejected and the measurements used to solve the least squares for the second time do not contain any bad data, we expect that the least squares estimates will fit all the data points as well as the least absolute value estimates.

When we rank the residuals, there are some cases in which the absolute values of two residuals are equal, but there is only one position for the corresponding measurement in the interpolated measurements set. In this case we implement a tie-breaking procedure. The two measurements that are equal constitute a tie. Then we have two least absolute value estimates that correspond to two different interpolated sets of measurements. Each set contains the same measurement (corresponding to the n smallest least squares residuals) with the only difference being that each set contains one of the two measurements that was involved in the tie. The cost functions of both estimates are then calculated and compared. The estimate with a smaller value of cost function is the unique least absolute value estimate, which is produced by the new technique and the tie-breaking procedure. If more than two residuals are equal, a tie-breaking procedure can be implemented, but there will be more than two cost functions.

In the following example, we will demonstrate a case in which the tie-breaking procedure must be implemented.

Example 2.9

Fit the data {(0, 2), (1, 5), (2, 7), (3, 7), (4, 11), (5, 15)}, with the line of form $y = ax + b$:

$$Z = \begin{bmatrix} 2 \\ 5 \\ 7 \\ 7 \\ 11 \\ 15 \end{bmatrix}, \quad H = \begin{bmatrix} 0 & 1 \\ 1 & 1 \\ 2 & 1 \\ 3 & 1 \\ 4 & 1 \\ 5 & 1 \end{bmatrix}, \quad \theta = \begin{bmatrix} a \\ b \end{bmatrix}$$

Step 1 Calculate the least error squares estimates θ^*:

$$\theta^* = \left[H^T H\right]^{-1} H^T Z = \begin{bmatrix} a \\ b \end{bmatrix} = \begin{bmatrix} 2.3714 \\ 1.9047 \end{bmatrix}$$

Step 2 Calculate the least error squares residuals:

$$\underline{v} = \underline{Z} - H\underline{\theta}^*$$

$$\underline{v} = \begin{bmatrix} 0.0952 \\ 0.7238 \\ 0.3524 \\ -2.0190 \\ -0.3905 \\ 1.2381 \end{bmatrix}$$

Step 3 Calculate the standard deviation:

$$\sigma = \sqrt{\frac{1}{m-n+1} \sum_{i=1}^{6} (v_i - v_{av})^2}$$

$$\sigma = 1.133$$

Step 4 Reject the measurement having residuals greater than the standard deviation value and for the new $\underline{Z}$ vector and H matrix. The points to be rejected are points 4 and 6.

$$\underline{Z} = \begin{bmatrix} 2 \\ 5 \\ 7 \\ 11 \end{bmatrix}, \quad H = \begin{bmatrix} 0 & 1 \\ 1 & 1 \\ 2 & 1 \\ 4 & 1 \end{bmatrix}$$

Step 5 Recalculate the least error squares estimates θ^*:

$$\theta^*_{\text{new}} = \begin{bmatrix} a \\ b \end{bmatrix} = \begin{bmatrix} 2.2 \\ 2.4 \end{bmatrix}$$

Step 6 Recalculate the least error squares residuals:

$$v_{\text{new}} = \begin{bmatrix} -0.4 \\ 0.4 \\ 0.2 \\ -0.2 \end{bmatrix}$$

Step 7 Rank the residuals in ascending order and select the two measurements corresponding to the smallest residuals. Here, the third and fourth measurements have the same absolute values for the residual. Thus, a tie-breaking procedure must be implemented. Also, the first and second measurements have equal absolute values for the residual. In this case four estimates must be calculated as

$$\hat{\theta}_1 = \begin{bmatrix} 2 & 1 \\ 0 & 1 \end{bmatrix}^{-1} \begin{bmatrix} 7 \\ 2 \end{bmatrix} = \begin{bmatrix} 2.5 \\ 2.0 \end{bmatrix}$$

$$\hat{\theta}_2 = \begin{bmatrix} 2 & 1 \\ 1 & 1 \end{bmatrix}^{-1} \begin{bmatrix} 7 \\ 5 \end{bmatrix} = \begin{bmatrix} 2 \\ 3 \end{bmatrix}$$

$$\hat{\theta}_3 = \begin{bmatrix} 4 & 1 \\ 0 & 1 \end{bmatrix}^{-1} \begin{bmatrix} 11 \\ 2 \end{bmatrix} = \begin{bmatrix} 2.25 \\ 2.00 \end{bmatrix}$$

$$\hat{\theta}_4 = \begin{bmatrix} 4 & 1 \\ 1 & 1 \end{bmatrix}^{-1} \begin{bmatrix} 11 \\ 5 \end{bmatrix} = \begin{bmatrix} 2 \\ 3 \end{bmatrix}$$

Step 8 Calculate the least absolute value residuals and calculate the cost function in each case:

$$J_1 = 1.50$$
$$J_2 = 1.00$$
$$J_3 = 1.25$$
$$J_4 = 1.00$$

The cost functions J_2 and J_4 are equal because $\hat{\theta}_2$ and $\hat{\theta}_4$ are identical. Therefore, the values of $\hat{\theta}_2$ and $\hat{\theta}_4$ are unique estimates produced by the new technique and the tie-breaking procedure.

When two cost functions are equal and the corresponding estimates are different, then both estimates are equally valid. Under such circumstances, the estimates produced by the new technique will not be unique. Extensive testing of the technique has demonstrated that this case rarely occurs and that the estimates are always unique.

2.5 Constrained LAV Estimation

The constrained state estimation problem can be handled by the Soliman and Christensen technique. If we have m measurements and ℓ constraints and $n > \ell$, the technique will interpolate at least n points of the given measurements. The constraints represent a good measurement so that the residual of the least squares solution for these constraints will be zero. The least absolute value technique must interpolate the ℓ constraints before interpolation of $(n-\ell)$ of the other measurements. The total number of the interpolated points will equal $(n-\ell+n=n)$. Thus, the new method will select directly the ℓ constraints and the $(n-\ell)$ measurements corresponding to the smallest LS residuals. Note that the number of constraints should be less than the number of unknowns; otherwise, the least absolute value will interpolate the n points from the constraints' equations only.

The solution technique may use the method proposed in the previous section for LS parameters estimation with constraints and then proceed in the same manner as the LAV technique to obtain the least absolute value optimal solution.

In the next example, we will apply the Soliman and Christensen technique for a problem that contains equality constraints.

Example 2.10

Fit the point $\{(-1, 2), (0, 5), (1, 6), (2, 13), (3, 20), (4, 29)\}$ with the polynomial of form $y = a_1 x^2 + a_2 x + a_3$, subject to the constraint $y(-2) = 5$:

$$\underline{Z} = \begin{bmatrix} 2 \\ 5 \\ 6 \\ 13 \\ 20 \\ 29 \end{bmatrix}, \quad H = \begin{bmatrix} 1 & -1 & 1 \\ 0 & 0 & 1 \\ 1 & 1 & 1 \\ 4 & 2 & 1 \\ 9 & 3 & 1 \\ 16 & 4 & 1 \end{bmatrix}, \quad \underline{\theta} = \begin{bmatrix} a_1 \\ a_2 \\ a_3 \end{bmatrix}$$

The equality constraint can be written as

$$C\underline{\theta} = \underline{d}, \quad C = [4 \quad -2 \quad 1], \quad \underline{d} = [5]$$

The constrained problem can be solved by least squares as mentioned earlier.

Step 1 Calculate the least error squares estimates θ^* by using

$$\theta^* = [H^T H]^{-1}\left[H^T Z - C^T\left[C[H^T H]^{-1}C^T\right]^{-1}\left[C[H^T H]^{-1}H^T Z - d\right]\right]$$

$$\theta^* = \begin{bmatrix} a_1 \\ a_2 \\ a_3 \end{bmatrix} = \begin{bmatrix} 1.136 \\ 1.796 \\ 4.049 \end{bmatrix}$$

Step 2 Calculate the least error squares residuals:

$$\underline{v} = \underline{Z} - H\underline{\theta}^*$$

$$\underline{v} = \begin{bmatrix} -1.388 \\ 0.951 \\ -0.982 \\ 0.812 \\ 0.355 \\ -0.415 \end{bmatrix}$$

Step 3 Calculate the standard deviation of the residuals

$$\sigma = \sqrt{\frac{1}{m-n+1}\sum_{i=1}^{\sigma}(v_i - v_{av})^2}$$

where $v_{av} = \frac{1}{m}\sum_{i=1}^{\sigma} v_i$

$$\sigma = 1.0796$$

Step 4 Reject measurements having residuals greater than the standard deviation and form the new $\underline{Z}$ vector and H matrix for the rest of the measurements. Measurement number 1 will be rejected.

$$\underline{Z} = \begin{bmatrix} 5 \\ 6 \\ 13 \\ 20 \\ 29 \end{bmatrix}, \quad H = \begin{bmatrix} 0 & 0 & 1 \\ 1 & 1 & 1 \\ 4 & 2 & 1 \\ 9 & 3 & 1 \\ 16 & 4 & 1 \end{bmatrix}$$

Step 5 Recalculate the least error squares estimates by using the same expression of step 1 as

$$\theta^*_{\text{new}} = \begin{bmatrix} a_1 \\ a_2 \\ a_3 \end{bmatrix} = \begin{bmatrix} 1.092 \\ 1.851 \\ 4.333 \end{bmatrix}$$

Step 6 Recalculate the least error squares residuals:

$$v_{\text{new}} = \begin{bmatrix} 0.666 \\ -1.277 \\ 0.594 \\ 0.282 \\ -0.215 \end{bmatrix}$$

Step 7 Rank the residuals and select the two measurements corresponding to the smallest residuals together with the constraint because the rank of H is 3. Form $\hat{Z}$ and $\hat{H}$. The smallest residuals are the fifth and the fourth residuals:

$$\hat{Z} = \begin{bmatrix} 29 \\ 20 \\ 5 \end{bmatrix}, \quad \hat{H} = \begin{bmatrix} 16 & 4 & 1 \\ 9 & 3 & 1 \\ 4 & -2 & 1 \end{bmatrix}$$

Step 8 Solve for the least absolute value estimates $\hat{\theta}$ by using

$$\hat{\theta} = \hat{H}^{-1}\hat{Z}$$

$$\hat{\theta} = \begin{bmatrix} a_1 \\ a_2 \\ a_3 \end{bmatrix} = \begin{bmatrix} 1 \\ 2 \\ 5 \end{bmatrix}$$

Step 9 Calculate the LAV residuals vector generated from this solution as

$$\hat{v}^T = [2.0 \quad 0.0 \quad 2.0 \quad 0.0 \quad 0.0 \quad 0.0], \quad \|\hat{v}\| = 4$$

2.6 Nonlinear Estimation Using the Soliman and Christensen Algorithm

The LAV cost function for nonlinear estimation is given by

$$J_1(\theta) = \sum_{i=1}^{m} |\Delta Z_i - H_i \Delta\theta_i|$$

where

$$\Delta Z_i = \Delta Z_i^\circ - H_i \hat{\theta}^\circ$$

To solve the nonlinear parameter estimation problem, we use the following steps:

Step 1 Assume an initial guess θ°.
Step 2 Linearize the state equations around the initial estimate θ°.

Step 3 Calculate the change in estimation parameter $\Delta\theta$ using the LES algorithm as

$$\Delta\theta^* = \left[H^T H\right]^{-1} H^T \Delta Z$$

Step 4 Update the state estimation parameters:

$$\theta_{\text{new}} = \theta_{\text{old}} + \Delta\theta^*$$

Step 5 If $\Delta\theta^*$ is less than a predetermined terminating criterion, go to step 6; otherwise, go to step 2.

Step 6 Calculate the least squares estimation residuals using

$$v_i = \Delta Z_i - H\Delta\theta_i^*, \quad i = 1, 2, \ldots, m$$

Step 7 Calculate the standard deviation of these residuals.

Step 8 Reject the outliers having residuals greater than the standard deviation.

Step 9 Repeat step 2 to step 6 for the remaining measurements.

Step 10 Rank the new residuals in ascending order.

Step 11 Select the n measurements that correspond to the smallest n residuals and form $\Delta\hat{Z}$ and $\hat{H}$.

Step 12 Calculate the change in least absolute value estimates using

$$\Delta\hat{\theta} = \hat{H}^{-1}\Delta Z$$

Step 13 Update the state estimation parameters:

$$\theta_{\text{new}} = \theta_{\text{old}} + \Delta\hat{\theta}$$

Step 14 If $\Delta\hat{\theta}$ is less than a prescribed terminating criterion, then terminate the iteration and go to step 15; otherwise, calculate the change in ΔZ vector and H matrix and then recalculate the new residuals and go to step 10.

Step 15 Print the least absolute value estimates.

The following example illustrates the preceding steps from the area of power system state estimation.

Example 2.11

The Soliman and Christensen LAV algorithm can be used in power system state estimation problems.

Consider the three-bus system given in Figure 2.1. Given a constant load of 1 p.u. (100 MW), assume that the voltages are constant at 1 p.u., and bus 3 is the slack bus. Find θ_1 and θ_2, based on the following power-flow measurements between buses M_{12}, M_{13}, and M_{32}:

$$M_{12} = 0.62\text{ p.u.}, \quad M_{13} = 0.06\text{ p.u.}, \quad \text{and} \quad M_{32} = 0.37\text{ p.u.}$$

Transmission line resistance is ignored.

The general expression for power flow from bus i to bus j is

$$P_{ij} = V_i V_j B_{ij} \sin\left(\theta_i - \theta_j\right)$$

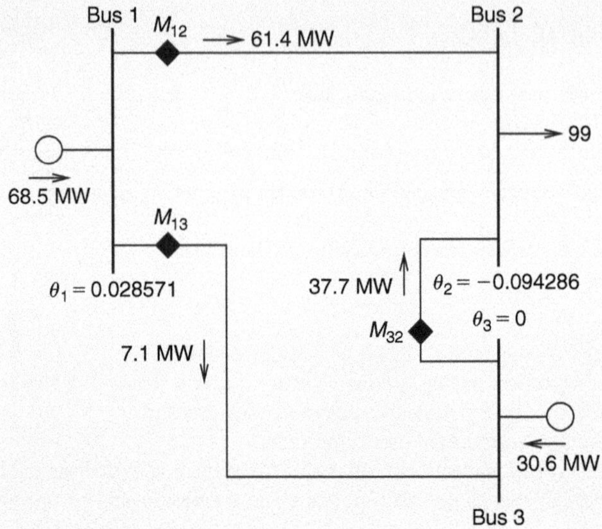

Figure 2.1 A three-bus power system.

If the voltages V_i and V_j are constant, then P_{ij} can be written using the first-order Taylor's series expansion as

$$\Delta P_{ij} = \frac{\partial P_{ij}}{\partial \theta_i}\Delta\theta_i + \frac{\partial P_{ij}}{\partial \theta_j}\Delta\theta_j$$

where

$$\Delta P_{ij} = P_{ij}^m(\theta_i, \theta_j) - P_{ij}^0(\theta_i, \theta_j)$$

$$\frac{\partial P_{ij}}{\partial \theta_i} = V_i V_j B_{ij}\cos(\theta_i - \theta_j)$$

$$\frac{\partial P_{ij}}{\partial \theta_j} = -V_i V_j B_{ij}\cos(\theta_i - \theta_j)$$

Then ΔP_{ij} can be written as

$$\Delta P_{ij} = [V_i V_j B_{ij}\cos(\theta_i - \theta_j)]\Delta\theta_i - [V_i V_j B_{ij}\cos(\theta_i - \theta_j)]\Delta\theta_j$$

and θ_1 and θ_2 are obtained using the following procedure:

1. Assume as an initial guess for θ_1 and θ_2 that $\theta_1^0 = \theta_2^0 = 0.0$.
2. Calculate ΔP_{ij} and the partial derivatives of P_{ij} for each measurement.
3. Solve the overdetermined equations for the LES solution of $\Delta\theta_i$ and $\Delta\theta_j$, taking into account the equality constraints.
4. Calculate the residuals and determine the two equations in two unknowns for which $\Delta\theta_1$ and $\Delta\theta_2$ can be solved using the Soliman and Christensen algorithm.

5. Update $\theta_1 = \theta_1^0 + \Delta\theta_1$ and $\theta_2 = \theta_2^0 + \Delta\theta_2$.
6. If the changes are small enough, convergence has been achieved; if not, return to step 2.

Performing the preceding steps yields

$$B = \begin{bmatrix} 0.62 \\ 0.06 \\ 0.37 \\ 1 \end{bmatrix}, \quad \underset{\text{constrained}}{A} = \begin{bmatrix} 5 & -4 \\ 2.5 & 0 \\ 0 & -4 \\ 5 & -9 \end{bmatrix}, \quad \underset{\text{constrained}}{\overline{\theta}} = \begin{bmatrix} \Delta\theta_1 \\ \Delta\theta_2 \end{bmatrix}$$

The first iteration (LAV solution) yields

$$\Delta\theta_1 = 0.029, \quad \Delta\theta_2 = -0.095$$

Then

$$\theta_1 = 0.029, \quad \theta_2 = -0.095$$

The second iteration yields

$$\Delta\theta_1 = 1.72\times10^{-4}, \quad \Delta\theta_2 = -1.51\times10^{-4}$$

and therefore,

$$\theta_1 = 0.029172, \quad \theta_2 = -0.095151 \text{ radians}$$

$\Delta\theta_1$ and $\Delta\theta_2$ are small, indicating that convergence has been achieved.

2.7 Leverage Points

Consider the simple linear regression model given by

$$\widetilde{Z} = H\widetilde{\theta} + \widetilde{\upsilon}$$

where, as defined earlier,

$$E[\widetilde{\upsilon}] = 0 \text{ and } E\left[\widetilde{\upsilon}^T \times \widetilde{\upsilon}\right] = R \text{ (diagonal)}$$

Let $Z = R^{-\frac{1}{2}} \times \widetilde{Z}$, $H = R^{-\frac{1}{2}}\widetilde{H}$, and $\upsilon R^{-\frac{1}{2}} \times \widetilde{\upsilon}$, then

$$Z = H\theta + \upsilon$$

where $E[\upsilon] = 0$, $E[\upsilon^T \upsilon] = I_m$; that is, the modified measurement error vector, υ, has unit covariance. Then the least error squares estimator for θ can be given by

$$\theta^* = \left[H^T H\right]^{-1} H^T Z$$

and the estimator for Z is

$$Z^* = H\left[H^T H\right]^{-1} H^T Z = K \times Z$$

Here, K is called the "hat" matrix. The diagonal elements of the matrix K can be expressed as

$$K_{ii} = h_i \left(H^T H\right)^{-1} h_i^T$$

where h_i is the ith row of the matrix H.

It can be shown that

$$0 \leq K_{ii} \leq 1$$

and the value of K_{ii} close to one corresponds to a leverage point. The residual vector of the measurements at the LES solution is

$$\gamma = Z - Z^* = (I - K) \times Z$$

For the leverage point, the residual will be very small, although it may be contaminated with a large error.

The residual covariance matrix can be written as

$$\mathrm{cov}(v) = (I - K)$$

If the measurement constitutes a leverage point (i.e., $K_{ii} \approx 1.0$), the $\mathrm{cov}(\gamma_i)$ will be near zero. In the case of a single leverage point, the normalized residual given by

$$\gamma_i^N = \frac{r_i}{\sqrt{1 - K_{ii}}}$$

will be inflated due to a small value of $(1 - K_{ii})$, and the normalized residual test will fail to identify the bad data when there are multiple leverage points in these measurements. Let us illustrate these facts via an example.

Example 2.12

Consider the model

$$Z_i = x_1 + h_i \times X_2, \quad i = 1, 2, \ldots, 7$$

where

$$Z_i = \begin{bmatrix} 0.5 \\ 1.0 \\ 1.5 \\ 2.0 \\ 2.5 \\ 1.5 \\ 1.5 \end{bmatrix}, \quad h_i = \begin{bmatrix} 1 \\ 2 \\ 3 \\ 4 \\ 5 \\ 15 \\ 16 \end{bmatrix}$$

The matrix H can be formulated as

$$H = \begin{bmatrix} 1 & 1 \\ 2 & 1 \\ 3 & 1 \\ 4 & 1 \\ 5 & 1 \\ 15 & 1 \\ 16 & 1 \end{bmatrix}$$

and the hat matrix K defined by

$$\underline{K} = H\left[H^T H\right]^{-1} H^T$$

has diagonal elements

$$\text{diag.}(K) = (0.276, 0.232, 0.197, 0.171, 0.153, 0.45, 0.523)$$

These elements indicate that the last two measurements are two leverage points because $K_{77} = 0.523$, $K_{66} = 0.450$, and their normalized residuals will be

$$\gamma_7^N = \frac{(1.5-8)}{\sqrt{1-0.523}} = -9.411$$

and

$$\gamma_6^N = \frac{(1.5-7.5)}{\sqrt{1-0.45}} = -8.090$$

In the literature

$$K_{ii} \geq 2\frac{n}{m}$$

is considered as an indication of a leverage point, where n is the number of parameters to be estimated and m is the number of measurements.

2.8 Comparison between LES Estimation and LAV Estimation Algorithms

After studying the main properties of least squares parameter estimation and the least absolute value parameter estimation technique, we can conclude the following:

The main difference in solution of the least squares and least absolute value parameter estimation technique is that the least absolute value estimate interpolates a subset of the available measurements, whereas the least squares parameter estimation technique interpolates the mean of the available measurements (i.e., the least squares may not fit any of the available measurements if the measurements set is contaminated with bad data).

If there is an erroneous measurement, the least squares estimator will give an optimal estimation when the error distribution is Gaussian or when there is a great redundancy in the available measurements. But if the measurements error follows another distribution and the sample of measurements is small, the solution of least squares will not be optimal. On the other hand, the least absolute value gives the optimal solution for any measurement's error distribution and a small sample of measurements. Thus, the least absolute value estimate will be the best when the error distribution is unknown and the sample of the measurements is small.

The least error squares technique is always biased by the presence of bad data. The least error squares technique does not detect or identify the bad data during the solution process, but it can be done separately after the least error squares solution. In the new least absolute value of the error estimate technique, the bad data are identified and rejected automatically during the solution process.

References

[1] G.S. Christensen, A.H. Rouhi, S.A. Soliman, A new technique for the unconstrained and constrained linear LAV parameter estimation, Can. J. Electr. Comput. Eng. 14 (1) (1989) 24–30.
[2] S.A. Soliman, G.S. Christensen, A new technique for curve fitting based on weight least absolute value estimation, J. Optim. Theory Appl. JOTA 62 (2) (1989) 281–299.
[3] S.A. Soliman, G.S. Christensen, A. Rouhi, A new technique for curve fitting based on minimum absolute deviation, Comput. Stat. Data. Anal. 6 (4) (1988) 341–351.
[4] E.J. Schlossmacher, An iterative technique for absolute deviations curve fitting, J. Am. Stat. Assoc. 68 (344) (1973) 857–869.
[5] P.D. Robers, A. Ben-Israel, An iterative programming algorithm for discrete linear L_1 approximation problems, J. Approximation Theory 2 (1969) 323–336.
[6] S.A. Soliman, G.S. Christensen, M.Y. Mohamed, Power system state estimation based on LAV, control and dynamic systems, in: C.T. Leondes (Ed.), Analysis and Control System Techniques for Electric Power Systems, vol. 44 (4), Academic Press, NewYork, 1991.
[7] I. Barrodale, F.D.K. Roberts, An improved algorithm for direct L_1 linear approximation, SIAM J. Numer. Anal. 10 (5) (1973) 839–849.
[8] V.A. Sposito, M.L. Hand, Using an approximate L_1 estimator communications in statistics, Simul. Comput. B6 (3) (1977) 236–268.
[9] I. Barrodale, F.D.K. Roberts, An efficient algorithm for discrete L_1 linear approximation with linear constraints, SIAM J. Numer. Anal. 15 (3) (1978) 603–611.
[10] G.F. McCormick, V.A. Sposito, Using the L_2–estimator in L_1–estimation, SIAM J. Numer. Anal. 13 (3) (1976) 337–343.
[11] W.D. Fisher, A note on curve fitting with minimum deviations by linear programming, J. Am. Stat. Assoc. 56 (1961) 359–362.
[12] I. Barrodale, A. Young, Algorithms for best L_1 and L_∞ linear approximations on a discrete set, Numerische Mathematik 8 (1966) 295–306.

3 Load Modeling for Short-Term Forecasting

3.1 Objectives

The objectives of this chapter are

- Proposing different models for short-term load forecasting.
- Introducing three models—namely, model A, model B, and model C—for winter and summer short-term load forecasting.
- Explaining different factors that affect these models for each season.
- Fitting each model to the different techniques that are used to estimate the parameters of the model—namely, least error squares (LES) and least absolute value (LAV) estimation techniques.

3.2 Introduction

In short-term load forecasting (STLF), the future load on a power system is predicted by extrapolating a predetermined relationship between the load and its influential variables—namely, time and/or weather. Determining this relationship is a two-stage process that requires (a) identifying the relationship between the load and the related variables, and (b) quantifying this relationship through the use of a suitable parameter estimation technique.

A prerequisite to the development of an accurate load-forecasting model is an in-depth understanding of the characteristics of the load to be modeled [2–6]. This knowledge of the load behavior is gained from experience with the load and through statistical analysis of past load data. Utilities with similar climatic and economic environments usually experience similar load behavior, and load models developed for one utility can usually be modified to suit another [44]. The load supplied by a power system is dynamic in nature and directly reflects the activities and conditions in the surrounding environment. This load can be separated into a standard or base load, a weather-dependent load, and a residual load. In the following sections, we review the characteristics of each of these components in turn [7,44].

3.3 Base Load

The base load results from the business and economic conditions of the service area, and is the largest component of total system load [44]. It accounts for approximately

DOI: 10.1016/B978-0-12-381543-9.00003-8

90% of total load and can be spectrally decomposed into four distinct components—namely:

1. A long-term component that reflects the economic growth of the area and is usually directly proportional to the growth of the national economy.
2. A seasonal component that results from changes in electricity demand from one season to another. In North America, this load pattern is characterized by midwinter and midsummer peaks interspaced by troughs occurring during the central spring and autumn seasons.
3. A weekly load cycle that results from the consumption pattern of one day of the week being characteristically different from the others. Weekly business cycles and repetitive local activities are the main reasons for this aspect of load behavior that is characterized by relatively constant midweek demands and smaller weekend loads.
4. A daily load cycle that results from the basic daily similarity of consumer activities. Low early-morning demand peaking at a mid-afternoon high usually characterizes this load cycle.

3.4 Weather-Dependent Load

Weather contributes significantly to the dynamics of a load, and much effort has been spent to find a viable relationship between the weather and the load so that an accurate load model could be developed [44]. The survey of the literature in Chapter 2 indicates that each utility has its own load model that depends on the weather of that load the utility is serving, and a load model for a utility does not necessarily suit (fit) the load of another utility.

The effects of weather on load are usually modeled by expressing the load as a linear regression of explanatory meteorological factors such as temperature, wind speed, humidity, etc. Although it is recognized that an extremely wide variety of explanatory weather variables is required to totally represent the effects of weather, studies have shown that a few basic meteorological factors usually account for most of the weather-dependent load.

The specific weather variables that are normally used to model weather-dependent load are dry bulb temperature, wind speed, humidity, and daylight illumination. The last is usually the least significant of these weather variables, and because its metering is difficult and costly, it is usually omitted from most models. The general effects of these weather variables on load are summarized next [7,44].

3.4.1 Temperature

In most load environments, dry bulb temperature is the most significant weather variable and usually accounts for the largest percentage of weather-dependent load. Deviations of temperature from the norm can result in major changes in the load pattern. These changes, however, do not occur immediately, but rather are delayed due to thermal storage in buildings.

The effects of temperature on load pattern are not uniform and are different from one utility to another and from one season to the next. A decrease in temperature below room temperature during the winter season means an increase in the heating load, but an increase in the temperature above room temperature during summer means increasing air conditioning load (increasing the cooling load).

Temperature effects are usually modeled by considering the load to be a function of the effective temperature or temperature deviation, rather than the actual temperature. This stems from the realization that the general effects of base temperature are already included in the seasonal load cycle, and only deviations from the norm will result in load changes.

In other words, each utility company designs its base load according to the normal temperature of the environment of that load, and any temperature deviation will lead to changes in the load.

3.4.2 Wind Speed

A factor that can contribute significantly to weather-dependent load is wind. Wind effects are especially prevalent during winter and are a direct consequence of the cooling power of the wind. The cooling effect of the wind depends on the wind speed and the dry bulb temperature. The heat loss from a building is proportional to the product of the square root of the wind speed and the temperature deviation from the comfort level of approximately 18°C. This effect is relatively small in postwinter seasons and, for simplicity, is usually included only in winter models [44].

Some researchers prefer to use the wind-chill factor as a means of representation of the wind in their models because wind-chill factor is often strongly correlated with winter load [7]. Others contend that wind-chill factor is only a measure of the discomfort level of the wind and temperature and, as such, is not a true index for gauging the resulting load response [7,44]. High wind-chills, however, have the psychological effect of causing people to turn up their thermostats.

3.4.3 Humidity

A weather variable that greatly influences air conditioning and other related cooling loads in summer is the level of humidity in the atmosphere. The effects of high humidity are generally noticeable only when the temperature is quite high, usually above room temperature. The humidity effect can be considered in the load model by representing it as a function of relative humidity, the temperature humidity index, or the dew-point temperature. The most common variable used in the literature is the humidity index.

The temperature humidity index is a measure of the discomfort level or equivalent heat stress in summer and depends on both the temperature and relative humidity, and normally shows greatest correlation with summer load and influences the load only above a predetermined cutoff temperature.

3.4.4 Illumination

Daytime illumination has a small effect on the load model, compared to the other two previously discussed factors. Surveying the literature shows that in most cases this factor is often omitted from most load models.

Low daytime illumination can cause an increase in daytime lighting load and advance the effects of nightfall, thereby altering the evening load pattern. This factor is influenced by such weather conditions as cloud cover, dust, fog, haze, etc., and is the measure for the level of luminous radiation received at ground level.

3.5 Residual Load

The residual load component occurs in load modeling and usually accounts for a small percentage of total load and results from irregularities in the behavior of the consuming public [44]. Abnormal consumer demands, though quite frequent in occurrence, are very difficult to model and predict, and are not accounted for in most load models.

The common factors of unpredictable load behavior range from public response to major television events, strikes, storms, disasters, time changes, etc.

3.6 Short-Term Load Models

Reviewing short-term load-forecasting methods indicates that the most important modeling techniques used can be classified in one of the following categories [7,9,12,16,50]:

1. Multiple linear regression.
2. General exponential smoothing.
3. Stochastic time series.
4. Expert systems approach.
5. State space model.

These models are classified on the basis of the name of the underlying mathematical technique used to estimate the parameters of the model. The preceding classifications are not unique, and the classification used with one utility is not necessarily suitable for another. However, one can combine these models or can use one model to initiate another model to predict certain parameters from past history. With unknown information about the load, these techniques can be combined to improve the accuracy of the forecast. Also, each model possesses distinct advantages and disadvantages compared to each other. In the following subsections, the first three methods are reviewed, whereas the last two methods are beyond the scope of this research.

3.6.1 Multiple Linear Regression

Multiple linear regression is the earliest technique of load-forecasting methods [44]. Here, load is expressed as a function of explanatory weather and nonweather variables that influence the load. The influential variables are identified on the basis of correlation analysis with load, and their significance is determined through statistical tests such as True and False tests.

Mathematically, the load model using this approach can be written as

$$y(t) = a_0 + \sum_{i=1}^{n} a_i x_i(t) + r(t) \tag{3.1}$$

where $y(t)$ is the load value at time t, $x_1(t), \ldots, x_n(t)$ are explanatory variables, $r(t)$ is the residual load at time t, and a_i are the regression parameters relating the load $y(t)$ to the explanatory variables. Previous analysis that uses this model treats a_i as a crisp number.

If the number of observations equals exactly the number of parameters to be estimated, then $r(t)$ is forced to zero. Equation (3.1) becomes

$$y(t) = a_0^* + \sum_{i=1}^{n} a_i^* x_i(t) \tag{3.2}$$

where the asterisk indicates the optimal estimated values of the parameters.

The multiple linear regression technique has found greatest application as an offline forecasting method and is generally unsuitable for online forecasting because it requires many external variables that are difficult to introduce into an online algorithm [44].

These models are relatively simple to apply but require extensive initial analysis to identify the regressors and their place in each model. Also, because the relationship between the load and weather variable is time specific, this model requires continuous reestimation of its parameters to perform accurately.

3.6.2 General Exponential Smoothing

In the general exponential smoothing technique, the load is modeled using a time-dependent fitting function that satisfies the relationship [7,44]

$$f(t) = Lf(t-1) \tag{3.3}$$

where $f(t)$ is the fitting function at time t, and L is a constant matrix called the transition matrix [44]. Mathematically, the model is expressed as

$$y(t) = \beta(t) f(t) + r(t) \tag{3.4}$$

where $y(t)$ = load at time t, $\beta(t)$ = coefficient vector at time t, and $r(t)$ = residual load or noise at time t.

The parameter vector is estimated from a data window of previous observations using the LES minimization technique. The estimated parameter vectors are obtained by minimizing the cost function

$$J = \sum_{j=0}^{N-1} w^j [y(N-j) - f(-j)\beta]^2 \tag{3.5}$$

where w is called the weighting factor, and $(1-w)$ is called the smoothing constant. The parameter vector that minimizes the cost function J can be written as

$$\hat{\beta}(N) = F^{-1}(N)h(N) \tag{3.6}$$

where

$$F(N) = \sum_{j=0}^{N-1} w^j f(-j)f^T(-j) \tag{3.7}$$

and

$$h(N) = \sum_{j=0}^{N-1} w^j f(-j)y(N-j) \tag{3.8}$$

The forecast at a lead time l is then given by

$$\hat{y}(N+1) = f(1)\hat{\beta}(N) \tag{3.9}$$

and the parameters of the forecasts can be updated using

$$\hat{\beta}(N+1) = L^T\hat{\beta}(N) + F^{-1}f(0)[y(N+1) - \hat{y}(N)] \tag{3.10}$$

and

$$\hat{y}(N+1+1) = f^T(1)\hat{\beta}(N+1) \tag{3.11}$$

This method can be used for both online and offline forecasting, although its recursive nature and generally poor long-range accuracy make it much more suitable for online forecasting. The low accuracy encountered for longer lead times stems from the fact that this technique cannot use related weather information, so this technique cannot account for weather-related load changes. Simplicity, recursiveness, and economical usage, however, make this technique a very attractive forecasting tool in practice.

3.6.3 Stochastic Time Series

In the stochastic time-series method, the load is modeled as the output of a linear filter driven by white noise [7,44]. Depending on the characteristics of the linear filter, different load models can be formulated.

The autoregressive (AR) and moving average (MA) processes are the two simplest forms of stochastic time series, and although neither of these processes is usually individually capable of accurately modeling the load, they form the basis for development of more complex processes.

In the autoregressive process, the current value of load is expressed linearly in terms of previous values and a random noise. The order of this process depends on the oldest previous value for which the load is regressed. The moving average process, on the other hand, expresses the load linearly in terms of current and previous values of a white noise series, and again the order of the series depends on the oldest previous value.

The autoregressive and moving average processes are usually combined to give the popular autoregressive moving average (ARMA) process, which has found widespread use in the power industry. In the ARMA process, the load at any instant is expressed as a linear combination of its past values and a white noise series. The order of this process is specified by the order of the AR and MA series included in its composition [1].

Time series used for AR, MA, or ARMA models are referred to as stationary processes when their means and covariances are stationary with respect to time. So if the process being modeled is nonstationary, it is first transformed to a stationary series before being modeled by the AR, MA, or ARMA process [1].

A nonstationary process is made into a stationary one using the method of differencing, and the order of a differencing process refers to the number of times the process has been differenced before achieving stationarity. Differenced processes modeled as AR, MA, or ARMA are now called integrated processes and are relabeled ARI, IMA, and ARIMA.

The autoregressive integrated moving average (ARIMA) process, like the ARMA process, is a very popular load-modeling technique that produces very accurate load forecasts. For longer lead times, however, a seasonal or periodic component must be included into these processes. This inclusion results in what is known as a seasonal process, and the abbreviations SARMA and SARIMA are now used [44].

The lack of weather input into time-series models usually limits their forecasting ability. Expressing these processes in transfer functions form makes it possible to add some weather information. This is usually limited to the singlemost influential variable—that is, temperature—which generally accounts for most of weather-induced load [1].

The popularity of the stochastic time-series approach in online forecasting stems mainly from the level of accuracy available and its ease of online implementation. The identification process of the time-series models is a major disadvantage because the process requires extensive analysis of raw load data through the use of range-mean, correlation, and autocorrelation analysis.

3.6.4 Qualities of Forecasting Models

The review of short-term load-forecasting methods indicates that depending on the forecasting technique employed, many different load models can be developed to predict the same load [8–11]. For these models to be considered good or efficient, however, their formulation must feature certain basic qualities, and their performance must be within tolerable limits.

The literature indicates that some of the preferred qualities in a load-forecasting algorithm include adaptiveness, recursiveness, economy, robustness, and accuracy [44].

3.6.4.1 Adaptiveness

The parameters of a short-term load-forecasting model are usually estimated from a fixed window of data and are accurate for only a specified period of time ahead. As the forecast period elapses and new measurements become available, the algorithm should be able to automatically update its data window and recompute its estimates.

3.6.4.2 Recursiveness

As new data such as weather and load measurements become available, the algorithm should be able to correct its forecasts and prediction for the next step.

3.6.4.3 Computational Economy

The pursuit of accuracy can lead to very complicated models that require the use of excessive computing facilities. A forecasting algorithm, however, should attempt to be computationally efficient with regards to execution time and care utilization.

3.6.4.4 Robustness

An algorithm should be robust to misspecification and erroneous data; that is, reasonable forecasts should be produced even if the model is predicting for conditions for which it was not designed, or even if its database is contaminated with bad or anomalous data.

3.6.4.5 Accuracy

The performance of a short-term load-forecasting algorithm depends largely on the forecasting lead time as well as on such factors as load behavior and model type.

For a model with a 24-hour prediction period, errors in the range of 2–3% are considered normal, whereas for models with lead times of 1 hour, the same error is considered large. Models with longer lead times than 24 hours show reduced accuracy, and for a lead time of one week, accuracies within 10% are to be expected.

3.7 Special Load-Forecasting Models

In short-term load forecasting, the future load on a power system is produced by extrapolating a predetermined relationship between the load and its influential variables—namely, time and/or weather information [13–15,44]. Determination of this relationship is a two-stage process that requires both

1. Identifying the relationship between the load and related variables.
2. Quantifying this relationship through the use of a suitable parameters estimation technique.

To study the effects of parameter estimation techniques on short-term load-forecasting accuracy, we need to identify and develop suitable load models that will allow for the application of these estimation techniques.

In the following sections, load models are developed for crisp parameters. These models will be used in both summer and winter forecasting modes, and as such, where applicable, winter and summer load formulations are included. In Chapter 5, fuzzy models are developed for winter and summer loads, and the techniques used to estimate these fuzzy parameters are discussed in Chapter 4. In this part of the chapter, crisp models are presented and discussed. These three models are developed in [44] for offline load models. The parameters are assumed to be crisp. Modifications will be carried out, if necessary, on these models for the fuzzy-type models, as will be seen in subsequent chapters.

The models will be referred to as A, B, and C, respectively. Model A is developed on the basis of multiple linear regression, whereas model B is developed on a harmonic basis; furthermore, model C is a hybrid that embodies properties of models A and B. These models are developed to forecast for 24 hours ahead [17–43,51–78].

3.7.1 Model A: Multiple Linear Regression Model

Model A expresses the load at any discrete time instant *t* as a function of a base load and a weather-dependent component [45–49]. The base load is assumed to be constant for each discrete time interval. The variable part of the load is weather dependent.

This model will be used for both winter and summer load forecast simulations, and because the relationship between load and weather differs significantly over these two seasons, a different load formulation will be required in each case. This will result in two load models—namely, a winter model and a summer model.

These models are based on the assumption that a common daily base load cycle is experienced by weekdays and that a constant but different base load cycle is experienced by weekend days (i.e., Saturday and Sunday). As such, two models are required to predict loads over a complete week (i.e., one for predicting weekday loads and one for predicting weekend loads). Correlation analysis of load and temperature deviations from the norm indicates that the load to be modeled depends on both immediate and previous values of temperature deviations. This correlation, however, is strongest for immediate values of temperature deviations and dies out in approximately 72 hours. The wind-chill and wind-cooling factors also display a similar relationship in winter, as does the temperature and humidity in summer. The wind-cooling factor, however, was selected in favor of wind-chill factor because it generally results in smaller prediction errors during forecast trial [44]. Based on early analyses, initial winter and summer models were formulated and tested in offline simulation mode. The two load model formulations described next were selected [44].

3.7.1.1 Winter Model

Mathematically, the load at any discrete instant *t*, where *t* varies from 1 to 24, can be expressed as

$$\begin{aligned}Y(t) = a_0(t) + a_1(t)T(t) + a_2(t)T^2(t) + a_3(t)T^3(t) + a_4(t)T(t-1) \\ + a_5(t)T(t-2) + a_6(t)T(t-3) + a_7(t)W(t) + a_8(t)W(t-1) \\ + a_9(t)W(t-2)\end{aligned} \tag{3.12}$$

where

$Y(t)$ = load at time t, $t = 1, 2, \ldots, 24$;
$T(t)$ = temperature deviation at time t;
$W(t)$ = wind-cooling factor at time t;
$a_0(t)$ = base load at time t;
$a_1(t), a_2(t), \ldots, a_9(t)$ are the regression parameters to be estimated at time t.

The temperature deviation at the instant t is calculated as the difference between the dry bulb temperature at time t and the average dry bulb temperature of the previous 20 weekdays' (four weeks') temperature measurements, corresponding to the same discrete instant; that is,

$$T(t) = T_d(t) - T_a(t) \tag{3.13}$$

where

$T_d(t)$ is the dry bulb temperature at time t, in °C;
$T_a(t)$ is the average dry bulb temperature at time t.

$$T_a(t) = [T_d(t-24) + T_d(t-48) + \cdots + T_d(t-480)]/20 \tag{3.14}$$

It should be noted that equations (3.13) and (3.14) refer to a database consisting only of weekday temperature recordings.

The wind-cooling factor is calculated from

$$W(t) = [18 - T_d(t)][V(t)]^{1/2} \tag{3.15}$$

where $V(t)$ is the wind speed in km/h at time t.

3.7.1.2 Summer Model

The winter equivalent of the load model given by equation (3.12) can be modified for the summer model to become

$$\begin{aligned} Y(t) = {} & a_0(t) + a_1(t)T(t) + a_2(t)T^2(t) + a_3(t)T^3(t) + a_4(t)T(t-1) \\ & + a_5(t)T(t-2) + a_6(t)T(t-3) + a_7(t)H(t) + a_8(t)H(t-1) \\ & + a_9(t)H(t-2) \end{aligned} \tag{3.16}$$

where

$Y(t)$ = load at time t;
$T(t)$ = temperature deviation at time t;
$a_0(t)$ = base load at time t;
$a_1(t), a_2(t), \ldots, a_9(t)$ are the regression parameters to be estimated at time t.

The temperature deviation is calculated as for the winter model.

The humidity factor $H(t)$, which replaces the wind-cooling factor in the winter model, is given by

$$H(t) = 0.55T_d(t) + 0.2T_p(t) + 5.05 \tag{3.17}$$

where $T_p(t)$ is the dew-point temperature at time t, in °C.

The humidity factor $H(t)$ is set to zero if the dry bulb temperature is less than 25°C, because at temperatures less than room temperature, the humidity effects are negligible.

Equations (3.12) and (3.16) give the multiple linear regression models for the load in winter and summer days. As such, this model is required to estimate 24 parameters (24 sets of the parameters) to predict the next day's hourly load profile.

Equations (3.12) and (3.16) can be rewritten in compact form as

$$Y(t) = f^T(t)X(t) \tag{3.18}$$

where $f(t)$ is a fitting function given by

$$f(t) = \begin{bmatrix} 1 \\ T(t) \\ T^2(t) \\ T^3(t) \\ T(t-1) \\ T(t-2) \\ T(t-3) \\ W(t) \\ W(t-1) \\ W(t-2) \end{bmatrix} \tag{3.19}$$

in winter, and

$$f(t) = \begin{bmatrix} 1 \\ T(t) \\ T^2(t) \\ T^3(t) \\ T(t-1) \\ T(t-2) \\ T(t-3) \\ H(t) \\ H(t-1) \\ H(t-2) \end{bmatrix} \tag{3.20}$$

in summer. Moreover, $X(t)$ is the parameter's vector to be estimated and is given by

$$X(t) = \begin{bmatrix} a_0(t) \\ a_1(t) \\ \vdots \\ a_7(t) \\ a_8(t) \\ a_9(t) \end{bmatrix} \tag{3.21}$$

In this chapter, the parameter vector of equation (3.21) is assumed to be crisp (a vector with constant values at time t). In Chapter 5, this vector will be assumed to be fuzzy (a vector with a certain middle and certain spread).

The corresponding parameters $X(t)$ at any given discrete interval are estimated using the previous four weeks' worth of weekday data corresponding to the discrete instant. The overdetermined system of equations corresponding to the estimates at the instant t will read

$$\begin{bmatrix} y(t-24) \\ y(t-48) \\ y(t-960) \end{bmatrix} = \begin{bmatrix} f(t-24) \\ f(t-48) \\ f(t-960) \end{bmatrix} X(t) \tag{3.22}$$

Equation (3.22), which involves crisp parameters estimation, can be solved using an appropriate estimation technique. After the parameter vector $X(t)$ is estimated, it can be substituted into equation (3.12) or (3.16) to obtain the load prediction for time t.

3.7.2 Model B: Harmonics Model

The load type of model B is expressed as a function of a constant base load and a Fourier harmonic series. It was discovered from studying early load data that there is a presence of a weekly load cycle that is characterized by distinct daily periodicities.

In this model, however, the weekly cycle is accounted for by the use of a daily load model, of which the parameters are estimated seven times weekly. Because this load does not take weather into consideration, a single load model will suffice for both winter and summer simulations.

Therefore, the load at any time t is

$$y(t) = a_0 + \sum_{i=1}^{N} [a_i \sin(i\omega t) + b_i \cos(i\omega t)] \tag{3.23}$$

$$y(t) = a_0 + \sum_{i=1}^{N} c_i \sin(i\omega t + \phi_i) \tag{3.24}$$

where

$$c_i = \sqrt{a_i^2 + b_i^2}$$

$$\tan \phi_i = b_i / a_i$$

Equation (3.23) is the most suitable form to model the load because it is a linear equation in the parameters to be estimated. In equation (3.23),

$y(t)$ = the load at time t;
N = number of harmonics to be chosen;
$\omega = 2\pi/24$;
a_0 = constant base load for each day of the week;
a_i, b_i, $i = 1, \ldots, N$ are the parameters corresponding to the harmonics in the load composition.

To predict the hourly load profile for any day of the week, we set up an over-determined system of equations using data from the previous four weeks corresponding to the day in question.

Equation (3.23) can be rewritten as

$$y(t) = f^T(t)x \tag{3.25}$$

where

$$f(t) = \begin{bmatrix} 1 \\ \sin \omega t \\ \cos \omega t \\ \vdots \\ \sin N\omega t \\ \cos N\omega t \end{bmatrix} \tag{3.26}$$

and

$$X = \begin{bmatrix} a_0 \\ a_1 \\ b_1 \\ \vdots \\ a_N \\ b_N \end{bmatrix} \tag{3.27}$$

The overdetermined system of equations can now be written as

$$\begin{bmatrix} y(t-168) \\ \vdots \\ y(t-192) \\ y(t-336) \\ \vdots \\ y(t-672) \end{bmatrix} = \begin{bmatrix} f^T(t-168) \\ \vdots \\ f^T(t-192) \\ f^T(t-336) \\ \vdots \\ f^T(t-672) \end{bmatrix} [X] \tag{3.28}$$

Having obtained the parameter vector x, we can use equation (3.23) to forecast for the next 24 hours.

3.7.3 Model C: Hybrid Model

Model C consists of the sum of a time-varying base load and a weather-dependent load. This model is developed to eliminate the disadvantages of the previous two models, A and B.

Model A has the advantage of being weather responsive but suffers the disadvantages of requiring (a) 24 separate parameter estimates to predict the next day's load, and (b) the use of weekdays and weekends, both with winter and summer formulations.

Model B requires the use of a single model formulation; therefore, it estimates a single parameter's vector to predict the next day's load. However, it suffers the disadvantage of being weather insensitive.

Models A and B are combined to form model C to obtain a computationally efficient and weather-sensitive model. This new model will eliminate the use of separate weekday and weekend models, as is the case with model A. Also, by limiting the weather input to temperature only, a single load model could be used for both winter and summer load forecast simulations. The main disadvantage of model C is its assumption of a constant relationship between load and weather for all times of the day. However, if there is a set of parameters for every hour, the model becomes computationally inefficient.

Mathematically, load model C can be expressed at any discrete time instant as

$$\begin{aligned} y(t) = a_0 + \sum_{i=1}^{N} [a_i \sin(i\omega t) + b_i \cos(i\omega t)] + c_0 T(t) + c_1 T(t-1) + c_2 T(t-2) \\ + c_3 T(t-3) \end{aligned} \tag{3.29}$$

where $T(t)$ is the temperature deviation at time t, and is given by

$$T(t) = T_d(t) - T_c(t) \tag{3.30}$$

where $T_c(t)$ is the average dry bulb temperature for the discrete instant t, calculated from the previous 28 daily temperature measurements corresponding to the discrete instant; that is,

$$T_c(t) = [T_d(t-24) + \cdots + T_d(t-672)]/28 \tag{3.31}$$

Equation (3.29) can be rewritten in vector form as

$$y(t) = f^T(t)X \tag{3.32}$$

where

$$f^T(t) = [1 \sin \omega t \cos \omega t \ldots \sin N\omega t \cos N\omega t \; T(t) \; T(t-1) \ldots T(t-3)] \tag{3.33}$$

and

$$X^T = [\,a_0 \quad a_1 \quad b_1 \quad \ldots \quad a_N \quad b_N \quad c_0 \quad c_1 \quad \ldots \quad c_3\,] \tag{3.34}$$

and the parameter's vector X can be estimated as for model B; that is, from the system of equations given by

$$\begin{bmatrix} y(t-168) \\ \vdots \\ y(t-192) \\ y(t-336) \\ \vdots \\ y(t-672) \end{bmatrix} = \begin{bmatrix} f^T(t-168) \\ \vdots \\ f^T(t-192) \\ f^T(t-336) \\ \vdots \\ f^T(t-672) \end{bmatrix} [X] \tag{3.35}$$

The next day's forecast can then be done by substituting for X and the predicted values of temperature deviation into equation (3.29).

References

[1] M.A. Abu-El-Magd, N.K. Sinha, Short-term load demand modeling and forecasting: a review, IEEE Trans. Syst. Man Cybern. 12 (3) (1982) 370–382.

[2] M.L. Willis, A.E. Schauer, J.E.D. Northcote, T.D. Vismor, Forecasting distribution system loads using curve shape clustering, IEEE Trans. Power Apparatus Syst. 102 (4) (1983) 893–901.

[3] M.L. Willis, R.W. Powell, D.L. Wall, Load transfer coupling regression curve fitting for distribution load forecasting, IEEE Trans. Power Apparatus Syst. 103 (5) (1984) 1070–1076.

[4] M.L. Willis, J.E.D. Northcote-Green, Comparison tests of fourteen distribution load forecasting methods, IEEE Trans. Power Apparatus Syst. 103 (6) (1984) 1190–1197.

[5] S. Rahman, R. Bhatnager, An expert system based algorithm for short term load forecast, IEEE Trans. Power Syst. 3 (2) (1988) 392–399.
[6] Q.C. Lu, W.M. Grady, M.M. Crawford, G.M. Anderson, An adaptive nonlinear predictor with orthogonal escalator structure for short-term load forecasting, IEEE Trans. Power Syst. 4 (1) (1989) 158–164.
[7] I. Moghram, S. Rahman, Analysis and evaluation of five short-term load forecasting techniques, IEEE Trans. Power Syst. 4 (4) (1989) 1484–1491.
[8] N.F. Hubele, C.-S. Cheng, Identification of seasonal short-term load forecasting models using statistical decision functions, IEEE Trans. Power Syst. 5 (1) (1990) 40–45.
[9] M.E. El-Hawary, G.A.N. Mbamalu, Short-term power system load forecasting using the iteratively reweighted least squares algorithm, Electric Power Syst. Res. 19 (1990) 11–22.
[10] S. Rahman, Formulation and analysis of a rule-based short-term load forecasting algorithm, IEEE Proc. 78 (5) (1990) 805–816.
[11] K.-L. Ho, Y.-Y. Hsu, C.-C. Lising, T.-S. Lai, Short term load forecasting of Taiwan power system using knowledge based expert system, IEEE Trans. Power Syst. 5 (4) (1990) 1214–1221.
[12] A.D. Papalexopoulos, T.C. Hesterberg, A regression-based approach to short-term load forecasting, IEEE Trans. Power Syst. 5 (4) (1990) 1535–1550.
[13] J.H. Park, Y.M. Park, K.Y. Lea, Composite modeling for adaptive short-term load forecasting, IEEE Trans. Power Syst. 6 (2) (1991) 450–457.
[14] Y.-Y. Msu, C.-C. Yang, Design of artificial neural networks for short-term load forecasting. Part I: self-organizing feature maps for day type identification peak load and valley load forecasting, IEE Proc. Gener. Transm. Distrib.-C 138 (5) (1991) 407–413.
[15] Y.-Y. Msu, C.-C. Yang, Design of artificial neural networks for short-term load forecasting. Part II: multilayer feedforward networks for peak load and valley load forecasting, IEE Proc. Gener. Transm. Distrib.-C 138 (5) (1991) 414–418.
[16] W.M. Grady, L.A. Groce, T.M. Huebner, Q.C. Lu, M.M. Crawford, Enhancement, implementation and performance of an adaptive short-term load forecasting algorithm, IEEE Trans. Power Syst. 6 (4) (1991) 1404–1410.
[17] K.Y. Lee, Y.T. Cha, J.H. Park, Short-term load forecasting using an artificial neural network, IEEE Trans. Power Syst. 7 (1) (1992) 124–132.
[18] K.-L. Ho, Y.-Y. Msu, C.-C. Yang, Short-term load forecasting using a multilayer neural network with an adaptive learning algorithm, IEEE Trans. Power Syst. 7 (1) (1992) 141–149.
[19] T.M. Peng, N.F. Huebele, G.G. Karady, Advancement in the application of neural networks for short-term load forecasting, IEEE Trans. Power Syst. 7 (1) (1992) 250–257.
[20] S. Rahman, I. Drezga, Identification of a standard for comparing short-term load forecasting techniques, Electric Power Syst. Res. 25 (1992) 149–158.
[21] S.T. Chen, D.C. Yu, A.R. Moghaddamjo, Weather sensitive short-term load forecasting using non-fully connected artificial neural network, IEEE Trans. Power Syst. 7 (3) (1992) 1098–1105.
[22] Y.-Y. Hsu, K.L. Ho, Fuzzy expert systems: an application to short-term load forecasting, IEE Proc.-C 139 (6) (1992) 471–477.
[23] C.N. Lu, H.T. Wu, S. Vemuri, Neural network based short term load forecasting, IEEE Trans. Power Syst. 8 (1) (1993) 336–342.
[24] S. Rahman, O. Hazim, A generalized knowledge based short-term load forecasting technique, IEEE Trans. Power Syst. 8 (2) (1993) 508–514.
[25] J.-L. Chen, R. Tsai, S.-S. Liang, A distributed problem solving system for short-term load forecasting, Electric Power Syst. Res. 26 (1993) 219–224.

[26] M. Djukanovic, B. Babic, D.J. Sobajic, Y.-H. Pao, Unsupervised/supervised learning concept for 24-hour load forecasting, IEE Proc. Gener. Transm. Distrib.-C 140 (4) (1993) 311–318.
[27] A. Asar, J.R. McDonald, A specification of neural network applications in the load forecasting problem, IEEE Trans. Control Syst. Technol. 2 (2) (1994) 135–141.
[28] D.K. Ranaweara, N.F. Hubele, A.D. Pagalexopoules, Application of radical basis function neural network model for short-term load forecasting, IEE Proc. Gener. Transm. Distrib. 142 (1) (1995) 45–50.
[29] O. Mohamed, D. Park, R. Merchant, T. Dinh, C. Tong, A. Azeem, et al., Practical experiences with an adaptive neural network short-term load forecasting system, IEEE Trans. Power Syst. 10 (1) (1995) 254–265.
[30] J.-F. Chen, W.-M. Wang, C.-M. Huang, Analysis of an adaptive time-series autoregressive moving-average (ARMA) model for short-term load forecasting, Electric Power Syst. Res. 34 (1995) 187–196.
[31] D. Srinivasan, A.C. Liew, C.S. Chang, Applications of fuzzy systems in power systems, Electric Power Syst. Res. J. 35 (1995) 39–43.
[32] A.A. El-Keib, X. Ma, H. Ma, Advancement of statistical based modeling techniques for short term load forecasting, Electric Power Syst. Res. J. 35 (1995) 51–58.
[33] K.-H. Kim, J.-K. Park, K.-J. Hwang, S.-H. Kim, Implementation of hybrid short-term load forecasting system using artificial neural networks and fuzzy expert systems, IEEE Trans. Power Syst. 10 (3) (1995) 1534–1539.
[34] A.G. Bakirtzis, J.B. Theocharis, S.J. Kiartzis, K.J. Satsios, Short term load forecasting using fuzzy neural networks, IEEE Trans. Power Syst. 10 (3) (1995) 1518–1524.
[35] J.A. Momoh, K. Tomsovic, Overview and literature survey of fuzzy set theory in power systems, IEEE Trans. Power Syst. 10 (3) (1995) 1676–1690.
[36] H. Mori, H. Kobayashi, Optimal fuzzy inference for short term load forecasting, IEEE Trans. Power Syst. 11 (1) (1996) 390–396.
[37] H.-T. Yang, C.-M. Haung, C.-L. Haung, Identifications of ARMAX model for short term load forecasting: an evolutionary programming approach, IEEE Trans. Power Syst. 11 (1) (1996) 403–408.
[38] A.G. Bakirtzis, V. Petrldls, S.J. Klartzls, M.C. Alexladls, A.H. Malssis, A neural network short term load forecasting model for the Greek power system, IEEE Trans. Power Syst. 11 (2) (1996) 858–863.
[39] K. Liu, S. Subbaratan, R.R. Shoults, M.T. Manry, C. Kwan, F.L. Lewis, et al., Comparison of very short-term load forecasting techniques, IEEE Trans. Power Syst. 11 (2) (1996) 877–882.
[40] T.W.S. Chow, C.T. Leung, Nonlinear autoregressive integrated neural network model for short term load forecasting, IEE Proc. Gener. Transm. Distrib. 143 (5) (1996) 500–506.
[41] T.W.S. Chow, C.T. Leung, Neural network based short-term load forecasting using weather compensation, IEEE Trans. Power Syst. 11 (4) (1996) 1736–1742.
[42] R. Lamedica, A. Prudenzi, M. Sforna, M. Caciotta, V.O. Cencellli, A neural network based technique for short-term forecasting of anomalous load periods, IEEE Trans. Power Syst. 11 (4) (1996) 1749–1756.
[43] S. Sargunaraj, D.P.S. Gupta, S. Devi, Short-term load forecasting for demand side management, IEE Proc. Gener. Transm. Distrib. 144 (1) (1997) 68–74.
[44] S.A. Soliman, S. Persaud, K. El-Nagar, M.E. El-Hawary, Application of least absolute value parameter estimation based on linear programming to short-term load forecasting, Electr. Power Energy Syst. 19 (3) (1997) 209–216.

[45] S.R. Huang, Short-term load forecasting using threshold autoregressive models, Online no. 19971144, IEE Proc., Gener. Transm. Distrib. 144 (1997) 477.
[46] O. Hyde, P.F. Hodnett, An adaptive automated procedure for short-term electricity load forecasting, IEEE Trans. Power Syst. 12 (1) (1997) 84–94.
[47] P.K. Dash, H.P. Satpathy, A.C. Liew, S. Rahman, A real-time short-term load forecasting system using functional link network, IEEE Trans. Power Syst. 12 (1997) 675–680.
[48] A. Khotanzad, R. Afkhami-Rohani, T.-L. Lu, A. Abaye, M. Davis, D.J. Maratukulam, ANNSTLF—a neural-network basic electric load forecasting system, IEEE Trans. Neural Netw. 8 (4) (1997) 835–846.
[49] M.H. Choueiki, C.A. Mount-Campbell, S.C. Ahalt, Building a quasi optimal neural network to solve short-term load forecasting problem, IEEE Trans. Power Syst. 12 (4) (1997) 1432–1439.
[50] A.S. Al-Fuhaid, M.A. EL-Sayed, M.S. Mahmoud, Cascaded artificial neural networks for short-term load forecasting, IEEE Trans. Power Syst. 12 (4) (1997) 1524–1529.
[51] M.H. Choueiki, C.A. Mount-Campbell, S.C. Ahalt, Implementing a weighted least squares procedure in training a neural network to solve the short-term load forecasting problem, IEEE Trans. Power Syst. 12 (4) (1997) 1689–1694.
[52] J. Vermaak, E.C. Botha, Recurrent neural networks for short-term load forecasting, IEEE Trans. Power Syst. 13 (1) (1998) 126–132.
[53] H.-T. Yang, C.-M. Haung, A new short-term load forecasting approach using self-organizing fuzzy ARMAX models, IEEE Trans. Power Syst. 13 (1) (1998) 217–225.
[54] S.E. Papadakis, J.B. Theocharis, S.J. Kiartzis, A.G. Bakertzis, A novel approach to short-term load forecasting using fuzzy neural networks, IEEE Trans. Power Syst. 13 (2) (1998) 480–492.
[55] W. Charytoniuk, M.S. Chen, P. Van Olinda, Nonparametric regression based short-term load forecasting, IEEE Trans. Power Syst. 13 (3) (1998) 725–730.
[56] I. Drezga, S. Rahman, Input variable selection for ANN-based short-term load forecasting, IEEE Trans. Power Syst. 13 (4) (1998) 1238–1244.
[57] M. Daneshdoost, M. Lotfalian, G. Bumroonggit, J.P. Ngoy, Neural network with fuzzy set-based classification for short-term load forecasting, IEEE Trans. Power Syst. 13 (4) (1998) 1386–1391.
[58] A. Khotanzad, R. Afkhami-Rohani, D.J. Maratukulam, ANNSTLF—artificial neural network short-term load forecasting—generation three, IEEE Trans. Power Syst. 13 (4) (1998) 1413–1422.
[59] A.P. Douglas, A.M. Breipohl, F.N. Lee, R. Adapa, The impacts of temperature forecast uncertainty on bayesian load forecasting, IEEE Trans. Power Syst. 13 (4) (1998) 1507–1513.
[60] R. Aggarwal, Y. Song, Artificial neural networks in power systems—part 3: examples of applications in power systems, Tutorial: ANNs in power systems, Power Eng. J. 12 (1998) 279–287.
[61] P.A. Mastorocostas, J.B. Theocharis, A.G. Bakirtzis, Fuzzy modeling for short-term load forecasting using the orthogonal least squares method, IEEE Trans. Power Syst. 14 (1) (1999) 29–36.
[62] H. Yoo, R.L. Pimmel, Short-term load forecasting using a self-supervised adaptive neural network, IEEE Trans. Power Syst. 14 (2) (1999) 779–784.
[63] H.L. Willis, L.A. Finley, M.J. Buri, Forecasting electric demand of distribution system in rural and sparsely populated regions, IEEE Trans. Power Syst. 10 (4) (1995) 2008–2013.
[64] P.H. Henault, R.B. Eastvedt, J. Peschon, L.P. Hajdu, Power system long term planning in the presence of uncertainty, IEEE Trans. Power Apparatus Syst. PAS-89 (1970) 156–164.

[65] G.S. Christensen, A. Rouhi, S.A. Soliman, A new technique for unconstrained and constrained LAV parameter estimation, Can. J. Electr. Comp. Eng. 14 (1) (1989) 24–30.
[66] Ministry of Electricity and Energy, Egyptian Electricity Authority, Load and energy forecast for the period 1996/1997 to 2009/2010, Report, 1998.
[67] H.K. Temraz, K.M. El-Nagar, M.M.A. Salama, Applications of non-iterative least absolute value estimation for forecasting annual peak electric power demand, Can. J. Electr. Comp. Eng. 23 (4) (1998) 141–146.
[68] D. Srinivasan, T.S. Swee, C.S. Cheng, E.K. Chan, Parallel neural network-fuzzy expert system strategy for short-term load forecasting: system implementation and performance evaluation, IEEE Trans. Power Syst. 14 (3) (1999) 1100–1106.
[69] I. Drezga, S. Rahman, Short-term load forecasting with local ANN predictors, IEEE Trans. Power Syst. 14 (3) (1999) 844–850.
[70] A.A. Ding, Neural-network prediction with noisy predictors, IEEE Trans. Neural Netw. 10 (5) (1999) 1196–1203.
[71] H.C. Wu, C. Lu, Automatic fuzzy model identification for short term load forecast, IEE Proc. Gener. Transm. Distrib. 146 (5) (1999) 477–482.
[72] W. Charytoniuk, M.S. Chen, Very short-term load forecasting using artificial neural networks, IEEE Trans. Power Syst. 15 (1) (2000) 263–268.
[73] K.H. Kim, H.S. Youn, Y.C. Kang, Short-term forecasting for special days in anomalous load conditions using neural networks and fuzzy inference method, IEEE Trans. Power Syst. 15 (2) (2000) 559–565.
[74] P.A. da Silva, L.S. Moulin, Confidence intervals for neural network based short-term load forecasting, IEEE Trans. Power Syst. 15 (4) (2000) 1191–1196.
[75] S.A. Villalba, C.A. Bel, Hybrid demand model for load estimation and short-term load forecasting in distribution electric systems, IEEE Trans. Power Syst. 15 (2) (2000) 764–769.
[76] R.H. Liang, C.C. Cheng, Combined regression-fuzzy approach for short-term load forecasting, IEE Proc. Gener. Transm. Distrib. 147 (4) (2000) 261–266.
[77] H.S. Hippert, C.E. Pedreira, R.C. Souza, Neural networks for short-term load forecasting: a review and evaluation, IEEE Trans. Power Syst. 16 (1) (2001) 44–55.
[78] M. Huang, H.T. Yang, Evolving wavelet-based networks for short-term load forecasting, IEE Proc. Gener. Transm. Distrib. 148 (3) (2001) 222–228.

4 Fuzzy Regression Systems and Fuzzy Linear Models

4.1 Objectives

The objectives of this chapter are

- Introducing principal concepts and mathematical notions of fuzzy set theory, a theory of classes of objects with nonsharp boundaries.
- Reviewing fuzzy sets as a generalization of classical crisp sets by extending the range of the membership function (or characteristic function) from [0, 1] to all real numbers in the interval [0, 1].
- Introducing a number of notions of fuzzy sets, such as representation support, α-cuts, convexity, and fuzzy numbers. We also discuss the resolution principle, which can be used to expand a fuzzy set in terms of its α-cuts.
- Introducing fuzzy mathematical programming and fuzzy multiple-objective decision making. We first introduce the required knowledge behind fuzzy set theory and fuzzy mathematics.
- Introducing fuzzy linear regression. The first part of this discussion describes how to estimate the fuzzy regression coefficients when the set of measurements available is crisp, whereas in the second part the fuzzy regression coefficients are estimated when the available set of measurements is a fuzzy set with a certain middle and spread.
- Introducing some simple examples for fuzzy linear regression.

4.2 Fuzzy Fundamentals

Human beings make tools for their use and also want to control the tools as they desire. A feedback concept is very important in being able to achieve control over these tools. As modern plants with many inputs and outputs become more and more complex, any description of a modern control system requires a large number of equations. Since about 1960, modern control theory has been developed to cope with the increased complexity of modern plants. The most recent developments may be said to be in the direction of optimal control of both deterministic and stochastic systems, as well as the adaptive and learning control of time-variant complex systems. These developments have been accelerated through the use of digital computers.

Modern plants are designed for efficient analysis and production by human beings. We are now confronted with the need to control living cells, which are nonlinear, complex, time variant, and mysterious. They cannot be mastered easily through classical or control theory or even modern artificial intelligence (AI) employing a

DOI: 10.1016/B978-0-12-381543-9.00004-X

powerful digital computer. So we are faced with many problems, and our problems can be seen in terms of decisions, management, and predictions. Solutions can be seen in terms of faster access to more information and of increased aid in analyzing, understanding, and utilizing the information that is not available. These two elements, a large amount of information coupled with a large amount of uncertainty, taken together constitute the basis for many of our problems today: complexity. How do we manage to cope with complexity as well as we do, and how could we manage to cope better? These are the reasons for introducing fuzzy notations because the fuzzy sets method is very useful for handling uncertainties and is essential for the knowledge acquisition of human experts. First, we have to know what *fuzzy* means? Fuzzy essentially means vague or imprecise information.

Everyday language provides one example of the way vagueness is used and propagated; for example, consider driving a car or describing the weather or classifying a person's age. So using the term *fuzzy* is one way engineers describe the operation of a system by means of fuzzy variables and terms. To solve any control problem, we might have a variable. This variable is a crisp set in the conventional control method; that is, it has a definite value and a certain boundary in such a way it can be defined by two groups:

1. Members, or those that certainly belong in the set inside the boundary.
2. Nonmembers, or those that certainly don't belong.

But sometimes collections and categories have boundaries that seem vague, and the transition from member to nonmember appears gradual rather than abrupt. These collections and categories are what we call *fuzzy sets*. Thus, fuzzy sets are a generalization of conventional set theory. Every fuzzy set can be represented by a membership function, and there is no unique membership. A function for any fuzzy set, a membership function, exhibits a continuous curve changing from 0 to 1 or vice versa, and this transition region represents a fuzzy boundary of the term.

For a computer language, we can define *fuzzy logic* as a method of easily representing analog processes with continuous phenomena that are not easily broken down into discrete segments, and the concepts involved are difficult to model sometimes. In conclusion, we can use the term *fuzzy* when

1. One or more of the control variables are continuous.
2. A mathematical model of the process does not exist, or it exists but is too difficult to encode.
3. A mathematical model is too complex to be evaluated fast enough for real-time operation.
4. A mathematical model involves too much memory on the designated chip architecture.
5. An expert is available who can specify the rules underlying the system behavior and the fuzzy sets that represent the characteristics of each variable.
6. A system has uncertainties in either its inputs or definition.

On the other hand, for systems in which conventional control equations and methods are already optimal or entirely adequate, we should avoid using fuzzy logic. One of the advantages of fuzzy logic is that we can implement systems too complex, too nonlinear, or with too much uncertainty to implement using traditional techniques. We also can implement and modify systems more quickly and squeeze additional capability from existing designs. Finally, fuzzy logic is simple to describe and verify. Before we introduce fuzzy models, however, we need to know some definitions:

- **Singleton:** A deterministic word of term or value (e.g., *male* or *female*, *dead* or *alive*, *80°C*, *30 Kg*). These deterministic words and numerical values have neither flexibility nor intervals. So a numerical value to be substituted into a mathematical equation representing a scientific law is a *singleton*.
- **Fuzzy number:** A fuzzy linguistic term that includes imprecise numerical value (e.g., "around 80°C," "bigger than 25").
- **Fuzzy set:** A fuzzy linguistic term that can be regarded as a set of singletons; the grades of it are not only [1] but also range from zero to one [0, 1]. Alternatively, it is a set that allows partial membership states. Whether ordinary or crisp, sets have only two membership states: inclusion and exclusion (member and nonmember). Fuzzy sets allow a degree of membership as well. Fuzzy sets are defined by labels and membership functions, and every fuzzy set has an infinite number of membership functions (μFs) that may represent it.
- **Fuzzy linguistic terms:** Elements that are ordered are fuzzy intervals, and the membership function is a bandwidth of this fuzzy linguistic term. Elements of fuzzy linguistic terms such as "robust gentleman" and "beautiful lady" are discrete and also disordered. This type of term cannot be defined by a continuous membership function, but defined by vectors.
- **Characteristic function:** This is comprised of a singleton, an interval, and a fuzzy linguistic term.
- **Control variable:** A variable that appears in the premise of a rule and controls the state of the solution variables.
- **Defuzzification:** The process of converting an output fuzzy set for a solution variable into a single value that can be used as output.
- **Overlap:** The degree to which the domain of one fuzzy set overlaps with that of another.
- **Solution fuzzy set:** A temporary fuzzy set created by the fuzzy model to resolve the value of a corresponding solution variable. When all the rules have been fired, the solution fuzzy set is defuzzified into the actual solution variable.
- **Solution variable:** The variable of which the value the fuzzy logic system is meant to find.
- **Fuzzy model:** The components of conventional and fuzzy systems are quite alike, differing mainly in that fuzzy systems contain "fuzzifers," which convert inputs into their fuzzy representations, and "defuzzifiers," which convert the output of the fuzzy process logic into "crisp" (numerically precise) solution variables.

In a fuzzy system, the values of a fuzzified input execute all the values in the knowledge repository that have the fuzzified input as part of the premise. This process generates a new fuzzy set representing each output or solution variable. Defuzzification creates a value for the output variable from that new fuzzy set. For physical systems, the output value is often used to adjust the setting of an actuator that, in turn, adjusts the states of the physical systems. The change is picked up by the sensors, and the entire process starts again. Finally, we can say that there are four steps to follow to design a fuzzy model.

Step 1 Define the Model Function and Operational Characteristics

The goal of the first step in designing a fuzzy model is to establish the architectural characteristics of a system and also to define the specific operating properties of the proposed fuzzy system. The fuzzy system designer's task lies in defining what information (data point) flows into the system, what basic operations are performed on the data, and what data elements are output from the system. Even if lacking a mathematical model of the system process, the designer must have a deep understanding of these three phenomena. This step is also the time to define exactly where the fuzzy subsystem fits into the total system architecture, which provides a clear picture of how inputs and outputs flow to and from the subsystem. Then the designer can estimate the number and ranges of inputs and outputs that will be required. This step also reinforces the input process–output design step.

Step 2 Define the Control Surfaces

Each control and solution variable in the fuzzy model is decomposed into a set of fuzzy regions. These regions are given a unique name, called labels, within the domain of the variable. Finally, a fuzzy set that semantically represents the concept associated with the label is created. Some rules of thumb help in defining fuzzy sets:

- First, the number of labels associated with a variable should generally be an odd number from 5 to 9.
- Second, each label should overlap somewhat with its neighbors. To get a smooth stable surface fuzzy controller, the overlap should be between 10% and 50% of the neighboring space, and the sum of vertical points of the overlap should always be less than one.
- Third, the density of the fuzzy sets should be highest around the optimal control point of the system and should decrease as the distance from that point increases.

Step 3 Define the Behavior of the Control Surfaces

The third step in designing a fuzzy model involves writing the rules that tie the input values to the output model properties. These rules are expressed in English-like language with syntax like the following:

If <*fuzzy proposition*>, then <*fuzzy proposition*>

That is, the IF, THEN rule, where fuzzy propositions are "x is y" or "x is not y." x is a scalar variable, and y is a fuzzy set associated with that variable. Generally, the number of rules a system requires is simply related to the number of control variables.

Step 4 Select a Method of Defuzzification

The fourth step in designing a fuzzy model is finding a way to convert an output fuzzy set into a crisp solution variable. The two most common ways are

- The composite maximum
- The composite momentary cancroids

Once the fuzzy model has been constructed, the process of solution and protocycling begins. The model is compared against known test cases to validate the results. When the results are not as desired, changes are made either to the fuzzy set descriptions or to the mappings encoded in the rules.

4.3 Fuzzy Sets and Membership

Fuzzy set theory is developed to improve the oversimplified model, thereby developing a more robust and flexible model to solve real-world complex systems involving human aspects [1,2]. Furthermore, it helps the decision maker not only to consider the existing alternatives under given constraints (optimize a given system), but also to develop new alternatives (design a system). Fuzzy set theory has been applied in many fields, such as operations research, management science, control theory, artificial intelligence/expert system, human behavior, etc.

4.3.1 Membership Functions

A classical (crisp or hard) set is a collection of distinct objects, defined in such a manner as to separate the elements of a given universe of discourse into two groups: those that belong (members) and those that do not belong (nonmembers). The transition of an element between membership and nonmembership in a given set in the universe is abrupt and well defined. The crisp set can be defined by the so-called characteristic function. Let U be a universe of discourse, the characteristic function of a crisp.

4.3.2 Basic Terminology and Definitions

Let X be a classical set of objects, called the universe, of which the generic elements are denoted by x [2]. The membership in a crisp subset of X is often viewed as a characteristic function μ_A from X to $\{0, 1\}$ such that

$$\begin{aligned} \mu_A(x) &= 1 \quad \text{if and only if } x \in \mathrm{A} \\ &= 0 \quad \text{otherwise} \end{aligned} \tag{4.1}$$

where $\{0, 1\}$ is called a valuation set.

If the valuation set is allowed to be the real interval [0, 1], $\widetilde{A}$ is called a fuzzy set proposed by Zadeh [2], and $\mu_A(x)$ is the degree of membership of x in $\widetilde{A}$. The closer the value of $\mu_A(x)$ is to 1, the more x belongs to $\widetilde{A}$ [2]. Therefore, $\widetilde{A}$ is completely characterized by the set of ordered pairs:

$$\widetilde{A} = \{(x, \mu_A(x)) \mid x \in X\} \tag{4.2}$$

It is worth noting that the characteristic function can be either a membership function or a possibility distribution. In this study, if the membership function is preferred, then the characteristic function will be denoted as $\mu_A(x)$. On the other hand, if the possibility distribution is preferred, the characteristic function will be specified as $\pi(x)$. Along with the expression of equation (4.2), Zadeh [2] also proposed the following notations. When X is a finite set $\{x_1, x_2, \ldots, x_n\}$, a fuzzy set $\widetilde{A}$ is then expressed as

$$\widetilde{A} = \mu_A(x_1)/x_1 + \cdots + \mu_A(x_n)/x_n = \sum_i \mu_A(x_i)/x_i \tag{4.3}$$

When X is not a finite set, A then can be written as

$$A = \int_X \mu_A(x)/x \tag{4.4}$$

Sometimes, we might need only objects of a fuzzy set but not its characteristic function to transfer a fuzzy set. To do so, we must consider two concepts: support and α-level cut.

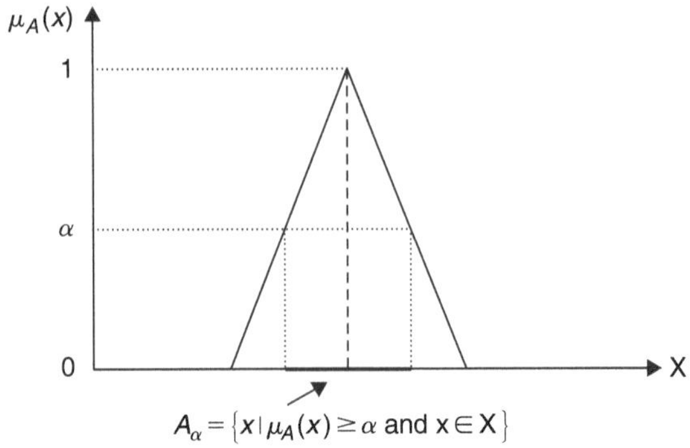

Figure 4.1 The α-level set (α-cut) of a fuzzy set A.

4.3.3 Support of a Fuzzy Set

The support of a fuzzy set A is the crisp set of all $x \in U$ such that $(x) > 0$ [1,2]. That is,

$$\text{supp}(A) = \{x \in U | \mu_A > 0\} \tag{4.5}$$

The α-level set (α-cut) of a fuzzy set A is a crisp subset of X and is denoted by Figure 4.1. An α-cut of a fuzzy set $\tilde{A}$ is a crisp set A, which contains all the elements of the universe U that have a membership grade in $\tilde{A}$ greater than or equal to α. That is,

$$A_\alpha = \{x | \mu_A(x) \geq \alpha \text{ and } x \in X\} \tag{4.6}$$

If $A_\alpha = \{x | \mu_A(x) > \alpha\}$, then A_α is called a strong α-cut of a given fuzzy set A or is called a level set of A. That is,

$$\prod A = \{\alpha | \mu_A(x) = \alpha, \text{ for some } x \in \mathsf{U}\} \tag{4.7}$$

4.3.4 Normality

A fuzzy set A is normal if and only if $\text{Sup}_x \, \mu_A(x) = 1$; that is, the supreme of $\mu_A(x)$ over X is unity. A fuzzy set is subnormal if it is not normal. A nonempty subnormal fuzzy set can be

$$A_\alpha = \{x | \mu_A(x) \geq \alpha \text{ and } x \in X\}$$

normalized by dividing each $\mu_A(x)$ by the factor $\text{Sup}_x \mu_A(x)$. A fuzzy set is empty if and only if $\mu_A(x) = 0 \text{ for } \forall x \in X) \forall x \in X$.

4.3.5 Convexity and Concavity

A fuzzy set A in X is convex if and only if for every pair of point x^1 and x^2 in X, the membership function of A satisfies the inequality

$$\mu_A(\partial x^1 + (1 - \partial)x^2) \geq \min(\mu_A(x^1), \mu_A(x^2)) \tag{4.8}$$

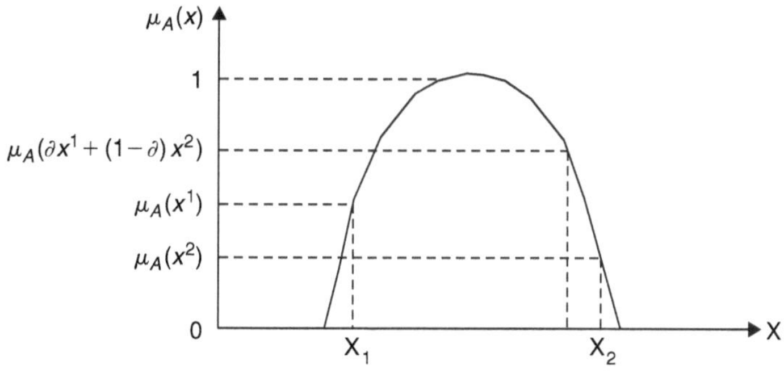

Figure 4.2 A convex fuzzy set.

where $\partial \in [0,1]$ (see Figure 4.2). Alternatively, a fuzzy set is convex if all α-level sets are convex.

Dually, A is concave if its complement A^c is convex. It is easy to show that if A and B are convex, so is $A \cap B$. Dually, if A and B are concave, so is $A \cup B$.

4.3.6 Basic Operation

This section provides a summary of some basic set-theoretic operations that are useful in fuzzy mathematical programming and fuzzy multiple-objective decision making. These operations are based on the definitions from Bellman and Zadeh [1].

1. **Inclusion**
   Let A and B be two fuzzy subsets of X. Then A is included in B if and only if

$$A_\alpha = \{x|\mu_A(x) \geq \alpha \,\text{and}\, x \in X\}$$
$$\mu_A(x) \leq \mu_B(x) \text{ for } \forall x \in X \tag{4.9}$$

2. **Equality**
   A and B are called equal if and only if

$$\mu_A(x) = \mu_B(x) \text{ for } \forall x \in X \tag{4.10}$$

3. **Complementation**
   A and B are complementary if and only if

$$\mu_A(x) = 1 - \mu_B(x) \text{ for } \forall x \in X \tag{4.11}$$

4. **Intersection**
   The intersection of A and B may be denoted by $A \cap B$, which is the largest fuzzy subset contained in both fuzzy subsets A and B. When the min operator is used to express the logical "and," its corresponding membership is then characterized by

$$\begin{aligned} \mu_{A\cap B}(x) &= \min(\mu_A(x), \mu_B(x)) \text{ for } \forall x \in X \\ &= \mu_A(x) \wedge \mu_B(x) \end{aligned} \tag{4.12}$$

   where $\wedge$ is a conjunction.

5. Union

The union ($A \cup B$) of A and B is dual to the notion of intersection. Thus, the union of A and B is defined as the smallest fuzzy set containing both A and B.

The membership function of A ∪ B is given by

$$\begin{aligned}\mu_{A\cup B}(x) &= \max(\mu_A(x), \mu_B(x)) \text{ for } \forall x \in X \\ &= \mu_A(x) \vee \mu_B(x)\end{aligned} \tag{4.13}$$

6. Algebraic Product

The algebraic product AB of A and B is characterized by the following membership function:

$$\mu_{A\cup B}(x) = \mu_A(x)\,\mu_B(x) \text{ for } \forall x \in X \tag{4.14}$$

7. Algebraic Sum

The algebraic sum $A \oplus B$ of A and B is characterized by the following membership function:

$$\mu_{A\oplus B}(x) = \mu_A(x) + \mu_B(x) - \mu_A(x)\mu_B(x) \tag{4.15}$$

8. Difference

The difference $A - B$ of A and B is characterized by

$$\mu_{A\cap B^c}(x) = \min(\mu_A(x), \mu_{B^c}(x)) \tag{4.16}$$

9. Fuzzy Arithmetic

a. Addition of Fuzzy Numbers

The addition of X and Y can be calculated by using α-level cut and max-min convolution.

***α*-level cut**. Using the concept of confidence intervals, the α-level sets of X and Y are $X_\alpha = [X_\alpha^L, X_\alpha^U]$ and $Y_\alpha = [Y_\alpha^L, Y_\alpha^U]$, where the result, Z, of the addition is

$$Z_\alpha = X_\alpha(+)Y_\alpha = [X_\alpha^L + Y_\alpha^L, X_\alpha^U + Y_\alpha^U] \tag{4.17}$$

for every $\alpha \in [0, 1]$.

Max-min convolution. The addition of the fuzzy numbers X and Y is represented as

$$Z(z) = \max_{z=x+y}[\min[\mu_X(x), \mu_Y(y)]] \tag{4.18}$$

b. Subtraction of Fuzzy Numbers

***α*-level cut**. The subtraction of the fuzzy numbers X and Y in the α-level cut representation is

$$Z_\alpha = X_\alpha(-)Y_\alpha = [X_\alpha^L - Y_\alpha^U, X_\alpha^U - Y_\alpha^L] \tag{4.19}$$

for every $\alpha \in [0,1]$.

Max-min convolution. The subtraction of the fuzzy numbers X and Y is represented as

$$\begin{aligned}\mu_Z(Z) = &\max_{z=x-y}\{[\mu_x(x), \mu_Y(y)]\} \\ &\max_{z=x+y}\{[\mu_x(x), \mu_Y(-y)]\} \\ &\max_{z=x+y}\{[\mu_x(x), \mu_{-Y}(y)]\}\end{aligned} \tag{4.20}$$

c. Multiplication of Fuzzy Numbers

***α*-level cut**. The multiplication of the fuzzy numbers *X* and *Y* in the *α*-level cut representation is

$$Z_\alpha = X_\alpha(.)Y_\alpha = \left[X_\alpha^L y_\alpha^L,\ X_\alpha^U Y_\alpha^U\right] \tag{4.21}$$

for every $\alpha \in [0,1]$.

Max-min convolution. The multiplication of the fuzzy numbers *X* and *Y* is represented by Kaufmann and Gupta [2] in the following procedure as

1. Find Z^1 (the peak of the fuzzy number *Z*) such that $\mu_Z(z^1) = 1$; then calculate the left and right legs.
2. The left leg of $\mu_Z(z)$ is defined as

$$\mu_z(z) = \max_{xy \le z}\{\min[\mu_x(x), \mu_Y(y)]\} \tag{4.22}$$

3. The right leg of $\mu_Z(z)$ is defined as

$$\mu_z(z) = \max_{xy \ge z}\{\min[\mu_x(x), \mu_Y(y)]\} \tag{4.23}$$

d. Division of Fuzzy Numbers

***α*-level cut**. The division is represented as follows:

$$Z_\alpha = X_\alpha(:)Y_\alpha = \left[x_\alpha^L / y_\alpha^U, x_\alpha^U / y_\alpha^L\right] \tag{4.24}$$

Max-min convolution. As defined earlier, we must find the peak and then the left and right legs:

1. The peak $Z = X\ (:)\ Y$ is used.
2. The left leg is presented as

$$\begin{aligned} \mu_z(z) = & \max_{x/y \le z}\{\min[\mu_x(x), \mu_Y(y)]\} \\ & \max_{xy \le z}\{\min[\mu_x(x), \mu_Y(1/y)]\} \\ & \max_{xy \le z}\{\min[\mu_x(x), \mu_{1/Y}(y)]\} \end{aligned} \tag{4.25}$$

3. The right leg is presented as

$$\begin{aligned} \mu_z(z) = & \max_{x/y \ge z}\{\min[\mu_x(x), \mu_Y(y)]\} \\ & \max_{xy \ge z}\{\min[\mu_x(x), \mu_Y(1/y)]\} \\ & \max_{xy \ge z}\{\min[\mu_x(x), \mu_{1/Y}(y)]\} \end{aligned} \tag{4.26}$$

10. *LR*-Type Fuzzy Number

A fuzzy number is defined to be of the *LR* type if there are reference functions *L* and *R* and positive scalars, as shown in Figure 4.3, *α* (left spread), *β* (right spread), and *m* (mean), such that

$$\mu_M(x) = \left\{ \begin{array}{l} L\left(\dfrac{m-x}{\alpha}\right) \text{for } x \le m \\ R\left(\dfrac{x-m}{\beta}\right) \text{for } x \ge m \end{array} \right\} \tag{4.27}$$

As the spread increases, M becomes fuzzier and fuzzier. Symbolically, we write

$$M = (m, \alpha\beta)_{LR} \tag{4.28}$$

11. **Interval Arithmetic**
Interval arithmetic is normally used with uncertain data obtained from different instruments. If we enclose those values obtained in a closed interval on the real line R—that is, this uncertain value is inside an interval of confidence R—$x \in [a_1, a_2]$, where $a_1 \leq a_2$.
12. **Triangular and Trapezoidal Fuzzy Numbers**
Triangular and trapezoidal fuzzy numbers are considered among the most important and useful tools in solving possibility mathematical programming problems. Tables 4.1 and 4.2 show all the formulas used in the *LR* representation of fuzzy numbers and interval arithmetic methods.

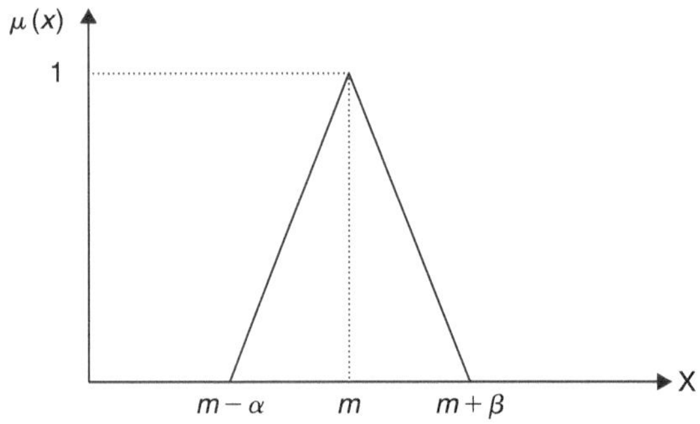

Figure 4.3 *LR*-type fuzzy number.

Table 4.1 Fuzzy Arithmetic on Triangular *LR* Representation of Fuzzy Numbers; $X = (x, \alpha, \beta)$ & $Y = (y, r, \delta)$

Image of Y:	$-Y = (-y, \delta, r)\ -Y = (-y, \delta, r)$
Inverse of Y:	$Y^{-1} = (y^{-1}, \delta y^{-2}, r y^{-2})$
Addition:	$X\,(+)\,Y = (x + y, \alpha + r, \beta + \delta)$
Subtraction:	$X\,(-)\,Y = X(+) - Y = (x - y, \alpha + \delta, \beta + r)$
Multiplication:	$X > 0,\ Y > 0 : X\,(\cdot)\,Y = (xy, xr + y\alpha, x\delta + y\beta)$
	$X < 0,\ Y > 0 : X\,(\cdot)\,Y = (xy, y\alpha - x\delta, y\beta - xr)$
	$X < 0,\ Y < 0 : X\,(\cdot)\,Y = (xy, -x\delta - y\beta, -xr - y\alpha)$
Scalar Multiplication:	$a > 0, a \in R : a\,(\cdot)\,X = (ax, a\alpha, a\beta)$
	$a < 0, a \in R : a\,(\cdot)\,X = (ax, -a\beta, -a\alpha)$
Division:	$X > 0,\ Y > 0 : X\,(:)\,Y = (x/y, (x\delta + y\alpha)/y^2, (xr + y\beta)/y^2)$
	$X < 0,\ Y > 0 : X\,(:)\,Y = (x/y, (y\alpha - xr)/y^2, (y\beta - x\delta)/y^2)$
	$X < 0,\ Y < 0 : X\,(:)\,Y = (x/y, (-xr - y\beta)/y^2, (-x\delta - y\alpha)/y^2)$

Table 4.2 Fuzzy Interval Arithmetic on Triangular Fuzzy Numbers; $X = (x^m, x^p, x^o)$ & $Y = (y^m, y^p, y^o)$

Image of Y:	$-Y = (-y^m, -y^o, -y^p)$
Inverse of Y:	$Y^{-1} = (1/y^m, 1/y^o, 1/y^p)$
Addition:	$X\,(+)\,Y = (x^m + y^m, x^p + y^p, x^o + y^o)$
Subtraction:	$X\,(-)\,Y = X(+) - Y = (x^m - y^m, x^p - y^o, x^o - y^p)$
Multiplication:	$X > 0,\ Y > 0 : X\,(\cdot)\,Y = (x^m y^m, x^p y^p, x^o y^o)$
	$X < 0,\ Y > 0 : X\,(\cdot)\,Y = (x^m y^m, x^p y^o, x^o y^p)$
	$X < 0,\ Y < 0 : X\,(\cdot)\,Y = (x^m y^m, x^o y^o, x^p y^p)$
Scalar Multiplication:	$a > 0, a \in R : a(\cdot)X = (ax^m, ax^p, ax^o)$
	$a < 0, a \in R : a(\cdot)X = (ax^m, ax^o, ax^p)$
Division:	$X > 0,\ Y > 0 : X\,(:)\,Y = (x^m/y^m, x^p/y^o, x^o/y^p)$
	$X < 0,\ Y > 0 : X\,(:)\,Y = (x^m/y^m, x^o/y^o, x^p/y^p)$
	$X < 0,\ Y < 0 : X\,(:)\,Y = (x^m/y^m, x^o/y^p, x^p/y^o)$

4.4 Fuzzy Linear Estimation

The fuzzy parameters linear estimation model or fuzzy regression model can be described by the following equation [3–13]:

$$Y = f(x, A) = A_1 x_1 + A_2 x_2 + A_3 x_3 + \cdots + A_n x_n \tag{4.29}$$

At any observation j; $j = 1, 2, \ldots, m$, equation (4.29) can be rewritten as

$$Y_j = f(x, A) = A_1 x_{1j} + A_2 x_{2j} + A_3 x_{3j} + \cdots + A_n x_{nj} \tag{4.30}$$

In fuzzy regression, the difference between the observed and estimated values is assumed to be due to the ambiguity inherently present in the system. Therefore, the preceding fuzzy regression model is built in terms of the possibility and evaluates all observed values as possibilities that the system should contain. The model in equation (4.29) is named as a possible regression model. In this model Y_j is the observation measurement j. This output observation may be a nonfuzzy or fuzzy observation; A_i, $i = 1, 2, \ldots, n$ are the fuzzy parameters of the model in the form of (p_i, c_i), where p_i is the middle and c_i is the spread. Or, it may take the form of the *LR* type as (p_i, c_i^L, c_i^R), and x_{ij} is the input to the model $i = 1, 2, \ldots, n$ and $j = 1, 2, \ldots, m$. In this section, three cases for the output Y_j are studied.

4.4.1 Nonfuzzy Output ($Y_j = m_j$)

In the nonfuzzy output model, the output Y_j is a nonfuzzy observation, but the model coefficients A_i, $i = 1, 2, \ldots, n$ are fuzzy parameters either in the form of $A_i = (p_i, c_i)$ or, $A_i = (p_i, c_i^L, c_i^R)$, $i = 1, \ldots, n$ for the *LR* type and the input x_{ij} is a nonfuzzy input. The membership functions for each type of A_i are shown in Figures 4.4 and 4.5.

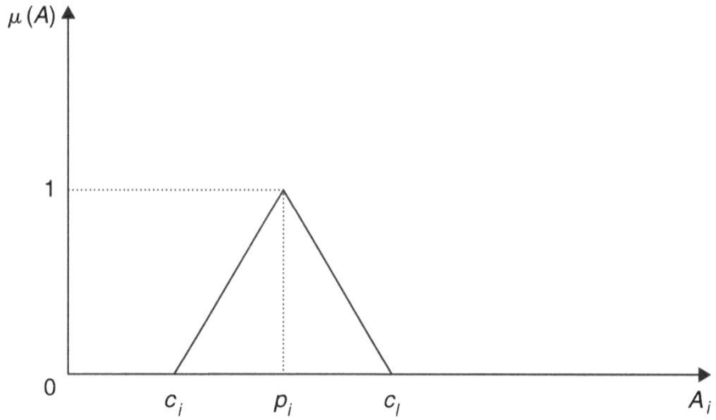

Figure 4.4 Membership functions of the fuzzy coefficients A_j.

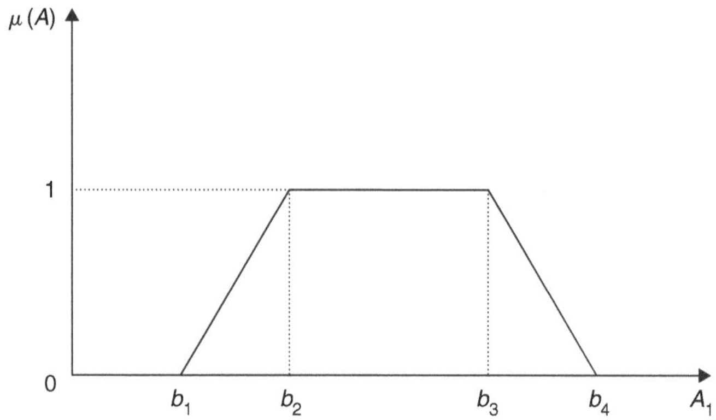

Figure 4.5 Trapezoidal membership function of A_j.

The equation that describes this membership can be written mathematically, for the triangular fuzzy number, as

$$\mu_A = \begin{cases} 1 - \dfrac{|p_j - a_i|}{c_j}; & p_j - c_j \le a_i \le p_j + c_j \\ 0 & \text{otherwise} \end{cases} \tag{4.31}$$

where the membership function of A_j of the *LR* type is assumed to be a trapezoidal function, as shown in Figure 4.5. Note that if $b_2 = b_3$, we obtain the triangular membership. In general, the membership function for the *LR* type can be described as

$$\mu_A = \begin{cases} L\left(p_j - \dfrac{x}{c_j^L}\right) & \text{for } x \le p_j \\ R\left(p_j - \dfrac{x}{c_j^R}\right) & \text{for } x \ge p_j \end{cases} \tag{4.32}$$

where p_i is the middle or the mean of A_j, c_j^L is the left spread, and c_i^R is the right spread.

Equation (4.29) can now be written as

$$Y_j = (p_1, c_1)x_{1j} + (p_2, c_2)x_{2j} + (p_n, c_n)x_{nj}, \qquad j = 1, 2, \ldots, m \tag{4.33}$$

for the first type of the fuzzy parameters and

$$Y_j = (p_1, c_1^L, c_1^R)x_{1j} + (p_2, c_2^L, c_2^R)x_{2j} + (p_n, c_n^L, c_n^R)x_{nj}, \qquad j = 1, 2, \ldots, m, \tag{4.34}$$

for the second type of the fuzzy parameters.

In the nonfuzzy output data regression described by equations (4.33) and (4.34), we seek to find the coefficients $A_i = (p_i, c_i)$ or $A_i = (P_i, c_i^L, c_i^R)$ that minimize the spread of the fuzzy output for all data sets. In mathematical form, this can be described as

Minimize

$$J_1 = \sum_{j=1}^{m} \sum_{i=1}^{n} |c_i x_{ij}| \tag{4.35}$$

such that the fuzzy regression model could contain all observed data in the estimated fuzzy numbers resulting from the model. This can be expressed mathematically as

$$y_j \geq \sum_{i=1}^{n} p_i x_{ij} - (1-\lambda) \sum_{i=1}^{n} c_i x_{ij} \tag{4.36}$$

$$y_j \leq \sum_{i=1}^{n} p_i x_{ij} + (1-\lambda) \sum_{i=1}^{n} c_i x_{ij} \tag{4.37}$$

Note that the first term on the right side of equations (4.36) and (4.37) represents the estimated middle of the fuzzy coefficients, whereas the second term represents the estimated spread of these coefficients and λ is the level of fuzziness and is specified by the user.

For the fuzzy coefficients of the *LR* type, the cost function to be minimized is

Minimize

$$J_1 = \sum_{j=1}^{m} \sum_{i=1}^{n} \left|(2m_j - 2p_j x_{ij} + c_i^L x_{ij} - c_i^R x_{ij})\right| \tag{4.38}$$

subject to satisfying the following two constraints on each data point

$$y_j \geq \sum_{i=1}^{n} p_i x_{ij} - (1-\lambda) \sum_{i=1}^{n} c_i^L x_{ji}, \qquad j = 1, \ldots, \ldots, m \tag{4.39}$$

$$y_j \leq \sum_{i=1}^{n} p_i x_{ij} + (1-\lambda) \sum_{i=1}^{n} c_i^L x_{ji}, \qquad j = 1, \ldots, \ldots, m \tag{4.40}$$

The problems formulated in equations (4.35), (4.36), and (4.37) and formulated in equations (4.38), (4.39), and (4.40) are linear optimization problems, which can be solved by the well-known linear programming–based simplex method. However, if the sum of the absolute value deviations in equations (4.35) and (4.38) is to be minimized, subject to satisfying the inequality constraints given by equations (4.36) and (4.37) and equations (4.39) and (4.40), then the problem turns out to be one of the least absolute value linear optimization problems and can be solved by using the software package RLAV available in the IMSL/STAT library.

4.4.2 Fuzzy Output Systems

If the output is fuzzy, in this case it may be represented by a fuzzy number in the form $Y_j = (m_j, \alpha_j)$ in the case of a triangular fuzzy number (TFN) or $Y_j = (m_j, \alpha_j^L, \alpha_j^R)$, $j = 1, 2, \ldots, m$ in the case of a trapezoidal membership function. For the TFN membership function, equation (4.30) can be written as

$$Y_j = (m_j, \alpha_j) = (p_1, c_1)x_{1j} + (p_2, c_2)x_{2j} + (p_n, c_n)x_{nj} \qquad j = 1, 2, \ldots, m \tag{4.41}$$

which can be rewritten as

$$(m_j, \alpha_j) = (p_1x_{1j} + p_2x_{2j} + \cdots + p_nx_{nj}, c_1x_{1j} + c_2x_{2j} + c_nx_{nj}), \quad j = 1, 2, \ldots, m \tag{4.42}$$

$$(m_j, \alpha_j) = \left(\sum_{i=1}^{n} p_ix_{ij}, \sum_{i=1}^{n} c_ix_{ij} \right) \tag{4.43}$$

Equation (4.43) is valid when

$$m_j = \sum_{i=1}^{n} p_ix_{ij} \qquad j = 1, 2, \ldots, m \tag{4.44}$$

$$\alpha_j = \sum_{i=1}^{n} c_ix_{ij} \qquad j = 1, 2, \ldots, m \tag{4.45}$$

The problem now turns out to be as follows: Given the fuzzy output $Y_j = (m_j, \alpha_j)$, the task is to find the fuzzy parameters (p_i, c_i), $i = 1, 2, \ldots, n$ that minimize the cost function given by

$$J_1(p_i, c_i) = \sum_{j=1}^{m} \left| \left(m_j - \sum_{i=1}^{n} p_ix_{ij} + \alpha_j - \sum_{i=1}^{n} c_ix_{ij} \right) \right| \tag{4.46}$$

subject to satisfying the following constraints on each measurement point:

$$m_j - (1-\lambda)\alpha_j \geq \sum_{i=1}^{n} p_i x_{ij} - \sum_{i=1}^{n} c_i x_{ij} \qquad j = 1, 2, \ldots, n \tag{4.47}$$

$$m_j + (1-\lambda)\alpha_j \leq \sum_{i=1}^{n} p_i x_{ij} + \sum_{i=1}^{n} c_i x_{ij} \qquad j = 1, 2, \ldots, n \tag{4.48}$$

If the fuzzy output is of the *LR* type, then equation (4.42) can be rewritten as

$$(m_j, \alpha_j^L, \beta_j^R) = \left(\sum_{i=1}^{n} p_i x_{ij}, \sum_{i=1}^{n} c_i^L x_{ij}, \sum_{i=1}^{n} c_i^R x_{ij}\right) \tag{4.49}$$

Equation (4.49) can be separated into the following equations:

$$m_j = \sum_{i=1}^{n} p_i x_{ij} \qquad j = 1, \ldots, \ldots, \ldots, m \tag{4.50}$$

$$\alpha_j^L = \sum_{i=1}^{n} c_i^L x_{ij} \qquad j = 1, \ldots, \ldots, \ldots, m \tag{4.51}$$

$$\beta_j^R = \sum_{i=1}^{n} c_i^R x_{ij} \qquad j = 1, \ldots, \ldots, \ldots, m \tag{4.52}$$

In this case, the objective function to be minimized is given as

$$J_1 = \frac{1}{4}\sum_{i=1}^{n} \left\| \left[4m_j - 4\sum_{i=1}^{n} p_i x_{xj} - \alpha_j^L + \sum_{i=1}^{n} c_i^L x_{xj} - \beta_j^R - \sum_{i=1}^{n} c_i^R x_{ij} \right] \right\| \tag{4.53}$$

subject to satisfying the following constraints:

$$m_j - (1-\lambda)c_j^L \geq \sum_{i=1}^{n} p_i x_{ij} - \sum_{i=1}^{n} c_i^L x_{ij}, \qquad j = 1, 2, \ldots, n \tag{4.54}$$

$$m_j + (1-\lambda)c_j^R \leq \sum_{i=1}^{n} p_i x_{ij} + \sum_{i=1}^{n} c_j^R x_{ij}, \qquad j = 1, 2, \ldots, n \tag{4.55}$$

Again, the problems formulated in equations (4.46), (4.47), and (4.48) and those formulated in equations (4.53), (4.54), and (4.55) for *LR* type are all linear optimization problems subjected to a set of linear constraints. These problems can be solved using the standard linear programming–based simplex method. However, if the objective functions are the minimization of the sum of the absolute value of the deviation, then the least absolute value optimization technique based on linear programming is used to solve the problems formulated here.

Example 4.1

Consider the data set shown in Table 4.3. The task here is to find a fuzzy model in the form of $y_j = A_0 + A_1x_{1j} + A_2x_{2j}$; $j = 1,\ldots,5$ that fits this set of data.

The output is nonfuzzy data. The objective function to be minimized is

$$J = c_0 + c_1\sum_{j=1}^{m} x_{ij} + c_2\sum_{j=1}^{m} x_{ij}$$
$$= c_0 + 3.68c_1 + 2.05c_2$$

subject to satisfying the following two constraints on each data point for $j = 1, 2, 5$:

$$y_j \geq (p_0 + p_1x_{1j} + p_2x_{2j}) - (1-\lambda)(c_0 + c_1x_{1j} + c_2x_{2j})$$

$$y_j \leq (p_0 + p_1x_{1j} + p_2x_{2j}) + (1-\lambda)(c_0 + c_1x_{1j} + c_2x_{2j})$$

The solution to the preceding linear programming problem using the simplex method is

$$A_0^* = (0.4391, 0.204), A_1^* = (0.0, 0.0), A_2^* = (6.963, 0.0)$$

With the cost function of $J = 0.204$, while the residual of each data point can be calculated as

$$Y_j^* = (0.4391,\ 0.204) + (6.963,\ 0.0)x_{2j}$$

the middle is

$$m_j^* = 0.4391 + 6.963x_{2j} \qquad j = 1,\ldots,\ldots,5$$

which gives a residual vector of

$$r^* = (-0.10208,\ -0.00986,\ 0.10199,\ -0.10200,\ 0.0683)^T$$

The preceding results are obtained when $\lambda = 0.5$ and the degree of fuzziness $= 0.5$.

Note that A_0^* is a fuzzy parameter because it has a spread of $c_0 = 0.204$, but the coefficients A_1^* and A_2^* are not fuzzy parameters.

Table 4.3 Five-Data Sample for Example 4.1

Y	x_1	x_2
3.54	0.84	0.46
4.05	0.65	0.52
4.51	0.76	0.57
2.63	0.70	0.30
1.90	0.73	0.20

Example 4.2

In a fuzzy regression, the output y is a TFN with c_j representing the error. The fuzzy output and the corresponding crisp input are given as:

(y_i, c_i)	x_i
(2.1, 0.2)	0.52
(4.6, 0.35)	1.36

where $\lambda = 0.4$ determines the fuzzy coefficients for a simple model $Y_j = A_0 + A_1 x_j$.

Because the output is a fuzzy number of TFN memberships, then the cost function to be minimized is

$$J = \sum_{j=1}^{2}(m_j - (p_0 + p_1 x_{1j}) + \alpha_j - (c_0 + c_1 x_{1j}))$$

$$= 7.25 - (p_0 + p_1 x_{1j})$$

Because the first term is a constant one, the cost function to be minimized is

$$J = -(p_0 + p_1 x_{1j} + c_0 + c_1 x_{1j})$$

subject to satisfying the following constraints:

$$m_j - 0.6\alpha_j \geq (p_0 + p_1 x_{1j}) - (c_0 + c_1 x_{1j}), \qquad j = 1, 2$$

$$m_j + 0.6\alpha_j \geq (p_0 + p_1 x_{1j}) + (c_0 + c_1 x_{1j}), \qquad j = 1, 2$$

Substituting the preceding data given we obtain

$$0.9 \geq p_0 + 0.52 p_1 - c_0 - 0.52 c_1$$

$$4.39 \geq p_0 + 1.36 p_1 - c_0 - 1.36 c_1$$

$$2.22 \leq p_0 + 0.52 p_1 + c_0 + 0.52 c_1$$

$$4.81 \leq p_0 + 1.36 p_1 + c_0 + 1.36 c_1$$

Using the linear programming–based simplex method approach, we obtain the following solution:

$$A_0^* = (2.855,\ 1.955),\ A_1^* = (0.0, 0.0)$$

where $J = 2.44$ and the residual vector of the inequality constraint is

$$r^* = (0.0,\ -3.49,\ -2.59,\ 0.00)^T$$

This indicates that the obtained solution is valid.

Example 4.3

This example is for the *LR* type; the solution obtained was based on minimum least absolute deviation. The data for the TFN fuzzy output are listed in Table 4.4.

The model required to fit these data points is in the form

$$Y_j = A_0 + A_1 x_j + A_1 x_j^2, \qquad j = 1, \ldots, \ldots, 16$$

The cost function to be minimized in this case is given as

$$J = 0.25\sum_{j=1}^{16}\left(4m_j - \alpha_j^L + \beta_j^R\right) - 0.25[(4p_0 + 96p_1 + 148.4p_2$$
$$- c_0^L - 24c_1^L - 96.6c_2^L + c_0^R + 24c_1^R + 96.6c_2^R)]$$

As mentioned earlier, if the first term of *J* is constant, then the cost function to be minimized is

$$J_1 = -0.25\left[(4p_0 + 96p_1 + 148.4p_2 - c_0^L - 24c_1^L - 96.6c_2^L + c_0^R + 24c_1^R + 96.6c_2^R)\right]$$

subject to satisfying the inequality constraints given as

$$m_j - (1-\lambda)c_j^L \geq \left(p_0 + p_1 x_{1j} + p_2 x_{2j}^2\right) - \left(c_0^L + c_1^L x_{1j} + c_2^L x_{2j}^2\right),$$
$$j = 1, \ldots, \ldots, 16$$

$$m_j + (1-\lambda)c_j^L \leq \left(p_0 + p_1 x_{1j} + p_2 x_{2j}^2\right) + \left(c_0^L + c_1^L x_{1j} + c_2^L x_{2j}^2\right),$$
$$j = 1, \ldots, \ldots, 16$$

Note that, the number of parameters to be estimated is 9 and the number of inequality constraints is 32. The solution to the fuzzy parameters for the proposed model has been found to be

$$A_0 = (12.75,\ 2.75,\ 0.0),\ \underline{A}_1 = (42.1,\ 0,\ 0),\ \text{and}$$
$$\underline{A}_2 = (140.794,\ 144.3,\ 0.0)$$

Table 4.4 Data for Example 4.3

No.	x_j	$Y_j = (m_j, \alpha_j^L, \beta_j^R)$	No.	x_j	$Y_j = (m_j, \alpha_j^L, \beta_j^R)$
1	0.0	(11.5, 3, 2.5)	9	1.6	(84., 15., 16.)
2	0.2	(24.8, 4.5, 4.)	10	1.8	(82., 15., 16.)
3	0.4	(40. 6., 7.)	11	2.0	(103.7, 16., 17.)
4	0.6	(45.2, 7., 7.)	12	2.2	(102.6, 16., 17.)
5	0.8	(49.1, 9. 9.)	13	2.4	(103.1, 16., 17.)
6	1.0	(70. 11. 12.)	14	2.6	(111., 17., 19.)
7	1.2	(70.9, 12. 12.)	15	2.8	(109., 17. 19.)
8	1.4	(80.1, 14. 15.)	16	3.0	(121.7, 18. 21.)

with an alarm from the linear program that this solution is not the unique solution. The model equation in this case can be written as

$$\underset{\sim}{Y} = (12.75, 2.75, 0.0) + (42.1, 0, 0)x + (140.794, 144.3, 0.0)x^2$$

This model satisfies all the constraints on the fuzzy optimization problem formulated previously. In the next section, we offer an example for the electrical load estimation.

Example 4.4

The fuzzy linear parameter estimation algorithm, proposed in the preceding sections, is implemented for determination of the required distribution system under uncertain conditions [18]. The uncertainty appears at input, at output, and in the nature of the system itself. Measured data are given in Table 4.5, where E_r is the yearly energy consumption, P_i is the installed capacity of electrical equipment at customers' sites, and P_r is the yearly peak load.

The task is to build a fuzzy linear model that relates the yearly energy consumption E_r and P_r in the form of

$$P_r = \underset{\sim}{A}_0 + \underset{\sim}{A}_r E_r$$

or

$$P_r = (p_0, c_0) + (p_1, c_1)E_r$$

Table 4.5 Power Measured at a Substation for Example 4.4

#	E_r(MWh)	P_i(kW)	P_r(kW)
1	21.79	125	30.
2	60.20	247	27.9
3	60.72	436	40.5
4	65.01	406	39.6
5	70.00	265	42.0
6	70.55	251	29.2
7	72.30	520	42.0
8	79.05	540	42.0
9	80.39	310	30.9
10	114.0	443	57.0
11	114.45	573	57.5
12	125.00	438	37.2
13	148.00	578	55.5
14	162.10	610	59.6

The cost function to be minimized is

$$J_1 = c_0 + \sum_{j=1}^{14} c_1 E_{ir}$$

subject to satisfying the following two constraints:

$$P_{ri} \geq (p_0 + p_1 E_{ir}) - (1. - \lambda)(c_0 + c_1 E_{ir}), \qquad i = 1, 2, \ldots, \ldots, 14$$

$$P_{ri} \leq (p_0 + p_1 E_{ir}) + (1. - \lambda)(c_0 + c_1 E_{ir}), \qquad I = 1, 2, \ldots, \ldots, 14$$

The solution to this linear optimization problem when $\lambda = 0.5$ is

$$P_r = (29.76, 21.85) + (0.14692, 0.0)E_r$$

with $J = 21.85$. Note that $\underset{\sim}{A}_0$ is a fuzzy number because it has a spread of 21.85, but $\underset{\sim}{A}_1$ is a crisp number. By using this model, we notice that P_r is fuzzy data having a constant spread of 21.85 along the whole measurement. In other words, the yearly peak load P_r is a fuzzy load having a TFN membership with a middle given in the table and a spread of 21.85 kW.

If λ is chosen to be zero, then the following solution is obtained:

$$P_r = [29.756, 10.925] + [0.14692, 0.0]E_r$$

with $J = 10.925$; that is, the fitted middle model does not change at both values of λ, but as the degree of fuzziness decreases, the spread decreases.

Another test is conducted such that when we model P_r by a second-order model with E_r, it has been shown that the first-order model mentioned previously is adequate to model such a load because the fuzzy coefficient of the second-order term equals zero.

Example 4.5

In Example 4.4, we stated that it is required to model P_r as a function of P_i in a first-order model as

$$P_r = f(P_i) \text{ or}$$

$$P_r = \underset{\sim}{A}_0 + \underset{\sim}{A}_1 P_i$$

The cost function to be minimized in this case, according to the data available in Table 4.5, is

$$J_1 = c_0 + 5742\,c_i$$

subject to satisfying the following two constraints on the measurement set:

$$P_{rj} \geq (p_0 + p_1 P_{ij}) - (1. - \lambda)(c_0 + c_1 P_{ij}), \qquad j = 1, 2, \ldots, \ldots, 14$$

$$P_{rj} \leq (p_0 + p_1 P_{ij}) + (1. - \lambda)(c_0 + c_1 P_{ij}), \qquad j = 1, 2, \ldots, \ldots, 14$$

The results obtained in this test for $\lambda = 0.5$ are

$$P_r = (25.78, 19.6813) + (0.0483, 0.0)P_i$$

with $J = 19.6813$. It has been found that such a model is adequate for these data, and a higher-order model gives zero fuzzy coefficients.

If the yearly peak load P_r is presented as a function of E_r and P_i as

$$P_r = \underline{A}_0 + \underline{A}_1 E_r + \underline{A}_2 P_i$$

then the cost function to be minimized in this case is

$$J = c_0 + 1243.56c_1 + 5742c_2$$

subject to satisfying

$$(P_r)_j \geq p_0 + p_1(E_r)_j + p_2(P_i)_j - (1 - \lambda)(c_0 + c_1(E_r)_j + c_2(P_i)_j),$$
$$j = 1, \ldots, \ldots, 14$$

$$(P_r)_j \leq p_0 + p_1(E_r)_j + p_2(P_i)_j + (1. - \lambda)(c_0 + c_1(E_r)_j + c_2(P_i)_j),$$
$$j = 1, \ldots, \ldots, 14$$

The solution to the preceding optimization problem at $\lambda = 0.5$ is

$$P_r = (25.51, 15.59) + (0.0027, 0.00)E_r + (0.0483, 0.0)P_i$$

with $J = 19.59$. Note that $\underline{A}_0$ is a fuzzy number with a spread of 19.59 and that P_r is a fuzzy number with a spread = 19.59.

Example 4.6

The yearly peak load in the preceding example is given as an *LR* type, as shown in Table 4.6 [18]. The task is to model this load as

$$P_s = \underline{A}_0 + A_r E_r$$

where $\underline{A}_0 = \left(p_0, c_0^L, c_0^R\right)$ and $\underline{A}_1 = \left(p_1, c_1^L, c_1^R\right)$.

Using the cost function defined in equation (4.53) and the constraints defined in equations (4.54) and (4.55), for this linear model, we obtain the following results:

$$P_s = (0.0, 0.0, 0.0) + (2.134, 1.974, 0.0)E_r$$

Now, the spread of P_r at a given measurement j, is

$$\left(c_s^L, c_s^R\right) = (1.974E_r,\ 0.0)_j \qquad j = 1, \ldots, \ldots, 14$$

while the middle is

$$(m_s)_j = (2.134E_r)_j \qquad j = 1, \ldots, \ldots, 14$$

Table 4.6 Samples of Measurements in SS for Example 4.6

#	E_r(MWh)	P_i(kW)	P_r(kW)
1	21.79	125	(30, 25, 33)
2	60.20	247	(27.9, 24, 30.5)
3	60.72	436	(40.5, 34.8, 44.9)
4	65.01	406	(39.6, 36.1, 43)
5	70.00	265	(42.0, 38, 45.7)
6	70.55	251	(29.2, 26, 33)
7	72.30	520	(42.0, 37.6, 45)
8	79.05	540	(42.0, 38.5, 46)
9	80.39	310	(30.9, 27, 34.5)
10	114.0	443	(57.0, 52.5, 60.9)
11	114.45	573	(57.5, 52.8, 61.4)
12	125.00	438	(37.2, 34.4, 40.8)
13	148.00	578	(55.5, 52, 59.8)
14	162.10	610	(59.6, 55.3, 64.7)

4.5 Fuzzy Short-Term Load Modeling

Most of the work on offline short-term load models available today assumes that the parameters of the model are constant crisp values [14–19]. This assumption is, to some extent, true, as long as there are no big changes in weather parameters from day to day. The load power is characterized by both uncertainty and ambiguity.

In this section, the load models used in Chapter 3 are reformulated to account for fuzziness of the load characteristics. In the first subsection the input is assumed to be crisp, whereas the load model parameters are expressed as fuzzy numbers having a certain middle and spreads. Three models are used in this section—namely, fuzzy load models A, B, and C. Fuzzy load model A is a multiple linear regression model. This model takes into account the weather parameters. Fuzzy load model B is a harmonic model and does not account for the weather parameters. Fuzzy load model C is a hybrid model that combines models A and B and takes into account the weather parameters.

In this section we assume that the input data are fuzzy numbers having certain middles and spreads. The parameters of the load model are fuzzy. The fuzzy numbers used for the fuzzy variables in this section are assumed to have a symmetrical triangular membership function.

The following system is considered:

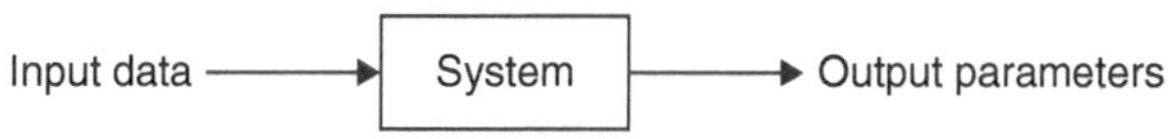

If the input data are crisp (nonfuzzy) and the system parameters A_i (i, $1, \ldots, n$) are crisp (nonfuzzy), then the output is also crisp (nonfuzzy) with an error deviation between the actual and the estimated or predicted values. If the input data are crisp (nonfuzzy) and the system parameters are fuzzy and follow a membership function (e.g., a triangular membership function), then the output is fuzzy and follows the same membership as in the system parameters. If the input data are fuzzy and the system parameters are fuzzy, then the output is fuzzy. The output will have some resemblance of shape to the membership function used.

The membership functions used in this section are triangular membership functions with fuzzy numbers having a certain middle and equal left and right spreads. The objective of the fuzzy parameters estimation is to minimize the spreads of the fuzzy parameters. If spreads of zero are attained, then the output is crisp with an error deviation from the actual value. If the spreads are minimized, then the output will follow the shape of a triangular membership function and the output value will be in a range between upper and lower values.

4.5.1 Multiple Fuzzy Linear Regression Model: Crisp Data

$$(Y_j(t) = m_j(t), \qquad j = 1, \ldots, m;\ t = 1, 2, \ldots, \ldots, 24)$$

The input data of the load model are assumed to be crisp values, whereas the load parameters are fuzzy. The load, in this model, can be expressed mathematically as

$$Y_j(t) = \underline{A}_0 + \sum_{i=1}^{n} \underline{A}_i x_{ij}(t), \qquad j = 1, \ldots, m \tag{4.56}$$

where

$Y_j(t)$ is the value of the load power at time t;
$\underline{A}_0$ is the fuzzy base load having a triangular membership with a middle p_0 and spread c_0, as shown in Figure 4.6(a);
$\underline{A}_i$ are the fuzzy coefficients having a triangular membership with a middle p_i and spread c_i, as shown in Figure 4.6(b).

Equation (4.56) can be rewritten as

$$Y_j(t) = (p_{yj}(t), c_{yj}(t)) = m_j(t) = (p_0, c_0) + \sum_{i=1}^{n} (p_i, c_i)\, x_{ij}(t) \tag{4.57}$$

As shown earlier in this chapter, for the output data described by equation (4.57), the coefficients $\underline{A}_0$ (p_0, c_0) and $\underline{A}_i$ (p_i, c_i) are to be found such that the spread of the fuzzy output is minimized for all data sets. In mathematical form, this can be described as
Minimize

$$J = \left| \sum_t \left\{ c_0 + \sum_{j=1}^{m} \sum_{i=1}^{n} c_i x_{ij}(t) \right\} \right| \tag{4.58}$$

where $t \in [0, t_F]$, t_F is the number of days for which data are taken at the hour in question. The fuzzy regression model in equation (4.58) contains all observed data in the estimated fuzzy numbers resulting from the model. This can be expressed mathematically as

$$y_j(t) \geq \left[p_0 + \sum_{i=1}^{n} p_i x_{ij}(t) \right] - (1 - \lambda) \left[c_0 + \sum_{i=1}^{n} c_i x_{ij}(t) \right]; \quad j = 1, \ldots, m \tag{4.59}$$

and

$$y_j(t) \leq \left[p_0 + \sum_{i=1}^{n} p_i x_{ij}(t) \right] + (1 - \lambda) \left[c_0 + \sum_{i=1}^{n} c_i x_{ij}(t) \right]; \quad j = 1, \ldots, m \tag{4.60}$$

Note that the first term on the right side of equations (4.59) and (4.60) represents the estimated middle of the fuzzy coefficients, and the second term represents the estimated spread of these coefficients. λ is the level of fuzziness and is specified by the user. As λ increases, the fuzziness of the output increases. In the preceding equations, m is the number of observations, and n is the number of fuzzy parameters used in the model.

In the following subsections, two multiple fuzzy linear regression models are developed. The first model can be used to predict the load during the winter season, whereas the second model can be used to predict the load during the summer season. The only difference between the two models is that the winter model considers the

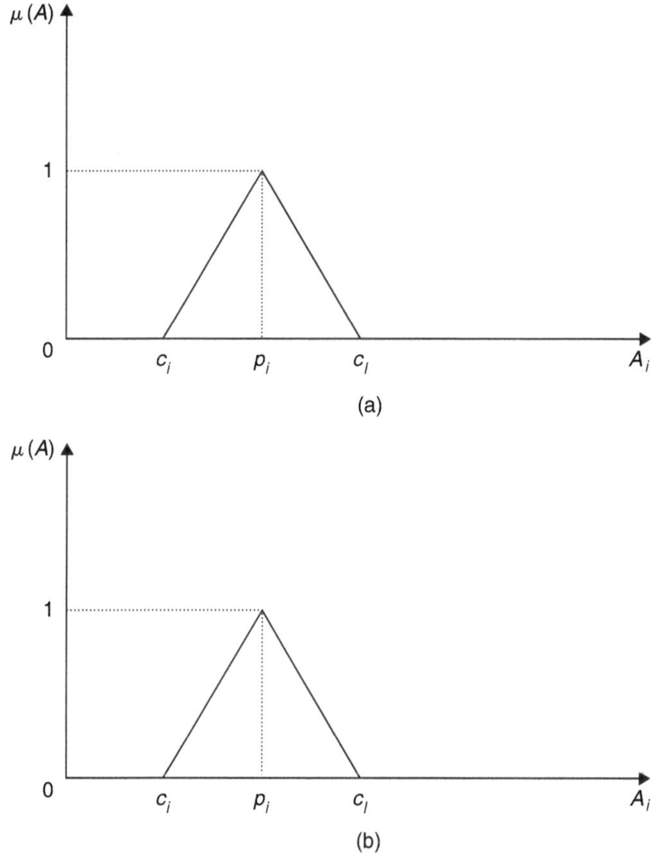

Figure 4.6 (a) Membership function of A_0; (b) membership function of A_I.

wind-cooling factor as an explanatory variable, and the summer model considers the humidity factor as an explanatory variable.

4.5.1.1 Fuzzy Load Model A: Winter Model

The fuzzy winter model, in Chapter 3, equation (3.12), can be rewritten in fuzzy form as

$$\begin{aligned} Y_j(t) = {} & \underline{A}_0 + \underline{A}_1 T_j(t) + \underline{A}_2 T_j^2(t) + \underline{A}_3 T_j^3(t) + \underline{A}_4 T_j(t-1) + \underline{A}_5 T_j(t-2) \\ & + \underline{A}_6 T_j(t-3) + \underline{A}_7 W_j(t) + \underline{A}_8 W_j(t-1) + \underline{A}_9 W_j(t-2);\ j = 1, \ldots, m \end{aligned} \tag{4.61}$$

where $Y_j(t)$ is the load power j; $j = 1, \ldots, m$ at time t; $t = 1, 2, \ldots, 24$ and is assumed to be given as nonfuzzy data. $T_j(t)$ is the jth temperature deviation from nominal at time t and is given by equation (3.13). $W_j(t)$ is the wind-cooling factor at time t and is given by equation (3.15), and $\underline{A}_0, \underline{A}_1, \ldots, \underline{A}_9$ are load model fuzzy coefficients having middles $p_0, p_1, \ldots, p_9$ and spreads $c_0, c_1, \ldots, c_9$.

Equation (4.61) can be rewritten as

$$\begin{aligned} Y_j(t) &= (p_0, c_0) + (p_1, c_1)T_j(t) + (p_2, c_2)\,T_j^2(t) + (p_3, c_3)\,T_j^3(t) \\ &+ (p_4, c_4)T_j(t-1) + (p_5, c_5)T_j(t-2) + (p_6, c_6)T_j(t-3) \\ &+ (p_7, c_7)W_j(t) + (p_8, c_8)W_j(t-1) + (p_9, c_9)W_j(t-2); \\ & j = 1, \ldots, m \end{aligned} \tag{4.62}$$

In fuzzy linear regression, the spreads of the fuzzy coefficients are to be minimized. This results in an objective function that can be expressed mathematically as

$$\begin{aligned} J = \sum_t \Bigg\{ c_0 + \sum_{j=1}^{m} \Big[& c_1 T_j(t) + c_2 T_j^2(t) + c_3 T_j^3(t) + c_4 T_j(t-1). \\ & + c_5 T_j(t-2) c_6 T_j(t-3) + c_7 W_j(t) + c_8 W_j(t)(t-1) + c_9 W_j(t-2) \Big] \Bigg\} \end{aligned} \tag{4.63}$$

where $t \in [0, t_F]$, t_F is the number of days for which data are taken at the hour in question. This is subject to satisfying the two inequality constraints on each load power given as

$$\begin{aligned} y_j(t) \geq\; & p_0 + p_1 T_j(t) + p_2 T_j^2(t) + p_3 T_j^3(t) + p_4 T_j(t-1) + p_5 T_j(t-2) \\ & + p_6 T_j(t-3) + p_7 W_j(t) + p_8 W_j(t-1) + p_9 W_j(t-2) - (1-\lambda)(c_0 + c_1 T_j(t) \\ & + c_2 T_j^2(t) + c_3 T_j^3(t) + c_4 T_j(t-1) + c_5 T_j(t-2) + c_6 T_j(t-3) \\ & + c_7 W_j(t) + c_8 W_j(t-1) + c_9 W_j(t-2)), \quad j = 1, 2, \ldots, m \end{aligned} \tag{4.64}$$

$$\begin{aligned} y_j(t) \leq\; & p_0 + p_1 T_j(t) + p_2 T_j^2(t) + p_3 T_j^3(t) + p_4 T_j(t-1) + p_5 T_j(t-2) \\ & + p_6 T_j(t-3) + p_7 W_j(t) + p_8 W_j(t-1) + p_9 W_j(t-2) \\ & + (1-l)(c_0 + c_1 T_j(t) + c_2 T_j^2(t) + c_3 T_j^3(t) + c_4 T_j(t-1) \\ & + c_5 T_j(t-2) + c_6 T_j(t-3) + c_7 W_j(t) + c_8 W_j(t-1) \\ & + c_9 W_j(t-2)), \; j = 1, 2, \ldots, m \end{aligned} \tag{4.65}$$

The optimization problem formulated in equations (4.63) to (4.65) is linear and can be solved using linear programming based on the simplex method available in the IMSL/STAT library.

Having identified the middle and spread of each coefficient, we can then obtain the fuzzy load model for the winter season using equation (4.61) or equation (4.62).

4.5.1.2 Fuzzy Load Model A: Summer Model

The summer fuzzy model for short-term load forecasting can be written as

$$\begin{aligned} Y(t) = {} & \underline{A}_0 + \underline{A}_1 T(t) + \underline{A}_2 T^2(t) + \underline{A}_3 T^3(t) + \underline{A}_4 T(t-1) + \underline{A}_5 T(t-2) \\ & + \underline{A}_6 T(t-3) + \underline{A}_7 H(t) + \underline{A}_8 H(t-1) + \underline{A}_9 H(t-2) \end{aligned} \tag{4.66}$$

where

$\underline{Y}(t)$ is the summer load power at time t;
$T(t)$ is the temperature deviation at time t given by equation (3.13) in Chapter 3;
$\underline{A}_0, \underline{A}_1, \ldots, \underline{A}_9$ are the fuzzy load coefficients having a certain middle $p_0, p_1, \ldots, p_9$ and certain spread $c_0, c_1, \ldots, c_9$ at time t;
$H(t)$ is the temperature humidity factor given by equation (3.17) in Chapter 3.

The summer load model stated in equation (4.66) takes into account the temperature deviation and the temperature humidity factor for each hour and at three and two hours before.

Equation (4.66) can be rewritten as

$$\begin{aligned} Y_j(t) = {} & (p_0, c_0) + (p_1, c_1)\, T_j(t) + (p_2, c_2)\, T_j^2(t) + (p_3, c_3)\, T_j^3(t) \\ & + (p_4, c_4)\, T_j(t-1) + (p_5, c_5)\, T_j(t-2) + (p_6, c_6)\, T_j(t-3) \\ & + (p_7, c_7)\, H_j(t) + (p_8, c_8)\, H_j(t-1) + (p_9, c_9)\, H_j(t-2) \end{aligned} \tag{4.67}$$

In fuzzy linear regression, the parameters $\underline{A}_i = (p_i, c_i)$, $i = 1, \ldots, \ldots, 9$ are to be found that minimize the spread of the fuzzy output for all data sets. This can be expressed mathematically as

Minimize

$$\begin{aligned} J = \Bigg| \sum_t \Bigg\{ c_0 + \sum_{j=1}^{m} \Big[& c_1 T_j(t) + c_2 T_j^2(t) + c_3 T_j^3(t) + c_4 T_j(t-1) \\ & + c_5 T_j(t-2) + c_6 T_j(t-3) + c_7 H_j(t) + c_8 H_j(t-1) + c_9 H_j(t-2) \Big] \Bigg\} \Bigg| \end{aligned} \tag{4.68}$$

where $t \in [0, t_F]$, t_F is the number of days for which data are taken at the hour in question. This is subject to satisfying the following inequality constraints at j; $j = 1, \ldots, m$:

$$\begin{aligned} y_j(t) \geq {} & p_0 + p_1 T_j(t) + p_2 T_j^2(t) + p_3 T_j^3(t) + p_4 T_j(t-1) + p_5 T_j(t-2) \\ & + p_6 T_j(t-3) + p_7 H_j(t) + p_8 H_j(t-1) + p_9 H_j(t-2) \\ & - (1-\lambda)\big[c_0 + c_1 T_j(t) + c_2 T_j^2(t) + c_3 T_j^3(t) + c_4 T_j(t-1) + c_5 T_j(t-2). \\ & + c_6 T_j(t-3) + c_7 H_j(t) + c_8 H_j(t-1) + c_9 H_j(t-2) \big], \quad j = 1, 2, \ldots, m \end{aligned} \tag{4.69}$$

$$\begin{aligned} y_j(t) \le\ & p_0 + p_1 T_j(t) + p_2 T_j^2(t) + p_3 T_j^3(t) + p_4 T_j(t-1) + p_5 T_j(t-2) \\ & + p_6 T_j(t-3) + p_7 H_j(t) + p_8 H_j(t-1) + p_9 H_j(t-2) \\ & - (1-\lambda)\big[c_0 + c_1 T_j(t) + c_2 T_j^2(t) + c_3 T_j^3(t) + c_4 T_j(t-1) + c_5 T_j(t-2). \\ & + c_6 T_j(t-3) + c_7 H_j(t) + c_8 H_j(t-1) \ + c_9 H_j(t-2)\big], \quad j = 1, 2, \ldots, m \end{aligned} \tag{4.70}$$

The problem formulated in equations (4.68) to (4.70) is linear and can be solved by the linear programming optimization package available in the IMSL/STAT library.

Having obtained the fuzzy parameters $\underline{A}_i = (p_i, c_i),\ i = 1, \ldots, 9,$ we can then predict the load for the next 24 hours using equation (4.67).

4.5.1.3 Fuzzy Load Model B

Fuzzy load model B is a harmonic decomposition model and does not account for weather conditions. It does not account for temperature deviation, wind-cooling factor, or humidity factor. Thus, this model can be used for both winter and summer simulations.

The fuzzy load at any time t, therefore, can be written as

$$Y(t) = \underline{A}_0 + \sum_{i=1}^{n} (\underline{A}_i \sin i\omega t + \underline{B}_i \cos i\omega t) \tag{4.71}$$

where

$\underline{Y}(t)$ is the load power at time t and it is assumed to have crisp values;
$\underline{A}_0$, $\underline{A}_i$, and $\underline{B}_i$ are fuzzy parameters having certain middles and spreads and are given as $\underline{A}_0 = (p_0, c_0)$, $\underline{A}_i = (p_i,\ c_i)$, and $\underline{B}_i = (a_i,\ b_i)$.

The model described in equation (4.71) can be rewritten as

$$Y(t) = (p_0,\ c_0) + \sum_{i=1}^{n} [(p_i, c_i) \sin i\omega t + (a_i, b_i) \cos i\omega t] \tag{4.72}$$

Note that the middles and the spreads are constants and are estimated seven times weekly.

The objective is to find the fuzzy parameters that minimize the spread of the load power. Mathematically, this can be written as

Minimize

$$J = \left| \sum_t \left\{ c_0 + \sum_{j=1}^{m} \sum_{i=1}^{n} \left[c_i x_{ij}(t) + b_i y_{ij}(t) \right] \right\} \right| \tag{4.73}$$

where

$x_{ij}(t) = (\sin i\omega t)_j, \qquad j = 1, \ldots, m;\ i = 1, \ldots, n;$
$y_{ij}(t) = (\cos i\omega t)_j, \qquad j = 1, \ldots, m;\ i = 1, \ldots, n;$

m, n are the number of observations and harmonics chosen in the model, respectively; $t \in [0, t_F]$, t_F is the number of days for which data are taken at the hour in question.

This is subject to satisfying the inequality constraints given by

$$y_j(t) \geq \left[p_0 + \sum_{i=1}^{n}(p_i \sin i\omega t + \alpha_i \cos i\omega t)\right]_j - (1-\lambda)\left[c_0 + \sum_{i=1}^{n}(c_i \sin i\omega t + b_i \cos i\omega t)_j\right] \tag{4.74}$$

$$y_j(t) \leq \left[p_0 + \sum_{i=1}^{n}(p_i \sin i\omega t + \alpha_i \cos i\omega t)\right]_j + (1-\lambda)\left[c_0 + \sum_{i=1}^{n}(c_i \sin i\omega t + b_i \cos i\omega t)_j\right] \tag{4.75}$$

The optimization problem formulated in equations (4.73) to (4.75) is a linear optimization problem and can be solved using the simplex method of linear programming.

Having obtained the fuzzy load parameters, we can then predict the load for the next 24 hours using equation (4.72).

4.5.1.4 Fuzzy Load Model C

Fuzzy load model C is a fuzzy hybrid model that takes into account weather-dependent components. The base load in the model is a time-varying function and takes the form of Fourier's coefficients. This model can be considered as a combination of fuzzy load model A and fuzzy load model B. Here, the weather input is limited only to temperature deviation, and the model is used for both winter and summer load forecast simulations.

The fuzzy load model in this case can be written mathematically as

$$Y_j(t) = \left\{\underline{A}_0 + \sum_{i=1}^{n}[\underline{A}_i \sin i\omega t + \underline{B}_i \cos i\omega t]\right\}_j + \left\{\underline{C}_0 T_j(t) + \underline{C}_1 T_j(t-1) + \underline{C}_2 T_j(t-2) + \underline{C}_3 T_j(t-3)\right\} \tag{4.76}$$

where

$\underline{A}_0$, $\underline{A}_i$, $\underline{B}_i$ and are the weather-independent fuzzy parameters having certain middles and certain spreads;

$\underline{C}_0$, $\underline{C}_1$, $\underline{C}_2$, and $\underline{C}_3$ are the temperature-dependent fuzzy parameters.

The terms in the first brace in equation (4.76) can be considered as the base load, which depends only on time, whereas the terms in the second brace are the temperature-dependent load terms.

Equation (4.76) can be rewritten as

$$Y(t) = (p_0, c_0) + \sum_{i=1}^{n}[(p_i, \alpha_i)x_i(t) + (b_i, \beta_i)\, y_i(t)] + [(\gamma_o, s_0)T_j(t) + (\gamma_1, s_1)\, T_j(t-1) + (\gamma_2, s_2)\, T_j(t-2) + (\gamma_3, s_3)\, T_j(t-3)] \tag{4.77}$$

In equation (4.77), the first letter in the parameter's brackets indicates the middle of that parameter, and the second letter indicates the spread of this parameter.

In fuzzy regression, the fuzzy model parameters are to be found to minimize the spread of the output. In mathematical form, this can be expressed as

$$J = \left| \sum_t \left\{ c_0 + \sum_{j=1}^{m}\sum_{i=1}^{n}[\alpha_i x_{ij}(t) + \beta_i y_{ij}(t)] + \sum_{j=1}^{m}[s_0 T_j(t) + s_1 T_j(t-1) + s_2 T_j(t-2) + s_3 T_j(t-3)] \right\} \right| \tag{4.78}$$

where $t \in [0, t_F]$, t_F is the number of days for which data are taken at the hour in question.

This is subject to satisfying the following two constraints on the output so that the fuzzy regression model could contain all the observed data j, $j = 1, \ldots, m$ in the estimated fuzzy numbers resulting from the model. This can be expressed mathematically as

$$y_j(t) \geq \left[p_0 + \sum_{i=1}^{n}(p_i x_{ij}(t) + b_i y_{ij}(t) + \gamma_0 T_j(t) + \gamma_1 T_j(t-1) + \gamma_2 T_j(t-2) + \gamma_3 T_j(t-3)) \right] - (1-\lambda)\left[c_0 + \sum_{i=1}^{n}(\alpha_i x_{ij}(t) + \beta_i y_{ij}(t)) + s_0 T_j(t) + s_1 T_j(t-1) + s_2 T_j(t-2) + s_3 T_j(t-3) \right], \quad j = 1, \ldots, m \tag{4.79}$$

$$y_j(t) \leq \left[p_0 + \sum_{i=1}^{n}(p_i x_{ij}(t) + b_i y_{ij}(t) + \gamma_0 T_j(t) + \gamma_1 T_j(t-1) + \gamma_2 T_j(t-2) + \gamma_3 T_j(t-3)) \right] + (1-\lambda)\left[c_0 + \sum_{i=1}^{n}(\alpha_i x_{ij}(t) + \beta_i y_{ij}(t)) + s_0 T_j(t) + s_1 T_j(t-1) + s_2 T_j(t-2) + s_3 T_j(t-3) \right], \quad j = 1, \ldots, m \tag{4.80}$$

The problem formulated in equations (4.78) to (4.80) is a linear optimization problem and can be solved using linear programming based on the simplex method explained earlier in this chapter. Having identified the fuzzy model parameters, we can predict the load for the next 24 hours using equation (4.77).

4.5.2 Multiple Fuzzy Linear Regression Model: Fuzzy Data

In Section 4.5, the load power data are assumed to be nonfuzzy, whereas the parameters of the load power are fuzzy. Different linear optimization problems were derived with different load models. In this section, the load data are assumed to be fuzzy power values having a certain middle and certain spread $\underline{Y}_j(t) = [m_j(t), \alpha_j(t)]$, where $m_j(t)$ is the middle of the load power at the time t in question during the observation j, and $\alpha_j(t)$ is the spread of the load power at time t and observation j. Using this formulation of fuzzy numbers means that a triangular membership function is assumed, as shown in Figures 4.6(a) and (b).

The fuzzy model for the load power can be expressed mathematically as

$$Y_j(t) = [m_j(t), \alpha_j(t)] = \underline{A}_0 + \sum_{i=1}^{n} \underline{A}_i x_{ij}(t), \qquad j = 1, \ldots, m \tag{4.81}$$

which can be rewritten as

$$[m_j(t), \alpha_j(t)] = (p_0, c_0) + \sum_{i=1}^{n} (p_i, c_i)\, x_{ij}(t), \qquad j = 1, \ldots, m \tag{4.82a}$$

Alternatively, it can be separated as

$$[m_j(t), \alpha_j(t)] = \left[\left\{p_0 + \sum_{i=1}^{n} p_i x_{ij}(t)\right\}, \left\{c_0 + \sum_{i=1}^{n} c_i x_{ij}(t)\right\}\right], \qquad j = 1, \ldots, m \tag{4.82b}$$

Equation (4.82b) is valid only when: Given two fuzzy numbers $M_1 = (m_1, \alpha_1, \beta_1)_{\mathrm{LR}}$ and $M_2 = (m_2, \alpha_2, \beta_2)_{\mathrm{LR}}$ in terms of *LR* functions [2] that follow triangular membership function, where:

m_1 and m_2 are the centers of the membership function;
α_1 and α_2 are left-side spreads;
β_1 and β_2 are right-side spreads.

Then

$$M_1(m_1, \alpha_1, \beta_1)_{\mathrm{LR}} + M_2(m_2, \alpha_2, \beta_2)_{\mathrm{LR}} = (m_s, \alpha_s, \beta_s)_{\mathrm{LR}}$$

where

$$m_s = m_1 + m_2$$

$$\alpha_s = \alpha_1 + \alpha_2$$

$$\beta_s = \beta_1 + \beta_2$$

The center of the sum is equal to the sum of the centers, and each spread of the sum is the sum of its respective spread.

$$m_j(t) = p_0 + \sum_{i=1}^{n} p_i x_{ij}(t), \qquad j = 1,\ldots,\ldots,m \tag{4.83}$$

$$\alpha_j(t) = c_0 + \sum_{i=1}^{n} c_i x_{ij}(t), \qquad j = 1,\ldots,\ldots,m \tag{4.84}$$

The problem turns out to be: Given the fuzzy load power at time t, $\underline{Y}_j(t) = [m_j(t), \alpha_j(t)]$, the task is to find the fuzzy parameters $\underline{A}_0$ and $\underline{A}_i$ that minimize the cost function given by

$$J = \left| \sum_t \left\{ \sum_{j=1}^{m} \left\{ m_j(t) - p_0 - \sum_{i=1}^{n} p_i x_{ij}(t) + \alpha_j(t) - c_0 - \sum_{i=1}^{n} c_i x_{ij}(t) \right\} \right\} \right| \tag{4.85}$$

where $t \in [0, t_F]$, and t_F is the number of days for which data are taken at the hour in question.

This is subject to satisfying the following constraints on each measurement point

$$m_j(t) - (1-\lambda)\,\alpha_j(t) \geq \left(p_0 + \sum_{i=1}^{n} x_{ij}(t)\right) - \left(c_0 + \sum_{i=1}^{n} c_i x_{ij}(t)\right), \quad j = 1,\ldots,m \tag{4.86}$$

$$m_j(t) + (1-\lambda)\,\alpha_j(t) \leq \left(p_0 + \sum_{i=1}^{n} x_{ij}(t)\right) - \left(c_0 + \sum_{i=1}^{n} c_i x_{ij}(t)\right), \quad j = 1,\ldots,m \tag{4.87}$$

The problem formulated in equations (4.85) to (4.87) is a linear optimization problem. This problem can be solved using linear programming. In the following subsections, we discuss two multiple linear regression models: one for winter and one for summer.

4.5.2.1 Model A: Fuzzy Winter Model

Two factors affect this fuzzy winter model. The first is temperature deviation. The more temperature deviation, the more load power is needed. The second factor is wind cooling. As the wind-cooling factor increases, the load power increases. The load power data in this model are assumed to be a fuzzy power, unlike the load model in equation (4.62), where the load power is assumed to be crisp (nonfuzzy). Equation (4.62) can be rewritten as

$$\begin{aligned} Y_j(t) = (m_j(t), \alpha_j(t)) = {} & (p_0, c_0) + (p_1, c_1)T_j(t) + (p_2, c_2)T_j^2(t) + (p_3, c_3)T_j^3(t) \\ & + (p_4, c_4)T_j(t-1) + (p_5, c_5)T_j(t-2) + (p_6, c_6)T_j(t-3) \\ & + (p_7, c_7)W_j(t) + (p_8, c_8)W_j(t-1) + (p_9, c_9)W_j(t-2) \end{aligned} \tag{4.88}$$

Equation (4.88) can be rewritten as

$$\begin{aligned} m_j(t) &= p_0 + p_1 T_j(t) + p_2 T_j^2(t) + p_3 T_j^3(t) + p_4 T_j(t-1) + p_5 T_j(t-2) \\ &\quad + p_6 T_j(t-3) + p_7 W_j(t) + p_8 W_j(t-1) + p_9 W_j(t-2), \\ &\quad j = 1, \ldots, m \end{aligned} \tag{4.89}$$

$$\begin{aligned} \alpha_j(t) &= c_0 + c_1 T_j(t) + c_2 T_j^2(t) + c_3 T_j^3(t) + c_4 T_j(t-1) + c_5 T_j(t-2) \\ &\quad + c_6 T_j(t-3) + c_7 W_j(t) + c_8 W_j(t-1) + c_9 W_j(t-2), \\ &\quad j = 1, \ldots, m \end{aligned} \tag{4.90}$$

Given the fuzzy load power ($m_j(t)$, $\alpha_j(t)$) at any time t, the task is to determine the middle and the spread of each parameter that minimizes the cost function

$$\begin{aligned} J = \Bigg| \sum_t \sum_{j=1}^{m} [m_j(t) - \{p_0 + p_1 T_j(t) + p_2 T_j^2(t) + p_3 T_j^3(t) + p_4 T_j(t-1) \\ + p_5 T_j(t-2) + p_6 T_j(t-3) + p_7 W_j(t) + p_8 W_j(t-1) + p_9 W_j(t-2)\} \\ + \alpha_j(t) - \{c_0 + c_1 T_j(t) + c_2 T_j^2(t) + c_3 T_j^3(t) + c_4 T_j(t-1) + c_5 T_j(t-2) \\ + c_6 T_j(t-3) + c_7 W_j(t) + c_8 W_j(t-1) + c_9 W_j(t-2)\}] \Bigg| \end{aligned} \tag{4.91}$$

where $t \in [0, t_F]$, and t_F is the number of days for which data are taken at the hour in question.

This is subject to satisfying the following two constraints at each measurement point:

$$\begin{aligned} m_j(t) - (1-\lambda)\,\alpha_j(t) \geq [(\text{RHS of equation 4.89}) - (\text{RHS of equation 4.90})], \\ j = 1, \ldots, m \end{aligned} \tag{4.92}$$

$$\begin{aligned} m_j(t) + (1-\lambda)\,\alpha_j(t) \leq [(\text{RHS of equation 4.89}) + (\text{RHS of equation 4.90})], \\ j = 1, \ldots, m \end{aligned} \tag{4.93}$$

where RHS stands for right-hand side.

The problem formulated in equations (4.91) to (4.93) is one of linear optimization. This problem can be solved using standard linear programming.

Having identified the fuzzy parameters of the fuzzy winter model, we can predict the load in a winter day. The middle of the load can be predicted at any hour t using equation (4.89), and the spread can be predicted using equation (4.90).

4.5.2.2 Model A: Fuzzy Summer Model

The load in the fuzzy summer model is a function of the temperature deviation and humidity factor. The load power and the load model parameters are assumed to be fuzzy numbers. Mathematically, this can be expressed as

$$Y_j(t) = (m_j(t),\, \alpha_j(t)) = \underline{A}_0 + \underline{A}_1 T_j(t) + \underline{A}_2 T_j^2(t) + \underline{A}_3 T_j^3(t) + \underline{A}_4 T_j(t-1) \\ + \underline{A}_5 T_j(t-2) + \underline{A}_6 T_j(t-3) + \underline{A}_7 H_j(t) + \underline{A}_8 H_j(t-1) + \underline{A}_9 H_j(t-2), \\ j = 1, \ldots, m \tag{4.94}$$

where

$\underline{Y}_j(t)$ is the fuzzy load power i; $i = 1, \ldots, m$, at time t. This power has a middle $m_j(t)$ and a spread $\alpha_j(t)$;
$\underline{A}_0, \underline{A}_1, \ldots, \underline{A}_9$ are the fuzzy load parameters at time t with certain middle $p_0, \ldots, p_9$ and certain spread $c_0, c_1, \ldots, c_9$;
$T_j(t)$ is the temperature deviation at time t, $j = 1, \ldots, m$;
$H_j(t)$ is the humidity factor given by equation (3.17).

Equation (4.94) can be rewritten as

$$Y(t) = (m_j(t),\, \alpha_j(t)) = (p_0,\, c_0) + (p_1, c_1)\, T_j(t) + (p_2,\, c_2)\, T^2(t) \\ + (p_3,\, c_3) T^3(t) + (p_4,\, c_4)\, T(t-1) + (p_5, c_5)\, T(t-2) \\ + (p_6,\, c_6)\, T(t-3) + (p_7,\, c_7)\, H(t) + (p_8,\, c_8) H(t-1) \\ + (p_9,\, c_9)\, H(t-2) \tag{4.95}$$

provided that the memberships for the fuzzy numbers are triangular memberships. Equation (4.91) can be rewritten as two equations:

$$m_j(t) = p_0 + p_1 T_j(t) + p_2 T_j^2(t) + p_3 T_j^3(t) + p_4 T_j(t-1) + p_5 T_j(t-2) \\ + p_6 T_j(t-3) + p_7 H_j(t) + p_8 H_j(t-1) + p_9 H_j(t-2), \\ j = 1, \ldots, m \tag{4.96}$$

$$\alpha_j(t) = c_0 + c_1 T_j(t) + c_2 T_j{}^2(t) + c_3 T_j{}^3(t) + c_4 T_j(t-1) + c_5 T_j(t-2) \\ + c_6 T_j(t-3) + c_7 H_j(t) + c_8 H_j(t-1) + c_9 H_j(t-2), \\ j = 1, \ldots, m \tag{4.97}$$

In the fuzzy optimization linear problem, the model fuzzy parameters are to be found to minimize the spread of the fuzzy load power. Mathematically, this can be expressed as

Minimize

$$J = \left| \sum_t \left\{ \sum_{j=1}^{m} \left[(m_j(t) - \text{RHS of equation 4.92}) + (\alpha_j(t) - \text{RHS of equation 4.93}) \right] \right\} \right| \tag{4.98}$$

where $t \in [0,\, t_F]$, and t_F is the number of days for which data are taken at the hour in question.

This is subject to satisfying the following constraints:

$$m_j(t) - (1-\lambda)\,\alpha_j(t) \geq [(\text{RHS of equation 4.92}) - (\text{RHS of equation 4.93})],\quad j=1,\ldots,m \tag{4.99}$$

$$m_j(t) + (1-\lambda)\,\alpha_j(t) \leq [(\text{RHS of equation 4.92}) + (\text{RHS of equation 4.93})],\quad j=1,\ldots,m \tag{4.100}$$

The optimization problem formulated in equations (4.98) to (4.100) is one of linear optimization and can be solved using linear programming.

Having obtained the fuzzy load parameters, we then can use equation (4.91) to predict the fuzzy load power at any hour t in question.

4.5.3 Fuzzy Load Model B

Fuzzy load model B does not account for weather conditions in the load; it can be expressed as

$$Y_j(t) = (m_j(t),\, \alpha_j(t)) = \underline{A}_0 \sum_{i=1}^{n} [(\underline{A}_{\mathrm{i}} \sin i\omega t + \underline{B}_{\mathrm{i}} \cos i\omega t]_j, \qquad j=1,\ldots,m \tag{4.101}$$

The only difference between equation (4.71) and (4.101) is the load power $Y_j(t)$ at time t. In (4.71) the load power is assumed to be a crisp value, whereas in (4.101) it is assumed to be a fuzzy value having a middle $m_j(t)$ and a spread $\alpha_j(t)$. Equation (4.101) can be rewritten as

$$(m_j(t),\, \alpha_j(t)) = (p_0,\, c_0) + \sum_{i=1}^{n} [(p_i,\, c_i)\, \sin i\omega t + (b_i,\, \beta_i)\, \cos i\omega t]_j \quad j=1,\ldots,m \tag{4.102}$$

which can be split into

$$m_j(t) = p_0 + \sum_{i=1}^{n} [(p_i \sin i\omega t + b_i \cos i\omega t)]_j, \qquad j=1,\ldots,m \tag{4.103}$$

$$\alpha_j(t) = c_0 + \sum_{i=1}^{n} [c_i \sin i\omega t + \beta_i \cos i\omega t]_j, \qquad j=1,\ldots,m \tag{4.104}$$

The task is to find the fuzzy load parameters that minimize the spread of the fuzzy load power. This can be expressed mathematically as

$$J = \left| \sum_t \left\{ \sum_{j=1}^{m} [(m_j(t) - \text{RHS of equation } 4.103 + (\alpha_j(t) - \text{RHS of equation } 4.104)] \right\} \right| \tag{4.105}$$

where $t \in [0,\ t_F]$, and t_F is the number of days for which data are taken at the hour in question.

This is subject to satisfying the following two constraints as

$$m_j(t) - (1-\lambda)\,\alpha_j(t) \geq [\text{RHS of equation } 4.103 - \text{RHS of equation } 4.104]; \quad j = 1, \ldots, m \tag{4.106}$$

$$m_j(t) + (1-\lambda)\,\alpha_j(t) \leq [\text{RHS of equation } 4.103 + \text{RHS of equation } 4.104]; \quad j = 1, \ldots, m \tag{4.107}$$

The problem formulated in equations (4.105) to (4.107) is one of linear optimization that can be solved using linear programming. Having identified the middle and the spread of fuzzy parameters, we then can use the harmonic load model described in equation (4.101) to predict the load at any hour t. Note that the load power obtained in this case is independent of the weather conditions and depends only on the hour in question.

The next model, model C, combines fuzzy load model A and fuzzy load model B. This model takes weather conditions into account.

4.5.4 Fuzzy Load Model C

Fuzzy load model A derived earlier has the advantage of being weather responsive; the fuzzy coefficients of this model depend on the weather conditions. These conditions include temperature deviation and cooling factor.

Fuzzy load model B is weather insensitive. The fuzzy coefficients of this model depend only on the time in question.

In this section, the two models A and B are combined into one fuzzy model, C. The resulting fuzzy load model C is weather sensitive. This fuzzy model is suitable for all weekdays and can be used for both winter and summer load-forecast simulations. Its main disadvantage is the assumption that the relation between load and weather is constant throughout the day.

The fuzzy model for the load in this case can be expressed mathematically as

$$\begin{aligned} Y_j(t) = (m_j(t),\ \alpha_j(t)) &= \left\{ \underline{A}_0 + \sum_{i=1}^{n} (\underline{A}_i \sin i\omega t + \underline{B}_i \cos i\omega t) \right\}_j \\ &+ \left\{ \underline{C}_0 T_j(t) + \underline{C}_1 T_j(t-1) + \underline{C}_2 T_j(t-2) + \underline{C}_3 T_j(t-3) \right\}_j, \\ & j = 1, \ldots, m \end{aligned} \tag{4.108}$$

where

$m_j(t)$, $\alpha_j(t)$ is the middle and spread of load power j, $j = 1, \ldots, m$ at time t;
$\underline{A}_0$, $\underline{A}_i$, and $\underline{B}_i$ are the weather-independent fuzzy parameters with certain middles and spreads;
$\underline{C}_0$, $\underline{C}_1$, $\underline{C}_2$, and $\underline{C}_3$ are the temperature-dependent fuzzy parameters with certain middles and spreads.

The left-hand side (LHS) of equation (4.108) is the fuzzy load power. The terms in the first bracket on the right-hand side (RHS) of equation (4.108) can be considered as the fuzzy base load, and it depends only on time, whereas the second bracket contains the temperature-dependent fuzzy load terms.

Equation (4.108) can be rewritten as

$$\begin{aligned}(m_j(t),\ \alpha_j(t)) = & \left\{(p_0,\ c_0) + \sum_{i=1}^{n}[(p_i,\ \theta_i)x_i(t) + (b_i,\ \beta_i)y_i(t)]\right\}_j \\ & + \{(\gamma_0, c_0^{'})T_j(t) + (\gamma_1, c_1)T_j(t-1) + (\gamma_2, c_2)T_j(t-2) \\ & + (\gamma_3, c_3)T_j(t-3)\}_j, \qquad j = 1, \ldots, m\end{aligned} \tag{4.109}$$

For simplicity, let

$$x_i(t) = \sin i\omega t, \qquad i = 1, \ldots, n \tag{4.110a}$$

$$y_i(t) = \cos i\omega t, \qquad i = 1, \ldots, n \tag{4.110b}$$

In equation (4.109), the first letter in all small brackets of the equations indicates the middle of the parameter, and the second letter indicates the spread of that parameter. A triangular membership is used for each parameter.

In the fuzzy model developed in equation (4.109), the task is to find the fuzzy model parameters to minimize the spread of the output. Mathematically, the fuzzy linear optimization problem can be expressed as

Minimize

$$\begin{aligned}J = \Bigg| \sum_t \Bigg\{ \sum_{j=1}^{m} m_j(t) - & \left\{p_0 + \sum_{i=1}^{n} [p_i x_i(t) + b_i y_i(t)]_j + \gamma_0 T_j(t) + \gamma_1 T_j(t-1)\right. \\ & \left. + \gamma_2 T_j(t-2) + \gamma_3 T_j(t-3)\right\} + \left\{\alpha_j(t) - \left[c_0 + \sum_{i=1}^{n}\left\{\theta_i x_i(t) + \beta_i y_i(t)\right\}_j\right.\right. \\ & \left.\left. + \underline{c}_0^{'} T_j(t) + c_1 T_j(t-1) + c_2 T_j(t-2) + c_3 T_j(t-3)\right\}\right]\Bigg\}\Bigg|\end{aligned} \tag{4.111}$$

where $t \in [0,\ t_F]$, and t_F is the number of days for which data are taken at the hour in question.

Subject to satisfying the following two constraints for each measurement point given as

$$\begin{aligned} m_j(t) - (1-\lambda)\,\alpha_j(t) \geq & \left\{ p_0 + \sum_{i=1}^{n} \{[p_i x_i(t) + b_i y_i(t)]\}_j \right. \\ & \left. + \gamma_0 T_j(t) + \gamma_1 T_j(t-1) + \gamma_2 T_j(t-2) + \gamma_3 T_j(t-3) \right\} \\ & - \left\{ c_0 + \sum_{i=1}^{n} \{\theta_i x_i(t) + \beta_i y_i(t) \right\}_j + c_0' T_j(t) + c_1 T_j(t-1) \\ & + c_2 T_j(t-2) + c_3 T_j(t-3), \qquad j = 1, \ldots, m \end{aligned} \tag{4.112}$$

$$\begin{aligned} m_j(t) + (1-\lambda)\,\alpha_j(t) \leq & \left\{ p_0 + \sum_{i=1}^{n} [p_i x_i(t) + b_i y_i(t)] + \gamma_0 T(t) + \gamma_1 T_j(t-1) \right. \\ & \left. + \gamma_2 T_j(t-2) + \gamma_3 T_j(t-3) \right\} \\ & + \left\{ c_0 + \sum_{i=1}^{n} \theta_i x_i(t) + \beta_i y_i(t) \right\}_j + c'_0 T_j(t) + c_1 T_j(t-1) \\ & + c_2 T_j(t-2) + c_3 T_j(t-3) \qquad j = 1, \ldots, m \end{aligned} \tag{4.113}$$

The problem formulated in equations (4.111) to (4.113) is one of linear optimization and can be solved by linear programming.

Having obtained the middle and spread of each fuzzy parameters, we can calculate the load power at any hour in question using equation (4.109).

4.6 Conclusion

This chapter presented a new formulation for fuzzy short-term load-forecasting models. In the first part of the chapter, the load power is considered given as crisp (non-fuzzy) data, while the load model parameters are fuzzy, having certain middles and spreads. The problem turns out to be one of linear optimization.

In the second part of the chapter, the load power is considered to be fuzzy power data having certain middles and spreads. Three different fuzzy models—A, B, and C—were developed, and new fuzzy equations were obtained. The resulting optimization problem is linear and can be solved using linear programming.

References

[1] T.J. Ross, Fuzzy Logic with Engineering Applications, McGraw-Hill, New York, 1995.
[2] M.E. El-Hawary, Electric Power Applications of Fuzzy Systems, IEEE Press, New York, 1998.

[3] J. Nazarka, W. Zalewski, An application ofthe fuzzy regression analysis to the electrical load estimation, Electrotechnical Conference MELECON '96, 3, 1563–1566, IEEE Catalog #96CH35884, May 13–16, 1996.
[4] H. Tanaka, S. Uejima, K. Asai, Linear regression analysis with fuzzy model, IEEE Trans. Syst. Man Cybern 12 (6) (1982) 903–907.
[5] P.T. Chang, E.S. Lee, Fuzzy least absolute deviations regression based on the ranking of fuzzy numbers, IEEE World Congr. Fuzzy Syst. IEEE Proc. 2 (1994) 1365–1369.
[6] J. Watada, Y. Yabuchi, Fuzzy robust regression analysis, IEEE World Congr. Fuzzy Syst. IEEE Proc. 2 (1994) 1370–1376.
[7] R. Alex, P.Z. Wang, A new resolution of fuzzy regression analysis, IEEE Int. Conf. Syst. Man Cybern. 2 (1998) 2019–2021.
[8] H. Ishibuchi, M. Nii, Fuzzy regression analysis by neural networks with non-symmetric fuzzy number weights, IEEE Int. Conf. Neural Networks 2 (1996) 1191–1196.
[9] H. Ishibuchi, M. Nii, Fuzzy regression analysis with non-symmetric fuzzy number coefficients and its neural network implementation, Proc. Fifth IEEE Int. Conf. Fuzzy Syst. 1 (1996) 318–324.
[10] S. Ghoshray, Fuzzy linear regression analysis by symmetric triangular fuzzy number coefficients, Proc. IEEE Int. Conf. Intell. Eng. Syst. (1997) 307–313.
[11] M.-S. Yang, C.-M. Ko, On cluster-wise fuzzy regression analysis, IEEE Trans. Syst. Man. Cybern. Part B Cybern. 27 (1) (1997) 1–13.
[12] H. Tanaka, H. Lee, Interval regression analysis by quadratic programming approach, IEEE Trans. Fuzzy Syst. 6 (4) (1998) 473–481.
[13] C.J. Huang, C.E. Lin, C.L. Haung, Fuzzy approach for generator maintenance scheduling, Electr. Power Syst. Res. 24 (1992) 31–38.
[14] I.M. Altas, A.M. Sharaf, A fuzzy logic power tracking controller for a photovoltic energy conversion scheme, Electr. Power Syst. Res. 25 (1992) 227–238.
[15] C.S. Chen, J.N. Sheen, Applying fuzzy mathematics to the evaluation of avoided cost for a load management program, Electr. Power Syst. Res. 26 (1993) 117–125.
[16] N. Rajakovic, S. Ruzic, Sensitivity analysis of an optimal short term hydro-thermal schedule, IEEE Trans. Power Syst. 8 (3) (1993) 1235–1241.
[17] R. Kenarangui, A. Seifi, Fuzzy power flow analysis, Electr. Power Syst. Res. 29 (1994) 105–109.
[18] J.Y. Fan, J.D. McDonald, A real-time implementation of short-term load forecasting for distribution power systems, IEEE Trans. Power Syst. 9 (2) (1994) 988–944.
[19] D. Srinivasan, M.A. Lee, Survey of hybrid fuzzy neural approach to electric load forecasting, IEEE Int. Conf. Syst. Man Cybern. 5 (1995) 4004–4008.

5 Dynamic State Estimation

5.1 Objectives

The objective of this chapter is to study the dynamic state estimation problem and its applications to electric power system analysis. Furthermore, the different approaches used to solve this dynamic estimation problem are also discussed in this chapter. After you finish reading this chapter, you should be familiar with

- The five fundamental components of an estimation problem, which are
  1. The variables to be estimated.
  2. The measurements or observations available.
  3. The mathematical model describing how the measurements are related to the variable of interest.
  4. The mathematical model of the uncertainties present.
  5. The performance evaluation criterion to judge which estimation algorithms are "best."
- Formulation of the dynamic state estimation problem.
- The Kalman filtering algorithm as a recursive filter used to solve this problem.
- The weighted least absolute value filter.
- Different problems that face Kalman filtering and weighted least absolute value filtering algorithms.

5.2 Discrete Time Systems

The problem of static estimation formulated in Chapter 2 uses all the available information (measurements) as one set of "snapshot measurements" to estimate the system states (parameters) [1]. In real time, especially for online applications, most of the system states are dynamic in nature, and there must be a mechanism for including dynamic effects of the system model and its inputs (both controlled and uncontrolled). Given the uncertainty in the measurement-based estimates, algorithms that optimally combine these estimates with model-based estimates can be formed.

The linear discrete systems can be modeled deterministically as

$$\underline{x}_k = \phi_{k-1}\underline{x}_{k-1} + \Gamma_{k-1}\underline{u}_{k-1} + \Lambda_{k-1}\underline{w}_{k-1}, \; k = 1, \ldots, K \tag{5.1}$$

with x_0 given. Here, x_k is the system state vector, u_k is the input vector, and w_k is the random disturbance vector, all evaluated at time instant k.

The dimensions of x, u, and w are $n \times 1$, $m \times 1$, and $s \times 1$, respectively. It is assumed that the parameter matrices ϕ_{k-1}, Γ_{k-1}, Λ_{k-1}, and u_{k-1} are known without error;

DOI: 10.1016/B978-0-12-381543-9.00005-1

however, the initial condition x_0 is a Gaussian random variable prescribed by its mean value and covariance matrix as

$$E(\underline{x}_0) = M_0 \tag{5.2}$$

$$E\left[(\underline{x}_0 - M_0)(\underline{x} - M_0)^T\right] = P_0 \tag{5.3}$$

and the disturbance input (sometimes called "process noise") is a zero-mean Gaussian random sequence:

$$\left.\begin{array}{l} E(\underline{w}_k) = 0 \\ E(\underline{w}_k\,\underline{w}_k^T) = Q_k \end{array}\right\} \text{ for } k = 0 \text{ to } K \tag{5.4}$$

The disturbance input covariance matrix, Q_k, could be symmetric, implying the existence of instantaneous cross-correlation of disturbances, but if Q_k is diagonal, then components of the disturbance input are uncorrelated with each other. If, in addition, $E\left(w_k\, w_{k-1}^T\right) = 0$ for nonzero values of ℓ, disturbances at one instant are completely uncorrelated with those at any other instant.

The expected value of the state of the kth instant is expressed as

$$\begin{aligned} E(\underline{x}_k) = \underline{M}_k &= E\left[\phi_{k-1}\underline{x}_{k-1} + \Gamma_{k-1}\underline{u}_{k-1} + \Lambda_{k-1}w_{k-1}\right] \\ &= \phi_{k-1}E(\underline{x}_{k-1}) + \Gamma_{k-1}\underline{u}_{k-1} + \Lambda_{k-1}E(\underline{w}_{k-1}) \\ &= \phi_{k-1}\underline{M}_{k-1} + \Gamma_{k-1}\underline{u}_{k-1} \end{aligned} \tag{5.5}$$

with given $\underline{M}_0$.

In this equation, known mean values are indistinguishable from deterministic values. The equation propagates the expected value of the state, with the understanding that random perturbations from the expected values are likely to occur.

The covariance of the state perturbation can be propagated in a similar way to obtain

$$\underline{P}_k = \phi_{k-1}P_{k-1}\phi_{k-1}^T + \Lambda_{k-1}Q_{k-1}\Lambda_{k-1}^T + \phi_{k-1}M_{k-1}\Lambda_{k-1}^T + \Lambda_{k-1}M_{k-1}^T\phi_{k-1}^T \tag{5.6}$$

where P_0 is given. The cross-correlation between state perturbation and disturbance input is given by

$$M_{k-1} = E\left[(x_{k-1} - M_{k-1})\, w_{k-1}^T\right] \tag{5.7}$$

It equals zero if w_k is a white noise sequence.

In the estimation problem, the measurements are assumed to be linear functions of the state. We suppose that the dynamic system can be represented by plant and measurement models as follows:

Plant

$$\underline{x}_k = \phi_{k-1}\underline{x}_{k-1} + \underline{w}_{k-1} \tag{5.8}$$

or

$$\underline{x}_{k+1} = \phi_k \underline{x}_k + \underline{w}_k$$

Measurement

$$\underline{Z}_k = H_k \underline{x}_k + \underline{v}_k \tag{5.9}$$

Plant noise

$$E\left[\underline{w}_k \underline{w}_k^T\right] = 0 \tag{5.10}$$

$$E\left[\underline{w}_k \underline{w}_i^T\right] = Q_k \tag{5.11}$$

Observation noise

$$E\left[\underline{v}_k \underline{v}_k^T\right] = 0 \tag{5.12}$$

$$E\left[\underline{v}_k \underline{v}_i^T\right] = R_k \tag{5.13}$$

The measurement and plant noises v_k and w_k are assumed to be a zero-mean Gaussian sequence, and the initial value $\underline{x}_0$ is a Gaussian variant with known mean $\underline{x}_0$ and known covariance matrix P_0.

5.3 Discrete Time-Optimal Filtering

The objective of this section is to find a "best" estimate of the state vector $\underline{x}_k$ in the discrete interval $0 \le k \le K$ [1–3]. The estimates $\hat{\underline{x}}_k$ are to be conditioned on all available information about a linear dynamic process and a set of measurements taken during the interval. It is assumed that the discrete interval of indices k corresponds to a sampling interval of time $0 \le t_k \le t_K$, where the time increment between successive samples, $t_k - t_{k-1}$, normally is constant.

Estimation problems fall into three categories, as shown in Figure 5.1. If the time at which the calculations are made, t_{calc}, is greater than $t_K (= t_{k_{max}})$, then all the measurements are available to estimate the value of each sample x_K at t_k, and an algorithm called a smoother can be applied to the data. If the time of calculation is the same as the time of estimate, t_k, then the computation of $\hat{x}_k$ must rely on past and present data only (i.e., on measurements in the "physically realizable" interval [0, t_k]). Such an estimation algorithm is called a filter, as it is meant to "filter out" the noise in the available signals. If an estimate of x_k at some future time—say t_K—is to be made at time t_{calc}, the estimation algorithm is called a predictor. The three general types of estimators for a discrete time system can be summarized as follows (see Figure 5.1) [2]:

- Smoothers use observations (measurements) beyond the time that the state of the dynamic system is to be estimated:

$$t_k < t_K,\ t_{calc} > t_K$$

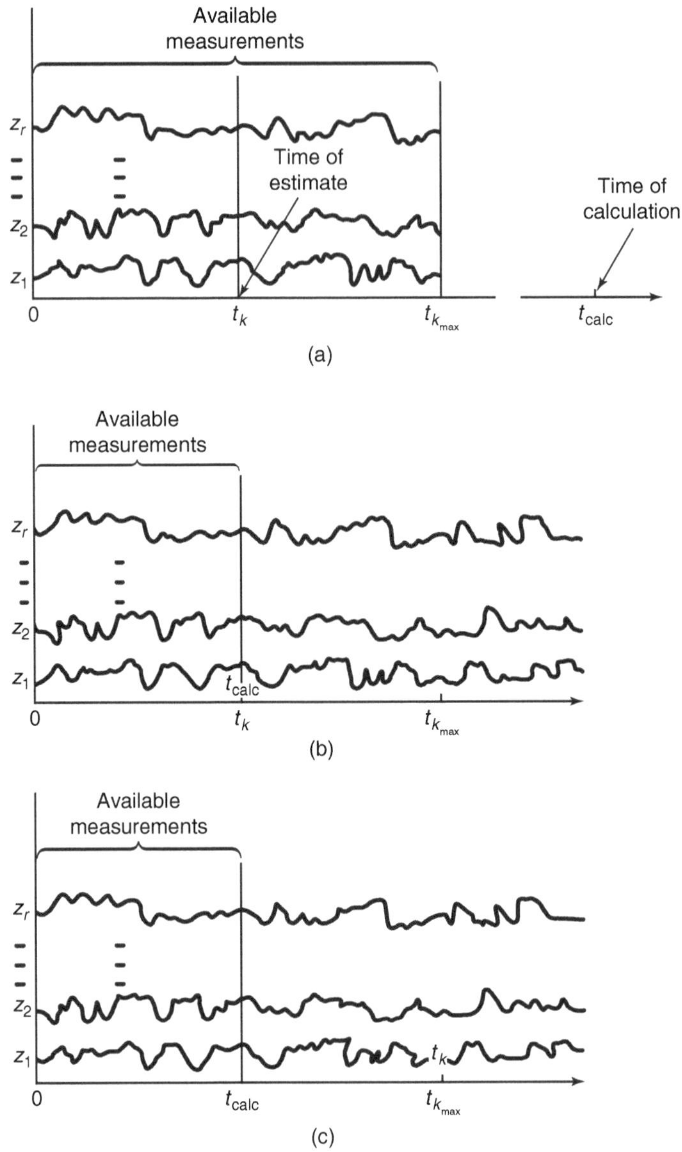

Figure 5.1 Classification of dynamic estimation problems: (a) smoothing problem; (b) filtering problem; and (c) prediction problem [1].

- Filters use observations up to and including the time that the state of the dynamic system is to be estimated:

$$t_k \leq t_K, \; t_{\text{calc}} \leq t_K$$

- Predictors use observations strictly prior to the time that the state of the dynamic system is to be estimated:

$$t_k > t_K \quad \text{and} \quad t_{\text{calc}} < t_K$$

Because the smoother can use more data than the filter or predictor, it usually can produce better estimates (except at the endpoint, $t_k = t_K$, where the estimators necessarily have the same information); however, the smoother can be applied only "after the fact." The filter and predictor can be implemented for real-time operation because they do not require future measurements.

5.3.1 Kalman Filter

A filter computes the state estimate $\hat{x}_k$, while in the process an optimal filter minimizes the spread of the estimate-error probability density [1–3]. This raises the peak of the function (or its likelihood) because the volume under the function is constant. (Recall that the cumulative probability in the range $[-\infty, +\infty]$ of all variables is one.) Therefore, for a Gaussian distribution, a minimum-variance estimator produces the same result as a maximum likelihood estimator.

A recursive optimal filter propagates the conditional probability density function from one sampling instant to the next, taking into account system dynamics and inputs. It also incorporates measurements and measurement error statistics in the estimate. Computing the weighting factors (or filter gains) that optimally combine measurements and extrapolations is an important intermediate step in the process.

We may take the mean (or expected) value of the conditional density function as the estimate of x_k and the covariance matrix as a measure of the spread (or uncertainty) in the estimate. The mean and covariance can be generated recursively in the intervals $0 \leq k \leq k_{\max}$ through the use of these five equations:

1. State estimate extrapolation (propagation).
2. Covariance estimate extrapolation (propagation).
3. Filter gain computation.
4. State estimate update.
5. Covariance estimate update.

The first two of these were introduced earlier, and the last three follow directly from results readily available in recursive weighted least squares estimation theory. Given the state estimate from a prior iteration, step 1 uses the dynamic process model to propagate the estimate of the state mean value to the next sampling instant without considering new measurements. Step 2 does the same thing for the state covariance matrix, assuming that "process noise" of known covariance is forcing the system. The result of step 2 enters the computation of the optimal filter gains. The filter gain computation of step 2 weights prior knowledge of measurement error covariance with state estimate covariance on a purely statistical basis. The actual measurements have no effect on the gain computation. These measurements correct the state estimate in step 4, adding the product of the gain matrix and the measurement residual to the state estimate propagated by step 1. A similar correction is made to the covariance estimate of step 5, accounting for the known covariance of measurement errors.

Before formulating the five equations of the estimator, we will review available knowledge of the system. The system's dynamic equation is

$$x_k = \phi_{k-1}x_{k-1} + \Gamma_{k-1}u_{k-1} + \Lambda_{k-1}w_{k-1} \tag{5.14}$$

with variables defined as before. ϕ_{k-1}, Γ_{k-1}, Λ_{k-1}, and u_{k-1} are known without error in the interval $0 \le k \le k_{\max}$; the expected values of the initial state and its covariance are known as

$$E(x_0) = \hat{x}_0 \tag{5.15}$$

$$E\left[(x_0 - \hat{x}_0)(x_0 - \hat{x}_0)^T\right] = P_0 \tag{5.16}$$

and the disturbance input is a zero-mean white Gaussian random sequence:

$$E(w_k) = 0 \tag{5.17}$$

$$E\left(w_k w_k^T\right) = Q_k^{'} \tag{5.18}$$

$$E\left(w_k w_j^T\right) = 0, \quad (j \neq k) \tag{5.19}$$

The observation vector is

$$z_k = H_k x_k + v_k \tag{5.20}$$

where H_k is known and the measurement error is a zero-mean white Gaussian random sequence that is uncorrelated with the disturbance input:

$$E(v_k) = 0 \tag{5.21}$$

$$E\left(v_k v_k^T\right) = R_k \tag{5.22}$$

$$E\left(v_k v_j^T\right) = 0, \quad (j \neq k) \tag{5.23}$$

$$E\left(v_k w_j^T\right) = 0 \quad (\text{all } j \text{ and } k) \tag{5.24}$$

The dynamic system is linear, and the recursive least squares estimator is linear; therefore, the principle of superposition applies. In the following equations, we must distinguish between estimates made before and after any updates occur; $\hat{x}_k(-)$ is the state estimate that results from the propagation equation alone (i.e., before the measurements are considered), and $\hat{x}_k(+)$ is the corrected state estimate that accounts for measurements. $P_k(-)$ and $P_k(+)$ are defined similarly. For completeness, the initial conditions should be treated as $\hat{x}_0(+)$ and $P_0(+)$.

The state estimate extrapolation equation is given by

$$\hat{x}_k(-) = \phi_{k-1}\hat{x}_{k-1}(+) + \Gamma_{k-1}u_{k-1} \tag{5.25}$$

The covariance estimate extrapolation equation is

$$\begin{aligned} P_k(-) &= \phi_{k-1}P_{k-1}(+)\phi_{k-1}^T + Q_{k-1} \\ &= \phi_{k-1}P_{k-1}(+)\phi_{k-1}^T + \Lambda_{k-1}Q'_{k-1}\Lambda_{k-1}^T \end{aligned} \tag{5.26}$$

The uncorrected mean value and covariance estimates are propagated from previously corrected estimates, with no modifications to the equations (other than the nomenclature). The recursive mean value estimator provides the *filter gain computation* as

$$K_k = P_k(-)H_k^T\left[H_kP_k(-)H_k^T + R_k\right]^{-1} \tag{5.27}$$

where $P_k(-)$ replaces P_{k-1} as the pre-update covariance estimate. Similarly, the *state estimate update* equation can be written as (with $x_k(-)$ replacing x_{k-1})

$$\hat{x}_k(+) = \hat{x}_k(-) + K_k\left[z_k - H_k\hat{x}_k(-)\right] \tag{5.28}$$

and the covariance estimate update is given by

$$P_k(+) = \left[P_k(-)^{-1} + H_k^TR_k^{-1}H_k\right]^{-1} \tag{5.29}$$

The functional relationships of the filter are illustrated by the block diagrams of Figure 5.2, where the recursive nature of the algorithm is clearly evident (note the "delay" blocks). Both the known inputs to the system and the observations "drive" the state estimator. The optimal gain K_k depends only on the relative strengths of the stochastic input and the measurement, and its computation is entirely separated from the two state equations. Thus, the deterministic input u_{k-1} and the measurement z_k have no effect on K_k. The filter gain may be computed ahead of time and stored as a function of k if the time variations in the system and its statistics were known; real-time implementation of the Kalman filter (KF) then would consist only of the computations shown in Figure 5.2(a) with the equations of Figure 5.2(b) computed before state estimation begins.

The five Kalman filter equations (equations 5.25 to 5.29) can be expressed as three equations, although this may not decrease the amount of computation needed. Substituting equation (5.26) in equation (5.29) provides a single *covariance estimation equation*:

$$P_k(+) = \left\{\left[\phi_{k-1}P_{k-1}(+)\phi_{k-1}^T + Q_{k-1}\right]^{-1} + H_k^TR_k^{-1}H_k\right\}^{-1} \tag{5.30}$$

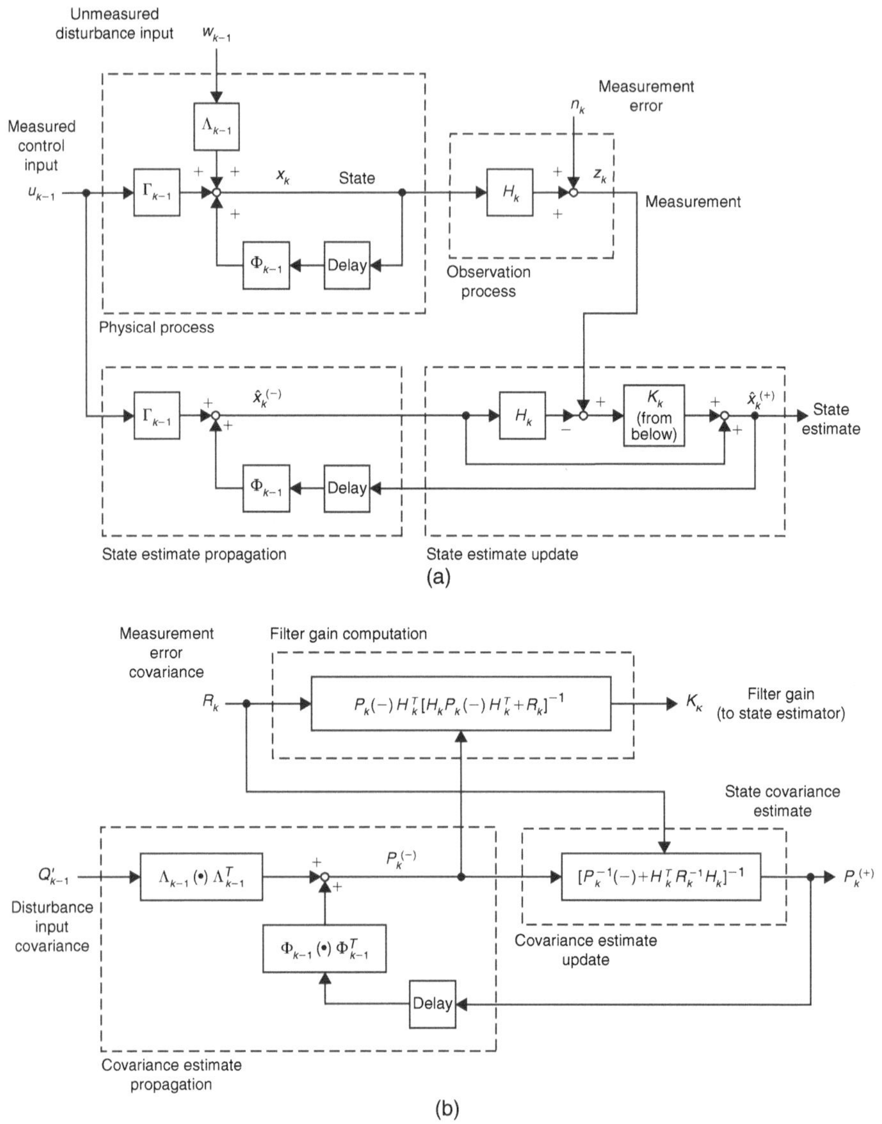

Figure 5.2 Discrete-time system and linear-optimal filter. (a) Dynamic system and state estimator. (b) Covariance estimator and gain computation.

The equation for K_k (equation (5.27)) required the intermediate variable $P_k(-)$, which does not appear in this solution; however, the gain computation can be reformulated in terms of $P_k(+)$. Equation (5.27) can be rearranged as follows:

$$\begin{aligned} K_k &= P_k(-)H_k^T\left[H_kP_k(-)H_k^T + R_k\right]^{-1} \\ &= P_k(-)H_k^T R_k^{-1}\left[I_r + H_kP_k(-)H_k^T R_k^{-1}\right]^{-1} \end{aligned} \tag{5.31}$$

$$K_k\left[I_r + H_kP_k(-)H_k^TR_k^{-1}\right] = P_k(-)H_k^TT_k^{-1} \tag{5.32}$$

$$\begin{aligned} K_k &= P_k(-)H_k^TR_k^{-1} - K_kH_kP_k(-)H_k^TR_k^{-1} \\ &= (I_n - K_kH_k)P_k(-)H_k^TR_k^{-1} \end{aligned} \tag{5.33}$$

From the matrix inversion lemma, equation (5.29) can be rewritten as

$$P_k(+) = P_k(-) - P_k(-)H_k^T\left[H_kP_k(-)H_k^T + R_k\right]^{-1}H_kP_k(-) \tag{5.34a}$$

$$= (I_n - K_kH_k)P_k(-) \tag{5.34b}$$

with K_k expressed in equation (5.27). (Equation (5.34b) implies that the update step reduces the magnitude of the covariance estimate, while the propagation step in equation (5.26) increases the magnitude.) Consequently, the *gain matrix equation* could be written in the alternate form

$$K_k = P_k(+)H_k^TR_k^{-1} \tag{5.35}$$

Substituting equation (5.25) in equation (5.28) provides the third equation, the *state estimator*:

$$\hat{x}_k(+) = \left[\phi_{k-1}\hat{x}_{k-1}(+) + \Gamma_{k-1}u_{k-1}\right] + K_k\left\{z_k - H_k\left[\phi_{k-1}\hat{x}_{k-1}(+) + \Gamma_{k-1}u_{k-1}\right]\right\} \tag{5.36}$$

The term in square brackets appears twice, but it is more efficient to compute it just once (i.e., to compute $\hat{x}_k(+)$ in two steps as done earlier).

In the following section, we offer some examples from power systems measurements to illustrate the application of the Kalman filtering algorithm. Before going further, we summarize in the following section the steps involved in the Kalman filtering algorithm.

The mathematical model for the system under consideration should be in the state form as

$$\underline{x}(k+1) = \phi(k)\underline{x}(k) + \underline{w}(k) \tag{5.37}$$

and the observation (measurement) of the process is assumed to occur at instant points of time in accordance with the relation

$$\underline{Z}(k) = H(k)\underline{x}(k) + \underline{v}(k) \tag{5.38}$$

Assume that we have a prior estimate, $\hat{\bar{x}}(k)$, and its error covariance $\overline{P}(k)$; then the general recursive Kalman filter equations are as follows:

1. Compute the Kalman filter gain, $\underline{K}(k)$, as follows:

$$\underline{K}(k) = \overline{P}(k)\underline{H}^T(k)\left[H(k)\overline{P}(k)H^T(k) + R_k\right]^{-1} \tag{5.39}$$

2. Compute the error covariance for the update estimate:

$$P(k) = \left[I - \underline{K}(k)H(k)\right]\overline{P}(k) \tag{5.40}$$

3. Update the estimate with the measurement $\underline{Z}(k)$ as follows:

$$\hat{\underline{x}}(k) = \hat{\overline{\underline{x}}}(k) + \underline{K}(k)\left[\underline{Z}(k) - H(k)\hat{\overline{\underline{x}}}(k)\right] \tag{5.41}$$

4. Project ahead the error covariance and the estimate:

$$\overline{P}(k+1) = \phi(k)P(k)\phi^T(k) + Q(k) \tag{5.42}$$

$$\hat{\overline{x}}(k+1) = \phi(k)\hat{\overline{x}}(k) \tag{5.43}$$

Example 5.1

The task in this example is to measure the amplitude, the phase angle, and the frequency of a noise-free voltage signal given by

$$v(t) = \hat{V}\cos(\omega_0 t + \phi) \tag{5.44}$$

where $\hat{V}$, ω_0, and ϕ are the amplitude, frequency, and phase angle of the signal, respectively.

The voltage signal can be written as

$$v(t) = \left(\hat{V}\cos\phi\right)\cos\omega_0 t - \left(\hat{V}\sin\phi\right)\sin\omega_0 t \tag{5.45}$$

Define the two states x_1 and x_2 by

$$x_1 = \hat{V}\cos\phi \tag{5.46}$$

and

$$x_2 = \hat{V}\sin\phi \tag{5.47}$$

Then equation (5.45) can be rewritten as

$$v(t) = x_1\cos\omega_0 t - x_2\sin\omega_0 t \tag{5.48}$$

The measurement at any time t includes the signal $v(t)$ plus noise. The term *noise* should include any non-60-Hz components, the measurement error, and the noise due to analog-to-digital (A/D) conversion. Thus, the discrete measurement $v(k)$ at any step k is

$$v(k) = \left[\cos(\omega_0 k\Delta T) - \sin(\omega_0 k\Delta T)\right]\begin{bmatrix} x_1 \\ x_2 \end{bmatrix} \tag{5.49}$$

which can be rewritten as

$$\underset{1\times1}{v(k)} = \underset{1\times2}{h(k)}\underset{2\times1}{x(k)} + \underset{1\times1}{v(k)} \tag{5.50}$$

For m samples, the preceding equation can be rewritten as

$$\underset{m\times1}{\underline{Z}(k)} = \underset{m\times2}{H(k)}\underset{2\times1}{x(k)} + \underset{m\times1}{\underline{v}(k)} \tag{5.51}$$

The state vector at step $(k+1)$ is related to the state vector at the prior step k by the state equation

$$\underline{x}(k+1) = \phi(k)\underline{x}(k) + \underline{\omega}(k) \tag{5.52}$$

where $\phi(k)$ is the state transition matrix and is chosen to be the identity matrix given by

$$\phi(k) = \begin{bmatrix} 1.0 & 0.0 \\ 0.0 & 1.0 \end{bmatrix} \tag{5.53}$$

Now, equations (5.51) and (5.52) are suitable for the Kalman filtering algorithm application. Having estimated the two states x_1 and x_2, we can calculate the voltage, amplitude, and phase angle as

$$\hat{V}^2(k) = \left[x_1^2(k) + x_2^2(k)\right] \tag{5.54}$$

and

$$\tan\phi(k) = \frac{x_2(k)}{x_1(k)} \tag{5.55}$$

The frequency deviation Δf can be calculated by taking the ratio of the change of ϕ over a certain number of time intervals as

$$\Delta f = \frac{1}{2\pi}\frac{\partial\phi}{\partial t} \tag{5.56}$$

The proposed models described earlier are implemented and used to calculate the steady-state voltage magnitude and frequency deviation and the rate of change of frequency for a voltage signal. The data consist of a sinusoidal waveform, the magnitude and phase of which may be selected. Here, we select the root mean square (rms) value of the signal to be 1 p.u., while its phase is 1.2 rad. A zero-mean Gaussian white noise was generated and superimposed on the waveform to present the noise measurement.

5.3.2 Initialization of the Kalman Filter

For offline application, it is necessary to initialize the recursive process of the Kalman filter, with an initial vector $\hat{\underline{x}}_0$ and its initial covariance matrix $\overline{P}_0$. Also, the system and measurement noise variances are needed.

A simple deterministic procedure can be implemented to compute the initial process vector, as well as its covariance matrix, using static least error squares estimates of previous measurements. Thus, in general, the initial process vector may be found as

$$\hat{\underline{x}}_0 = \left[H^T H\right]^{-1} H^T \underline{Z} \tag{5.57}$$

and the corresponding covariance error matrix is given by

$$\overline{P}_0 = \left[H^T H\right]^{-1} \tag{5.58}$$

For online applications, if the previous measurements are not available at the beginning of the estimation process, the initial process vector may be selected to be zero. The choice of the initial process vector does not need to be highly accurate. The effect of error introduced by the initial process vector diminishes very rapidly.

5.3.3 Divergence Problems in Kalman Filter

Because the discrete Kalman filter is recursive, the looping can be continued indefinitely, in principle at least [3]. There are practical limits, though, and under certain conditions divergence problems can arise. The three major problems can be summarized as round-off errors, modeling errors, and observability errors.

5.3.3.1 Round-off Errors

As the number of steps becomes large, the round-off error can lead to problems, especially in online applications, where computer constraints sometimes dictate fixed-point arithmetic. This always leads to difficulty when the dynamic range of the variables is large. There is no simple way to overcome such problems, and each problem should be treated or examined on its own merits. Fortunately, if the system is observable, the Kalman filter has a degree of natural stability. In this case a stable solution to the matrix P will normally exist. The following are some means to overcome the problem of the round-off error:

1. Avoid fixed-point arithmetic, if at all possible. Use double-precision arithmetic for offline analysis.
2. If possible, avoid deterministic processes in the filter modeling to avoid P approaching a semi-definite condition as the number of steps becomes large.
3. If measurement data are sparse, beware of propagating the P matrix in many tiny steps between measurements.
4. Make the P and $\overline{P}$ matrices symmetric with each recursive step because the covariance matrix must be symmetric.

5.3.3.2 Modeling Errors

The inaccurate modeling of the process being estimated may lead to a divergence problem. This problem is caused by the designer because the designer told the Kalman filter to use a process model that is actually far away from the real case, and the filter is trying to fit the wrong curve to the measurement data. This situation can also occur with nondeterministic as well as deterministic processes.

5.3.3.3 Observability Errors

A third kind of divergence may occur when the system is not observable. Physically, this means that one or more state variables (or linear combinations thereof) are hidden from the view of the observer (i.e., the measurement process). As a result, if the unobserved processes are unstable, the corresponding estimation errors will be similarly unstable. This problem is simply that the number of measurements available does not provide enough information to estimate all the state variables of the system. Note that the filter is still doing the best estimation possible under adverse conditions.

5.3.4 Soliman and Christensen Filter: Weighted Least Absolute Value Filter (WLAVF)

Consider a linear dynamic stochastic multistage discrete system [4,5] described by

$$\underline{x}(k+1) = \phi(k)\underline{x}(k) + \Gamma(k)\underline{u}(k) \tag{5.59}$$

The initial condition of $x(k)$, namely $x(0)$, is a Gaussian random vector, with the following statistics:

$$E\left[\underline{x}(0)\right] = \overline{\underline{x}}(0) \tag{5.60}$$

$$E\left[\left(\underline{x}(0) - \overline{\underline{x}}(0)(\underline{x}(0) - \overline{\underline{x}}(0))^T\right)\right] = M(0) \tag{5.61}$$

The input $u(k)$ is a Gaussian random vector sequence with these statistics

$$E\left[\underline{u}k\right] = \overline{\underline{u}}(0) \tag{5.62}$$

$$E\left[\left(\underline{u}(k) - \overline{\underline{u}}(k)(\underline{u}(k) - \overline{\underline{u}}(k))^T\right)\right] = V(k)\delta_{kj} \tag{5.63}$$

where δ_{kj} is the Kronecker delta. Furthermore, the input sequence $u(k)$ is independent of $x(0)$ so that

$$E\left[\left(\underline{u}(k) - \overline{\underline{u}}(k)(\underline{x}(0) - \overline{\underline{x}}(0))^T\right)\right] = 0 \tag{5.64}$$

The measurement equation, in this case, is given by

$$\underline{Z}(k) = H(k)\underline{x}(k) + \underline{v}(k) \tag{5.65}$$

where the measurement error $\underline{v}(k)$ is also a Gaussian sequence with the following statistics:

$$\left.\begin{array}{l} E\left[\underline{v}(k)\right] = 0 \\ E\left[\underline{v}(k)\underline{v}(j)\right] = R(k)\delta_{kj} \\ E\left[(\underline{u}(k) - \overline{\underline{u}}(k))\underline{v}^T(j)\right] = 0 \\ E\left[(\underline{x}(0) - \overline{\underline{x}}(0))\underline{v}^T(j)\right] = 0 \end{array}\right\} \tag{5.66}$$

The optimal estimate $\hat{\underline{x}}(k)$, based on least absolute value (LAV) criteria, of the state $\underline{x}(k)$ using the measurement $Z(j)$, $j \leq k$ can be obtained as follows:

$$\hat{\underline{x}}(k) = \overline{\underline{x}}(k) + \hat{\underline{K}}(k)\left[\hat{\underline{Z}}(k) - \hat{H}(k)\overline{\underline{x}}(k)\right] \tag{5.67}$$

where

$$\hat{\underline{K}}(k) = \left[\hat{R}(k)\hat{A}(k)\right]^{-1} \tag{5.68}$$

$$\hat{\underline{A}}(k) = \left[\hat{\underline{Q}}(k) + \hat{\underline{R}}^{-1}(k)\hat{H}(k)\right] \tag{5.69}$$

$$\hat{Q}(k) = \hat{C}M^{-1}(k) \tag{5.70}$$

or

$$\hat{\underline{K}}(k) = \left[\hat{H}(k) + \hat{R}(k)\hat{C}M^{-1}(k)\right]^{-1} \tag{5.71}$$

and

$$\overline{x}(k) = \phi(k-1)\hat{\underline{x}}(k-1) + \Gamma(k-1)\overline{\underline{u}}(k-1) \tag{5.72}$$

with $\overline{x}(0)$ specified. Furthermore,

$$M(k) = \left[\phi(k-1)\hat{P}(k-1)\phi^T(k-1) + \Gamma(k-1)v(k-1)\Gamma^T(k-1)\right] \tag{5.73}$$

where

$$\hat{P}(k) = \left[\hat{A}^T(k)\hat{B}(k)\hat{A}(k)\right]^{-1} \tag{5.74}$$

with

$$\hat{B}^{-1}() = \hat{R}^{-1}(k) + C\left[M^{-1}(k)\right]^T\hat{C}^T \tag{5.75}$$

Equations (5.67) and (5.72) can be combined in one equation as

$$\hat{x}(k) = \phi(k-1)\hat{x}(k-1) + \Gamma(k-1)\overline{u}(k-1) + \hat{K}(k)\left(Z(k) - \hat{H}(k)\overline{x}(k)\right) \tag{5.76}$$

Equation (5.76) describes the WLAVF for the state estimation of a linear multistage dynamic process. The filter is a combination of the model of the process and a correction term proportional to the difference between the actual measurement $Z(k)$ and the predicted measurement $\hat{H}(k)\overline{x}(k)$. The matrix $K(k)$ is referred to as the gain matrix, and $\left[Z(k) - \hat{H}(k)\overline{x}(k)\right]$ is sometimes referred to as the innovation.

The only difference between equation (5.76) and the well-known Kalman filter equations lies in the gain matrix $K(k)$, due the difference in the nature of the objective function used to derive the filter equations. In the Kalman filter, the objective function is the WLES objective function, but in the WLAV filter, the objective function used is in the WLAV sense, which is different from the WLES objective function.

It can be shown that the best WLAV approximation is superior to the best weighted least squares (WLS) approximation when estimating the true form of data that contain some inaccurate observations. Note that if there is only one measurement at a time, we can drop the hat (^) from the preceding equations, and there is no need to calculate the residual for each measurement.

In the following, we summarize the filter equations as [4,5]

1. State and Measurement Models

$$x(k+1) = \phi(k)x(k) + \Gamma(k)u(k)$$
$$Z(k) = H(k)x(k) + v(k)$$

with the following known model quantities:

$$\phi,\ H,\ \Gamma,\ R,\ M(0),\ V$$

2. Covariance Propagation Equations

$$M(k) = \phi(k-1)P(k-1)\phi^T(k-1) + \Gamma(k-1)V(k-1)\Gamma^T(k-1)$$
$$P(k) = \left[A^T(k)B(k)A(k)\right]^{-1}$$

where

$$B^{-1}(k) = R^{-1}(k) + C(M^{-1}(k))^T C^T$$
$$A(k) = \left[Q(k) + R^{-1}(k)H(k)\right]$$

and

$$Q(k) = CM^{-1}(k)$$

3. Estimation Equations

$$\hat{x}(k) = \phi(k-1)\hat{x}(k-1) + \Gamma(k-1)\overline{u}(k-1) + K(k)(Z(k) - H(k)\overline{x}(k))$$
$$\overline{x}(k) = \phi(k-1)\hat{x}(k-1) + \Gamma(k-1)\overline{u}(k-1)$$

where

$$K(k) = \left[H(k) + R(k)CM^{-1}(k)\right]^{-1}$$

A complete block diagram for the system dynamics and the filter is given in Figure 5.3.

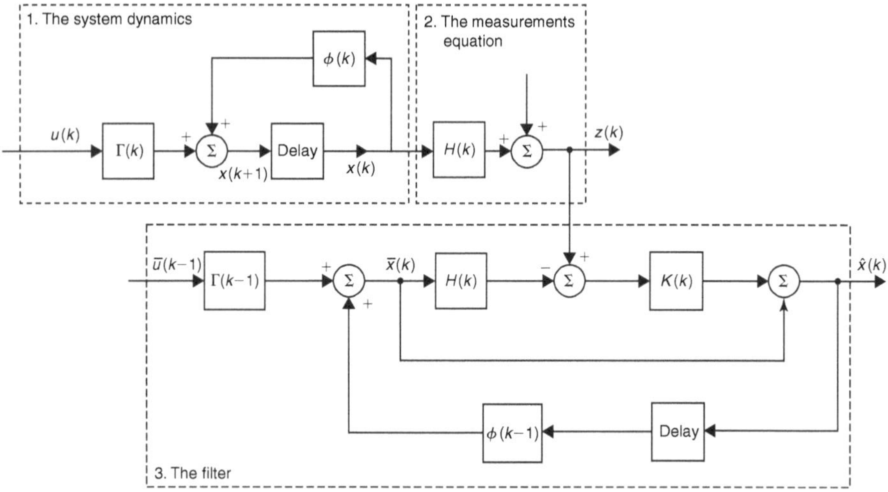

Figure 5.3 The system discrete filter.

5.3.4.1 Some Important Characteristics of the Filter

Following are some characteristics to be aware of:

1. $\bar{x}(k) = E[x(k)/\hat{x}(k-1)]$ is the expected value of $x(k)$ conditioned on the previous estimate. In other words, $\bar{x}(k)$ is the expected value of $x(k)$ before the measurement $Z(k)$ is made. On the other hand, we have $\hat{x}(k) = E[x(k)/Z(k)]$, the expected value of $x(k)$ conditioned on the present or current measurement $Z(k)$.
2. From (5.73) and (5.74), we notice that the propagation of the covariance matrices of the error in the estimate is independent of the measurement $Z(k)$. Therefore, these covariance matrices can be calculated beforehand and stored, if the parameters of the system and the observations measurement equations are known a priori.
3. The computation of the update estimate as $\hat{x}(k)$ involves only the current measurement and the error covariance after the measurement $Z(k)$ is made. Thus, computations can be carried out in real time.

In the following sections, we offer two examples to explain the main features of the filter. The first example is a first-order system, whereas the second example is a second-order system. Due to limitations of space, we limit ourselves to the final results. In Example 5.2, there is no noise in the state equation. Both methods give essentially the same results, but this might well change if noise is present in the state equation. However, this situation is not investigated here.

Example 5.2

The system state equation for this first-order system is given by

$$x(k+1) = 0.7x(k) + w(k)$$

The measurement equation is

$$Z(k+1) = 10x(k+1) + v(k+1)$$

The static properties of $x(k)$, $w(k)$, and $v(k)$ are

$$E[e(k)] = 0,\ E[w(k)w(j)] = V(k) = 0$$

$$E[x(0)] = \bar{x}(0) = 10,\ E\left[(x(0) - \bar{x}(0))^2\right] = 16 = M(0)$$

$$E[v(k)] = 0 \text{ and } E[v(k)v(j)] = R(k)\delta_{kj} = 100$$

Tables 5.1 and 5.2 give the optimal estimates of $x(k)$ for $k = 0, 1, 2, 3, 4$, when $Z(k)$ is 167.0, 95.9, 88.8, 55.3, and 15.3, by using a discrete KF and a proposed filter based on WLAV.

Table 5.1 Optimal Estimate of $x(k)$; Optimal Covariance Before and After Measurement and the Optimal Gain Using a Kalman Filter [4,5]

k	$z(k)$	$\hat{x}(k/k-1)$	$\hat{z}(k) = 10\hat{x}(k/k-1)$	$P(k/k-1)$	$K(k)$	$\hat{x}(k/k)$	$P(k/k)$
0	167.9	10	100	16.00	0.0941	16.39	0.9412
1	95.5	11.47	114.7	0.4612	0.0316	10.86	0.3156
2	88.8	7.606	76.06	0.1547	0.0134	7.77	0.1339
3	55.3	5.44	54.4	0.0656	0.0062	5.45	0.0616
4	15.3	5.81	38.1	0.0302	0.0029	5.74	0.0293

Table 5.2 Optimal Estimate of $x(k)$; Optimal Covariance Before and After Measurement and the Optimal Gain Using the Proposed Filter (WLAVF) [4,5]

K	$z(k)$	$\hat{x}(k)$	$\hat{z}(k) = 10\hat{x}(k)$	$P(k)$	$K(k)$	$\hat{x}(k)$	$M(k)$
1	167.9	7.00	70	2.655	0.04385	11.29	7.840
2	95.5	7.903	79.03	1.0318	0.01151	8.093	1.300
3	88.8	5.665	56.65	0.46	0.00487	5.819	0.5056
4	55.3	4.073	40.73	0.2163	0.00221	4.110	0.2256
5	15.3	2.877	28.77	0.1038	0.00105	2.863	0.1059

Example 5.3

This example provides a comparison between a Kalman filter and a new proposed algorithm for ship navigational fixes (a two-state system). We summarize these results in Table 5.3.

Table 5.3 Comparison between KF and WLAVF

k	**KF (Refer to [3] for List of Symbols)**	**WLAVF**
$k = 0$	$\hat{x}_0 = \overline{x}_0 = \begin{bmatrix} 0 \\ 10 \end{bmatrix}$	$\hat{x}_0 = \overline{x}_0 = \begin{bmatrix} 0 \\ 10 \end{bmatrix}$
	$P_0 = \begin{bmatrix} 2 & 0 \\ 0 & 3 \end{bmatrix}$	$M_0 = \begin{bmatrix} 2 & 0 \\ 0 & 3 \end{bmatrix}$
$k = 1$	$\hat{x}_1^- = \begin{bmatrix} 10 \\ 10 \end{bmatrix}$	$\overline{x}(1) = \begin{bmatrix} 10 \\ 10 \end{bmatrix}$
	$\overline{P}_1 = \begin{bmatrix} 5 & 3 \\ 3 & 4 \end{bmatrix}$	$M(1) = \begin{bmatrix} 5 & 3 \\ 3 & 4 \end{bmatrix}$
	$P_1 = \begin{bmatrix} 1.429 & 0.857 \\ 0.857 & 2.714 \end{bmatrix}$	$K(1) = \begin{bmatrix} 0.773 \\ 0.238 \end{bmatrix}$
	$\hat{x}_1 = \begin{bmatrix} 9.286 \\ 9.571 \end{bmatrix}$	$\hat{x}_1 = \begin{bmatrix} 9.227 \\ 9.762 \end{bmatrix}$
		$P(1) = \begin{bmatrix} -262.2 & 853.7 \\ 853.7 & -2753.1 \end{bmatrix}$
$k = 2$	$\hat{x}_2^- = \begin{bmatrix} 18.857 \\ 9.571 \end{bmatrix}$	$\overline{x}_2 = \begin{bmatrix} 18.989 \\ 9.7622 \end{bmatrix}$
	$\overline{P}_2 = \begin{bmatrix} 5.857 & 3.571 \\ 3.571 & 3.714 \end{bmatrix}$	$M(2) = -\begin{bmatrix} 1310.97 & 1902.44 \\ 1902.44 & 2755.1 \end{bmatrix}$
	$P_2 = \begin{bmatrix} 1.491 & 0.909 \\ 0.909 & 2.097 \end{bmatrix}$	$K(2) = \begin{bmatrix} 0.799 \\ -0.103 \end{bmatrix}$
	$\hat{x}_2 = \begin{bmatrix} 19.336 \\ 9.864 \end{bmatrix}$	$\hat{x}(2) = \begin{bmatrix} 19.4 \\ 9.710 \end{bmatrix}$
		$P(2) = \begin{bmatrix} 4.35 & 34.39 \\ 34.39 & 266.68 \end{bmatrix}$

Table 5.3 Comparison between KF and WLAVF *Continued*

k	KF (Refer to [3] for List of Symbols)	WLAVF
$k=3$	$\hat{\bar{x}}_3 = \begin{bmatrix} 29.2 \\ 9.864 \end{bmatrix}$	$\bar{x}_3 = \begin{bmatrix} 29.11 \\ 9.71 \end{bmatrix}$
	$\overline{P}_3 = \begin{bmatrix} 5.4 & 3 \\ 3 & 3.091 \end{bmatrix}$	$M(3) = \begin{bmatrix} 339.81 & 301.07 \\ 301.07 & 267.68 \end{bmatrix}$
	$P_3 = \begin{bmatrix} 1.46 & 0.811 \\ 0.811 & 1.875 \end{bmatrix}$	$K(3) = \begin{bmatrix} 1.052 \\ 0.0683 \end{bmatrix}$
	$\hat{x}_3 = \begin{bmatrix} 29.054 \\ 9.783 \end{bmatrix}$	$\hat{x}(3) = \begin{bmatrix} 29.0 \\ 9.703 \end{bmatrix}$

5.4 Recursive Least Error Squares

In this section, for reader convenience, we offer a short summary for the mathematical background of the algorithm used in this chapter: recursive least error squares (RLES).

In the recursive least error squares, assume that the k_1 measurement, z_1, is available the corresponding observation equation is

$$z_1 = H_1 x_1 + \zeta_1 \tag{5.77}$$

The best estimate of x_1 based on least error squares is

$$x_1 = \left[H_1^T H_1\right]^{-1} H_1^T z_1 \tag{5.78}$$

The previous estimates are "batch processing" algorithms, in that all measurements are processed together to provide the estimate of a constant vector. To use these equations when new measurements become available, we would need to append the new data to z and repeat the entire process.

If k_2 is a new set of measurements, z_2, the new observation equation is

$$z_2 = H_2 x + \zeta_2 \tag{5.79}$$

When we redefine k as time indexes, let the observation vector at time k have r components, and the recursive mean-value estimator is

$$\hat{x}_k = \hat{x}_{k-1} + K_k(z_k - H_k \hat{x}_{k-1}) \tag{5.80}$$

with

$$K_k = P_{k-1} H_k^T \left(H_k P_{k-1} H_k^T + I\right)^{-1} \tag{5.81}$$

and

$$P_k = \left(P_{k-1}^{-1} + H_k^T H_k\right)^{-1} \tag{5.82}$$

where K_k is an $(n \times r)$ gain matrix, and P_k is an $(n \times n)$ matrix that represents the estimation error at the kth sampling instant. Equations (5.80) to (5.82) are recursive equations that can be used to estimate the parameter x at each time step.

References

[1] R.F. Stengel, Stochastic Optimal Control, John Wiley & Sons Inc., New York, 1986.
[2] A.E. Bryson Jr., Y.C. Ho, Applied Optimal Control, Hemisphere, New York, 1975.
[3] F.L. Lewis, Optimal Estimation with an Introduction to Stochastic Control Theory, John Wiley & Sons Inc., New York, 1986.
[4] G.S. Christensen, S.A. Soliman, Optimal filtering for continuous linear dynamic systems based on WLAV approximations, Automatica 26 (2) (1990) 397–400.
[5] G.S. Christensen, S.A. Soliman, Optimal filter for discrete dynamic systems based on WLAV approximations, Automatica 26 (2) (1990) 401–404.

6 Load-Forecasting Results Using Static State Estimation

6.1 Objectives

The objectives of this chapter are

- Deriving three models on the basis that the load powers are crisp in nature.
- Viewing the results obtained for the crisp load power data for the different load models developed in Chapter 3.
- Performing a comparison between the two static least error squares (LES) and least absolute value (LAV) estimation techniques.
- Using the estimated parameters to predict a load using both techniques, where we compare between them for summer and winter.

6.2 Description of the Data

A big Canadian utility supplied the hourly load power data used in this study for the years 1994 and 1995, and the Atlantic Climate Center of Environment Canada supplied the hourly weather conditions for the same two years extracted from Environment Canada's archives. These data include hourly dry bulb temperatures, wind speed, and percentage humidity recorded at Shearwater Airport in Halifax, Nova Scotia. A standard record format has been adopted for climatological data. Each record consists of station identification, date (year, month, and day), and element number, followed by the data repeated 24 times. The element number identifies each data type and implies the units and decimal position. The element numbers are described in Table 6.1 as they appeared in the data from Environment Canada. If there are missing data for a certain hour, denoted by −999, the average value of the hour before and hour after is used to replace this missing point.

6.3 Offline Simulation (Static Load Forecasting Estimation)

In this section we use the three offline load models for crisp load power data that were developed in Chapter 3 to predict the next day's hourly load profile for selected periods for winter and summer of 1994.

DOI: 10.1016/B978-0-12-381543-9.00006-3

Table 6.1 Elements and Their Units

Element	Units	Description
78	0.1°C	Dry Bulb Temperature
76	km/h	Wind Speed
80	percent	Humidity

For each load model, the LES and LAV algorithms are used to estimate the load model parameters. The results for each load model parameter are given in table format, whereas the final forecasts for LES and LAV together with the actual load are given in the form of curves. Furthermore, the estimated parameters for each model are used to predict the load 24 hours ahead for the same time period.

The following abbreviations are used in this section:

z = Actual recorded load.
z_{LS} = Load forecast made from the least error squares.
z_{LAV} = Load forecasted using the least absolute value algorithm.

The percentage errors corresponding to the forecasted loads are given by

$$\varepsilon_L = \left[(z - z_{LS})/z\right] \times 100$$

and

$$\varepsilon_{LAV} = \left[(z - z_{LAV})/z\right] \times 100$$

where ε is positive (underestimation) if the actual value z is larger than the estimated value, and negative (overestimated) if the estimated value is greater than the actual value.

6.4 Model A Results

Model A was described in Chapter 3 for crisp load powers. It is a multiple linear regression model of which the parameters are constants during the hour considered. A parameter estimate from the available data is obtained for every hour.

An excessive volume of computations is associated with a single 24-hour load prediction. The days are chosen for the prediction process in a random way. The model is applied to different days in the same period of time (same time) for the same season. The parameters estimated for each of these days are not reported because the obtained predictions for more days are essentially the same.

Two approaches are applied. The first approach estimates a given parameter for every hour in question in the day. The days are chosen randomly. The second approach assumes the model parameters to be constants during the whole day studied. The estimated parameters in the two approaches are used to predict the load for one day ahead of a working day and a weekend day in summer and winter.

6.4.1 Model Parameters Estimation for Every Hour in a Summer Weekday (24 Sets)

Tables 6.2 and 6.3 give the estimated parameters for a summer weekday using the LES and LAV algorithms. Table 6.4 gives the estimated load and percentage errors in the estimates using the least error squares and least absolute value algorithms. Figure 6.1 gives a comparison between actual and estimated loads, and Figure 6.2 gives the errors in the estimated powers compared to actual load. From these tables and figures, we can note the following:

- LES estimates the actual load value with a maximum error of 9.1% (underestimated) at hour 24 and a minimum error of 0.1% (overestimated) at hour 1.
- Most error values are below 4% (19 hours). LAV estimates the actual load value with a maximum error of 10.6% (underestimated) at hour 23 and a minimum error of 0% at hours 3, 10, 13, 18, and 22.
- Because many error values are less then 4% for both algorithms, estimated power values during the day (even with wide variations in the weather data) are still acceptable. If the redundancy in the estimated parameters is increased, the errors in the LAV estimates will be decreased.

The estimated parameters during the 24 hours are used to predict the load 24 hours ahead. Table 6.5 gives the predicted load power 24 hours ahead using the estimated load parameters given in Tables 6.2 and 6.3. Figure 6.3 gives the predicted load for 24 hours ahead, and Figure 6.4 shows the error in the predicted load. Examining this table and these figures reveals the following:

- LES predicts the load 24 hours ahead with a maximum error of 10.7% (overpredicted) at hour 9 and a minimum error of 0% at hour 4. Most error values are below 4% (15 hours).
- LAV predicts the load 24 hours ahead with a maximum error of 39.5% (overpredicted) at hour 7 and a minimum error of 0.1% (overpredicted) at hour 21. For several hours, the errors are over 4%. LAV needs more data to decrease the error values.
- Because the levels of error in LES prediction are less than LAV prediction for the load, the LES-predicted value represents the load better than LAV prediction. The error levels in LAV prediction can be reduced if the redundancy in the estimated parameters is increased.

6.4.2 Estimation of Constant Model Parameters for Weekday (One Set)

In the second approach, the load parameters are assumed to be constants during the day in question where there is only one group of parameters instead of 24 groups. Table 6.6 gives the estimated load and percentage error for a summer weekday. Figures 6.5 and 6.6 show the estimated load and the error in the estimated load. Examining the table and figures reveals the following:

- LES estimates the load with a maximum error of 10.2% (overestimated) at hour 22 and a minimum error of 0.1% (overestimated) at hour 16. Most error values are below 4% (17 hours).

Table 6.2 Estimated Parameters for Summer Weekday Using LES

Hour	A_0	A_1	A_2	A_3	A_4	A_5	A_6	A_7	A_8	A_9
1.0	1485.72	−0.06	0.67	−0.03	−85.81	125.68	−26.89	−5021	145.44	−167.20
2.0	1087.29	−77.95	0.09	−0.16	0.05	91.86	0.37	133.90	9.98	−159.39
3.0	1430.12	69.93	0.16	−0.15	−51.12	0.89	4.8	−122.4	86.21	5.42
4.0	1329.39	−101.96	−0.54	−0.19	171.69	−47.67	0.27	172.15	−296.75	97.17
5.0	1524.05	3.91	0.21	−0.09	−141.83	154.46	6.87	34.12	215.41	−238.16
6.0	183.12	−209.76	2.83	−0.05	46.45	142.08	16.73	407.91	−33.91	−357.56
7.0	2126.05	18.25	1.47	0.14	16.32	−15.47	7.32	15.62	−54.59	−16.73
8.0	2255.80	357.16	3.53	0.36	−216.20	−33.64	−80.21	−695.73	490.80	158.28
9.0	1676.21	−10.22	21.37	1.49	−100.93	93.19	87.58	167.32	−215.71	9.87
10.0	675.81	92.49	9.87	0.76	−17.63	−147.9	69.54	−195.15	60.49	146.96
11.0	−13.12	352.99	−6.15	−0.89	−579.85	233.87	−5.45	−1114.21	1787.35	−631.76
12.0	906.51	37.41	−2.21	−0.3	77.14	−65.40	−49.86	−2.45	−486.20	501.33
13.0	2402.79	−160.45	0.2	0.05	56.89	63.64	66.33	321.47	−107.42	−266.44
14.0	1070.16	202.14	−1.96	−0.15	−359.94	155.71	−9.96	−393.21	729.93	−340.41
15.0	3054.72	−807.87	−6.31	−0.45	239.56	671.98	−58.39	910.25	316.52	−1302.35
16.0	3572.46	−233.72	−14.14	−0.88	−100.83	338.87	5.27	470.43	−151.57	−403.66
17.0	3634.89	−398.62	−11.87	−0.64	417.67	−141.99	115.88	718.68	−871.51	65.02
18.0	2637.68	35.35	0.41	−0.04	−286.61	235.38	34.42	2.97	621.69	−679.62
19.0	138.92	230.39	−2.47	−0.21	−13.45	−97.77	−125.81	−467.09	184.76	260.46
20.0	2270.39	−371.77	−9.7	−0.63	769.15	−352.97	−47.10	646.96	−1304.62	618.39
21.0	2196.61	52.98	−0.35	0.21	−140.15	140.63	−57.59	−70.79	286.13	−254.96
22.0	401.09	−395.79	0.83	−0.30	368.14	−38.64	73.38	906.26	−821.55	−46.64
23.0	2079.17	66.14	2.11	0.08	−44.17	−1.21	4.87	−25.46	29.71	−44.55
24.0	1664.7	85.43	0.4	−0.05	−54.52	7.19	−19.68	−143.52	123.10	−9.41

Table 6.3 Estimated Parameters for Summer Weekday Using LAV

Hour	A_0	A_1	A_2	A_3	A_4	A_5	A_6	A_7	A_8	A_9
1.0	3121.23	8.04	−3.64	0.23	−199.75	212.04	7.7	−65.25	467.30	−482.31
2.0	1468.80	−23.16	−1.02	−0.27	0.73	42.86	0.92	35.08	2.53	−66.94
3.0	2255.94	−339.70	−17.13	−1.57	332.40	5.29	−4.95	701.62	700.61	−51.41
4.0	1303.64	−59.84	0.03	−0.17	146.48	−61.04	0.20	92.20	−253.86	134.58
5.0	2325.94	−55.66	4.65	0.11	132.66	−66.12	35.54	289.02	−347.47	−2.51
6.0	−50.26	−189.05	2.17	−0.03	59.67	98.03	22.82	369.27	−57.10	−288.02
7.0	40313.83	3180.82	11.05	18.78	−2127.89	−2978.19	2410.72	−3911.42	2112.78	350.55
8.0	2065.19	400.32	2.95	0.47	−204.63	−99.20	−87.98	−773.30	516.02	219.04
9.0	5167.72	−158.56	39.85	2.56	−347.79	456.89	280.96	802.52	−471.99	−517.27
10.0	−2688.83	−660.06	19.79	1.08	1571.86	−1197.89	252.98	1122.56	−2374.36	1367.30
11.0	1078.60	140.32	−5.51	−0.72	−229.54	132.16	−28.66	−674.78	1060.20	−383.87
12.0	−379.85	64.34	−1.51	−0.39	291.02	−300.76	−74.02	−64.05	−1006.76	1139.02
13.0	2717.30	−216.48	−0.44	0.00	133.62	28.22	87.19	414.68	−192.20	−286.04
14.0	1535.7	−36.68	−18.33	−0.99	−284.97	271.08	−30.666	−124.18	491.70	−389.96
15.0	2736.55	−773.48	−5.98	−0.44	351.08	517.49	−57.37	895.09	46.65	−1005.28
16.0	211.73	114.98	−3.27	−0.12	−174.79	13.87	1.5	−358.85	416.69	−35.45
17.0	−835.93	−46.97	−7.14	−0.17	419.89	−524.28	41.55	−67.83	−946.26	1068.56
18.0	−23585.22	3485.4	−15.7	−0.41	675.49	−5438.4	699.05	−6189.0	−920.71	7850.56
19.0	3671.00	−201.02	−9.82	−0.49	−58.98	228.14	9.67	400.88	107.68	−586.85
20.0	2175.94	−159.33	−10.15	−0.93	584.64	−382.53	−18.69	191.58	−891.17	656.41
21.0	3323.57	391.81	−5.9	−0.01	−342.10	−124.78	61.03	−701.63	693.20	−81.93
22.0	794.65	−179.54	2.27	0.13	133.18	−20.76	55.26	525.48	−349.20	−161.09
23.0	1152.33	−206.01	3.58	0.35	261.59	−96.35	35.88	344.41	−370.59	18.70
24.0	1180.20	−442.01	3.12	−0.33	328.66	112.57	25.73	851.90	−300.86	−557.56

Table 6.4 Estimated Load and Percentage Error for a Summer Weekday, 24 Parameter Sets, Model A

Daily Hours	Actual Load (MW)	LES Estimate	LAV Estimate	% LES Error	% LAV Error
1	674.0	674.9	674.7	−0.1	−0.1
2	609.9	618.8	610.4	−1.5	−0.1
3	559.6	569.0	559.7	−1.7	0.0
4	537.7	550.7	538.0	−2.4	−0.1
5	536.8	516.9	501.3	3.7	6.6
6	535.6	548.6	536.5	−2.4	−0.2
7	545.6	555.7	546.4	−1.9	−0.1
8	574.4	600.7	598.3	−4.6	−4.2
9	668.9	689.7	695.8	−3.1	−4.0
10	787.3	778.5	786.9	1.1	0.0
11	875.1	844.6	849.2	3.5	3.0
12	909.4	894.5	893.9	1.6	1.7
13	925.0	928.2	924.9	−0.3	0.0
14	903.0	894.5	903.7	0.9	−0.1
15	876.0	855.4	846.8	2.3	3.3
16	848.7	869.2	870.8	−2.4	−2.6
17	848.3	820.7	829.0	3.3	2.3
18	884.8	896.7	884.6	−1.3	0.0
19	880.9	896.0	906.3	−1.7	−2.9
20	837.9	894.9	887.3	−6.8	−5.9
21	805.1	873.7	870.2	−8.5	−8.1
22	824.4	833.4	824.4	−1.1	0.0
23	876.2	797.2	783.4	9.0	10.6
24	815.6	741.6	739.4	9.1	9.3

- LAV estimates the load with a maximum error of 17% (overestimated) at hour 7 and a minimum error of 0% at hours 8 and 14. Error values under 4% are at 14 hours.
- Both LES and LAV estimations for the load show a range of errors as the 24 parameter sets. The estimations deviate from the actual load with an acceptable range of errors.

The performance of two approaches for a summer weekday, as explained, is also examined for a summer weekend.

6.4.3 Model Parameter Estimation for Every Hour in a Summer Weekend Day (24 Sets)

Table 6.7 gives the estimated load and percentage error parameters using LES and LAV techniques for a summer weekend day. Figures 6.7 and 6.8 give the estimated

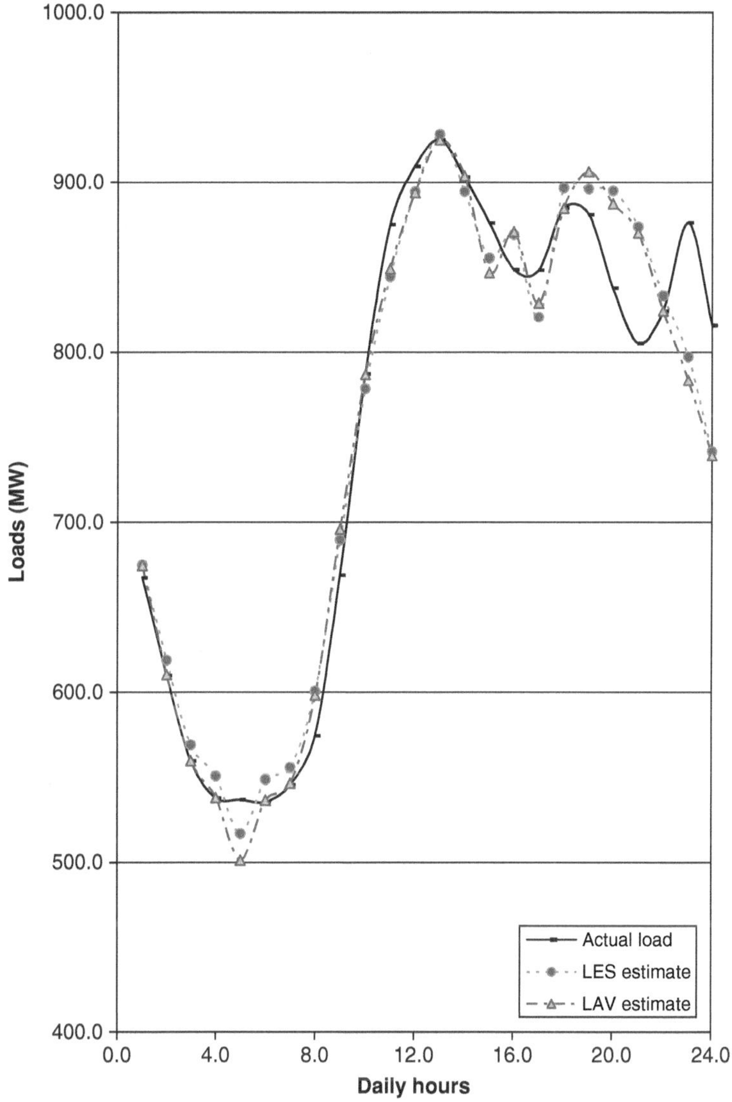

Figure 6.1 Estimated load for a summer weekday using 24 parameter sets, model A.

load and the percentage errors in this estimate using these parameters. Examining the table and figures reveals the following:

- LES estimates the load for a weekend day with a maximum error of 4.4% (overestimated) at hour 22 and a minimum error of 0.1% (overestimated) at hour 2. So, the estimated load values are good due to small error values.

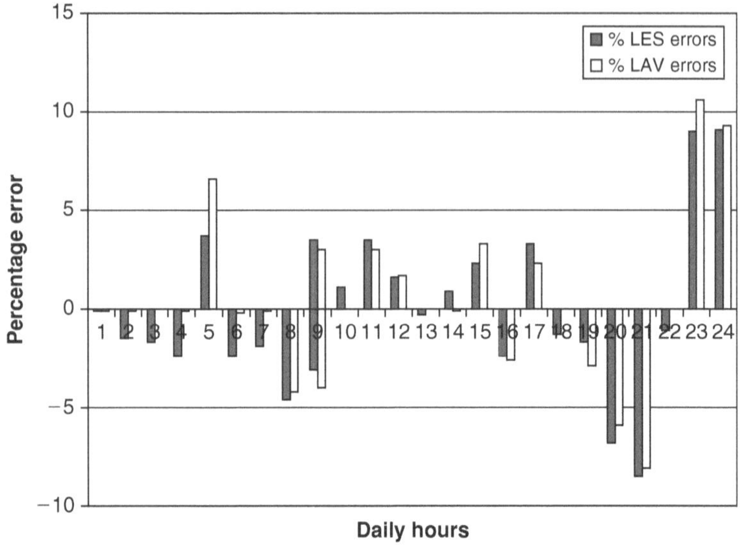

Figure 6.2 Estimated load error for a summer weekday using 24 parameter sets, model A.

- The LAV-estimated load value has a maximum error of 10.2% (underestimated) at hour 24 and a minimum error of 0% at hours 2 and 12. Because most of the rest of the error values are under 4% (either overestimated or underestimated), the estimated load value is good.

The 24 parameter sets are used to predict the load one week ahead. Table 6.8 and Figures 6.9 and 6.10 give the predicted load for a weekend ahead and the percentage error in this prediction. Examining the table and figures reveals the following:

- The maximum error in the LES-predicted load is 7.34% (overpredicted) at hour 20, and the minimum error is 0.1% (underpredicted) at hour 18. Most of the rest of the errors are less than 4% (overpredicted or underpredicted) in value.
- LAV predicts the load with a maximum error of 19.34% (overpredicted) at hour 24 and a minimum error of 0.32% (overpredicted) at hour 15. Most of the rest of the errors are less than 4% (overpredicted or underpredicted) in value.
- Due to the small values of errors, both LES and LAV give an acceptable load prediction.

6.4.4 Estimation of Constant Model Parameters for a Summer Weekend Day (One Set)

Table 6.9 gives the estimated parameters for a summer weekend day, and Table 6.10 and Figures 6.11 and 6.12 give the estimated load and percentage errors in this estimate using one set of parameters given in Table 6.9 for LES and LAV techniques. Examining the tables and figures reveals the following:

- Maximum error in LES load estimation is 15% (underestimated) at hour 8, while the minimum error is 0.1% (overestimated) at hour 12. Most of the rest of the errors (12 hours) are less than 4% (overestimated or underestimated) in value.

Table 6.5 Predicted Load and Percentage Error for Summer Weekday, 24 Parameter Sets, Model A

Daily Hours	Actual Load (MW)	LES Prediction	LAV Prediction	% LES Error	% LAV Error
1	681	686.8	583.9	−0.9	14.3
2	622.7	610	613.9	2	1.4
3	586.5	573.8	630.7	2.2	−7.5
4	569.7	569.5	564	0	1
5	572.3	566.2	556.8	1.1	2.7
6	569.7	573.6	570.2	−0.7	−0.1
7	594.2	573.2	828.8	3.5	−39.5
8	661.5	677.6	665.6	−2.4	−0.6
9	783.8	867.8	814.1	−10.7	−3.9
10	900.1	921.5	855.3	−2.4	5
11	979.6	945.1	938.4	3.5	4.2
12	1020	953.6	961.5	6.5	5.7
13	1047	966.5	944	7.7	9.8
14	1032	924.3	854.4	10.4	17.2
15	1016	1018	1020	−0.1	−0.3
16	1003	951.2	921.7	5.1	8.1
17	1011	923.2	869.7	8.7	14
18	1044	979	1007	6.2	3.5
19	1022	931.6	967.4	8.9	5.4
20	956.6	937.2	887	2	7.2
21	922.9	907.4	923.9	1.7	−0.1
22	957.2	1010	992.1	−5.5	−3.6
23	958.8	967.4	916.4	−0.9	4.4
24	874.5	872.7	860.1	0.2	1.6

- Maximum error in LAV load estimation is 15% (underestimated) at hour 8, while the minimum error is 0.1% (overestimated) at hour 12. Most of the rest of the errors (12 hours) are less than 4% (overestimated or underestimated) in value.
- Better estimated values can be obtained by using more data to reduce the errors.

The one set of parameters is used to predict the load one week ahead. Table 6.11 and Figures 6.13 and 6.14 give the obtained results. Examining the tables and figures reveals the following:

- Maximum error in LES load prediction is 9.4% (underpredicted) at hour 1, while the minimum error is 0.1% (overpredicted) at hour 12. Most of the rest of the errors (15 hours) are less than 4% (overpredicted or underpredicted).
- Maximum error in LAV load prediction is 9.5% (overpredicted) at hour 9, while the minimum error is 0.1% (overpredicted) at hour 12.
- For both techniques, LES and LAV, the predicted load can be represented better by using more data to reduce the errors.

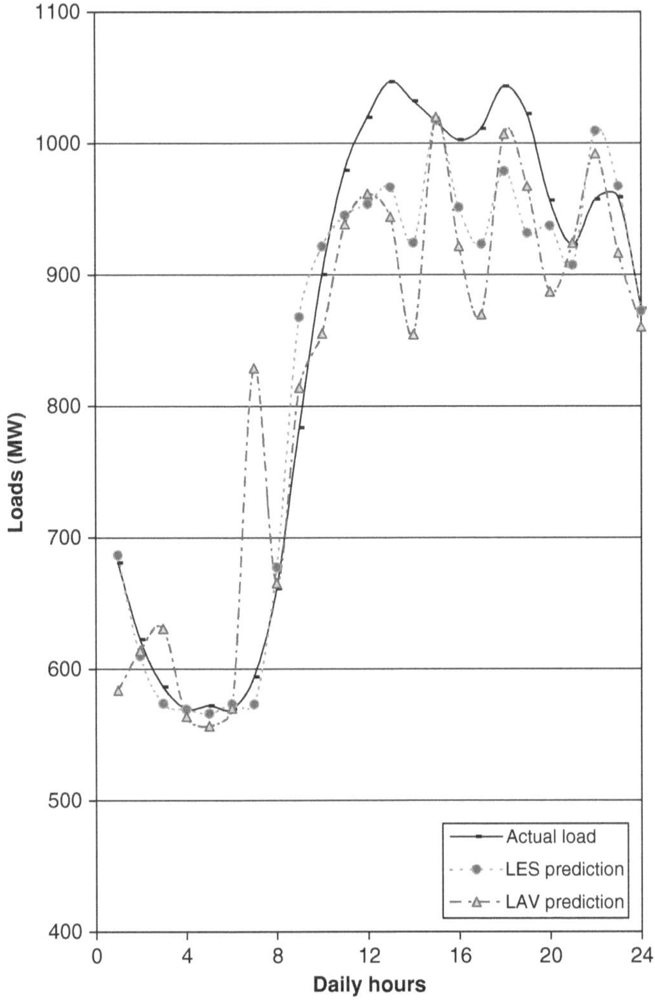

Figure 6.3 Predicted load for a summer weekday using 24 parameter sets, model A.

6.4.5 General Remarks for Summer Model A

The two approaches, LES and LAV, give acceptable predicted load values. There are errors involved between actual and estimated load and between actual and predicted load. To reduce errors, we need to use more quality data. The scope here is to show how LES and LAV algorithms are applied as predicting tools. Estimated parameter values obtained using 24 sets or one set of parameters produce estimated and predicted values deviating with errors from the actual value. The results obtained using 24 sets or one set contain error values. Thus, using one set of parameters will be more economical, requiring less effort and computing time.

These algorithms will be compared to the fuzzy algorithm later on in Chapter 7.

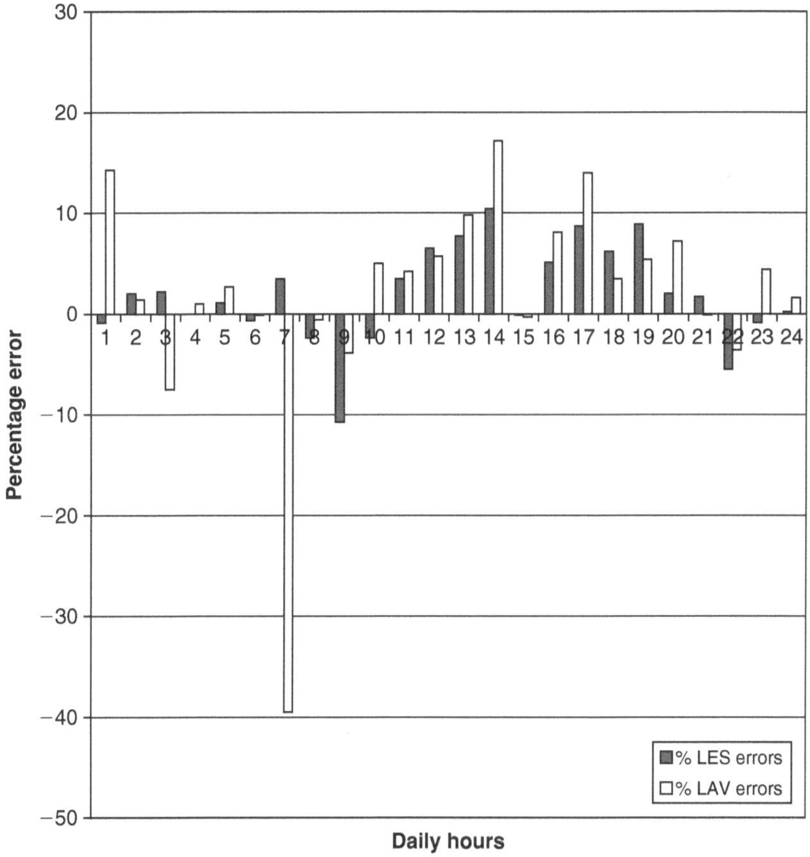

Figure 6.4 Predicted load error for a summer weekday using 24 parameter sets, model A.

6.4.6 Winter Predictions

Appendix 6.1 gives the results obtained for a winter weekday and winter weekend day using model A. The same arguments for the summer results can be made for the winter results. The estimated and predicted load values deviate from the actual load values. Tables 6.12a and 6.12b give a brief summary for estimated and predicted errors. The errors can be reduced by using more quality data so that the predicted values can resemble and predict the data as accurately as possible. The use of either 24 sets or one set of parameters gives high error values in the predicted load. Thus, using one set of parameters will be more economical and save time.

6.5 Model B

Model B is a weather-insensitive model that depends only on the hour (time) considered. The model proposed in Chapter 3 is used here to predict the load power for

Table 6.6 Estimated Load and Percentage Error for Summer Weekday, One Parameter Set, Model A

Daily Hours	Actual Load (MW)	LES Estimate	LAV Estimate	% LES Error	% LAV Error
1	674	691.9	679.6	−2.7	−0.8
2	609	624	624	−2.3	−5.9
3	559.7	572	582	−2.3	−3.98
4	537.7	554.4	540	−3.1	−0.4
5	536.8	567.5	561.7	−5.7	−4.6
6	535.6	536.4	533.9	−0.2	0.3
7	545.6	586.2	638.3	−7.4	−17
8	574.4	566.5	574.6	1.4	0
9	668.9	685.6	694.1	−2.5	−3.8
10	787.3	830.7	680.1	−5.5	13.6
11	875.1	787.5	828	10	5.4
12	909.4	878.7	859.2	3.4	5.5
13	925	915.2	902.2	1.1	2.5
14	903	933.3	903	−3.4	0
15	876	900.9	903.5	−2.8	−3.1
16	848.7	849.5	858.3	−0.1	−1.1
17	848.3	801.6	868.2	5.5	−2.3
18	884.8	851.9	871.2	3.7	−1.5
19	880.9	877.5	874.3	0.4	0.8
20	837.9	869.9	826.3	−3.8	1.4
21	805.1	841.9	827.2	−4.6	−2.7
22	824.4	908.9	723.4	−10.2	12.3
23	876.2	887.3	871.6	−1.3	0.5
24	815.6	820.1	756.6	−0.6	7.2

24 hours ahead for one working day and one weekend day in summer and winter. The model is applied to the data used in model A. Harmonics included are from the lowest number to 13. Nine harmonics are used in the static estimation process because it was found to produce the lowest error.

6.5.1 Summer Weekday

Table 6.13 gives the estimated load parameters for a summer weekday, and Table 6.14 and Figures 6.15 and 6.16 give the estimated load for a summer weekday and the percentage error in this estimate using the LES and LAV algorithms. Examining the tables and figures reveals the following:

- LES estimates the load with a maximum error of 26.2% (overestimated) at hour 2 and a minimum error of 2.5% (underestimated) at hour 18.
- LAV estimates the load with a maximum error of 87.4% (overestimated) at hour 3 and a minimum error of 0% at hour 24.

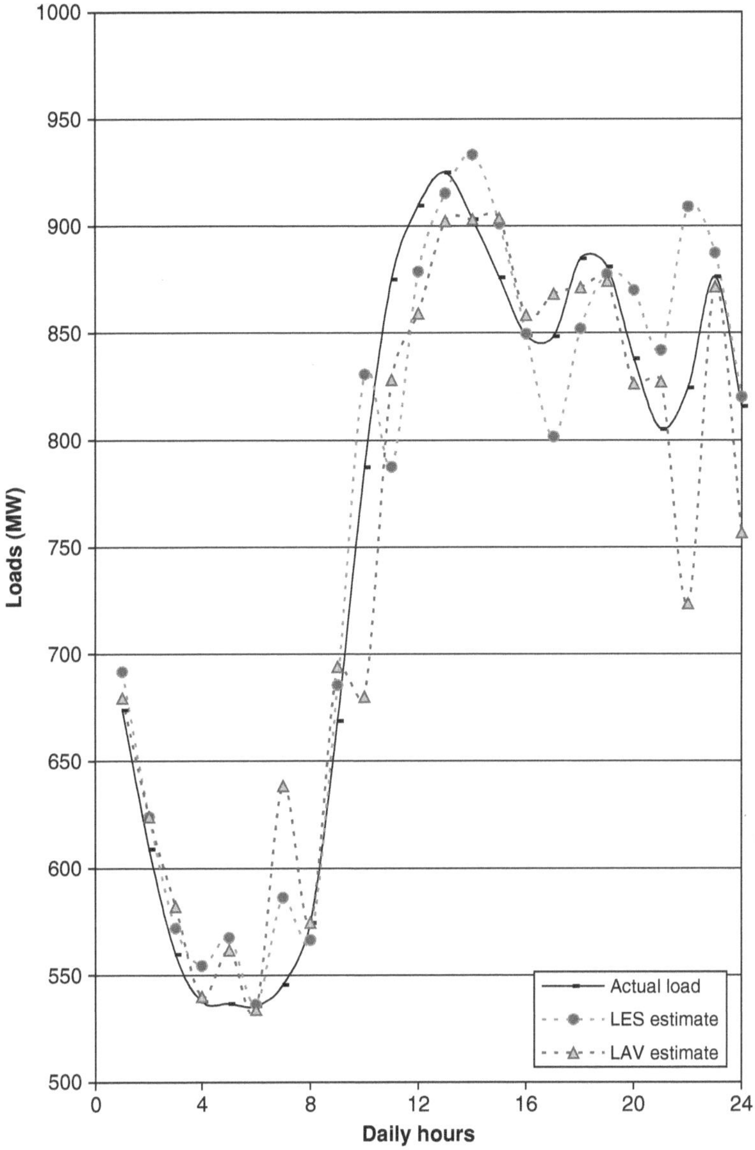

Figure 6.5 Estimated load for a summer weekday using one parameter set, model A.

- Estimated load error values of LAV from hour 4 to hour 22 are very small compared to the estimated load error values of LES (for example, at hour 4, LAV error is 0.05%, whereas LES error is 20.5%; and at hour 20, LAV error is 0.06%, whereas LES error is 14.8%). The LAV algorithm gives a better load estimate than the LES estimate for the hours 4 to 22. The large error values for some hours in LAV estimation are caused by bad data. More data can help screen these bad points and reduce error values.

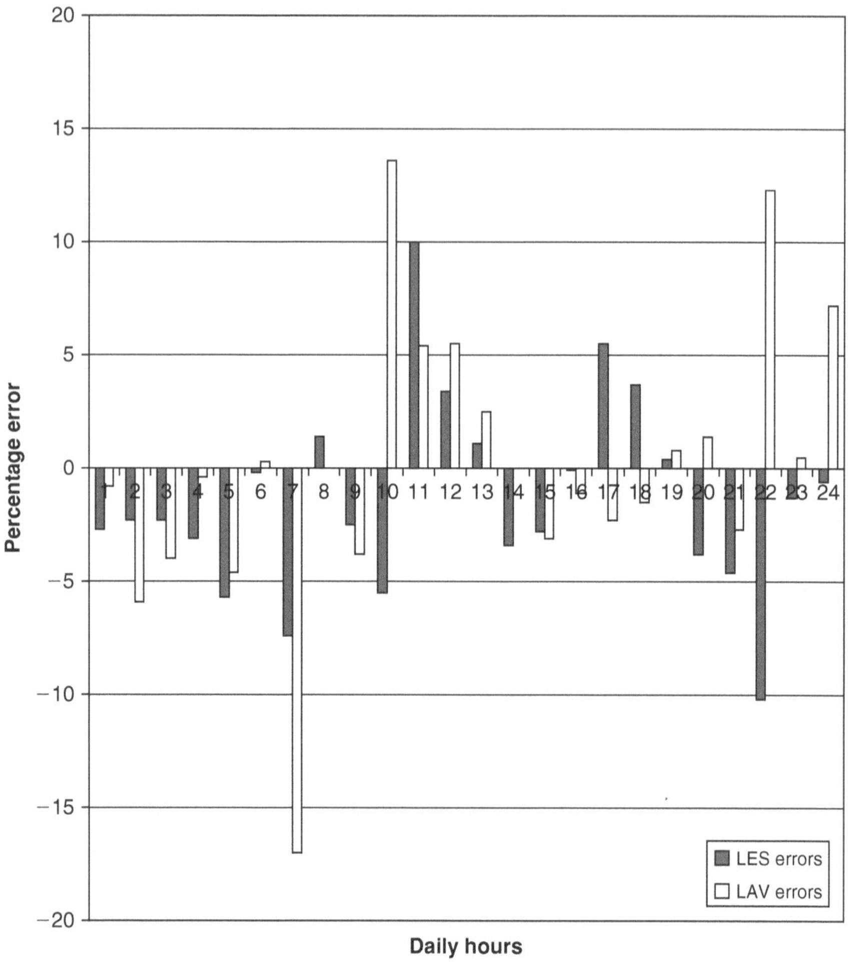

Figure 6.6 Predicted load error for a summer weekday using one parameter set, model A.

The parameters estimated for model B are used to predict the load 24 hours ahead for the same weekday, during the same season. Table 6.15 and Figures 6.17 and 6.18 show the predicted load and the error in this prediction. Examining the table and figures reveals the following:

- LES predicts the load with errors larger than 14% in 11 instances. The range of errors is from 27.54% (overpredicted) at hour 2 as the highest to 0.11% (underpredicted) at hour 12 as the lowest.
- LAV predicts the load with errors less than 0.6% in 19 instances. The error range is from the highest at 20.98% (overpredicted) to the lowest at 0.08% (overpredicted).
- LAV error values are less than LES error values in every hour except hour 23, where LES error is 16.46% (underpredicted) and LAV error is 18.16% (overpredicted). This indicates that in this case LAV gives a better representation than LES.

Table 6.7 Estimated Load and Percentage Error for a Summer Weekend Day, 24 Parameter Sets, Model A

Daily Hours	Actual Load (MW)	LES Estimate	LAV Estimate	% LES Error	% LAV Error
1	758.1	759.3	796.2	−0.2	−5
2	683.3	684.3	683.2	−0.1	0
3	640.8	629.3	660.6	1.8	−3.1
4	614.2	612.9	622.4	0.2	−1.3
5	597.3	607.6	598.2	−1.7	−0.2
6	586.8	597.3	618.7	−1.8	−5.4
7	590.3	568.3	568.5	3.7	3.7
8	601.3	613.9	592.5	−2.1	1.5
9	667.3	659.8	655.3	1.1	1.8
10	764.1	759.1	765.8	0.7	−0.2
11	848.8	843	825.8	0.7	2.7
12	885.7	884	885.8	0.2	0
13	907.7	901.3	903.7	0.7	0.4
14	897.2	904.6	986.6	−0.8	−10
15	869.5	857.4	903.1	1.4	−3.9
16	842.4	834.6	810.3	0.9	3.8
17	835.5	829.8	834.2	0.7	0.2
18	853.8	850.5	851.8	0.4	0.2
19	857.7	851.8	787.6	0.7	8.2
20	823.9	811.3	809.4	1.5	1.8
21	801.8	812.7	821.6	−1.4	−2.5
22	823.4	859.6	863.4	−4.4	−4.9
23	835.3	828.1	816.6	0.9	2.2
24	783.1	767.4	703.2	0.2	10.2

6.5.2 Summer Weekend Day

The same model, B, is used to forecast a weekend day load. Table 6.16 and Figures 6.19 and 6.20 give the estimated load and the percentage error in this estimate. Examining the table and figures reveals the following:

- LES estimates the load with a maximum error of 17.78% (overestimated) at hour 2 and a minimum error of 0% at hour 12.
- LAV estimates the load with a maximum error of 13.88% (overestimated) at hour 2 and a minimum error of 0.05% (overestimated) at hour 17 and of 0.01% (underestimated) at hour 24.
- The LAV-estimated load errors are less, with a wide margin, than the LES-estimated load errors in all the cases except for hour 1.
- Because LAV-estimated errors are less than 0.22% in 19 instances, LAV gives better estimates than LES.

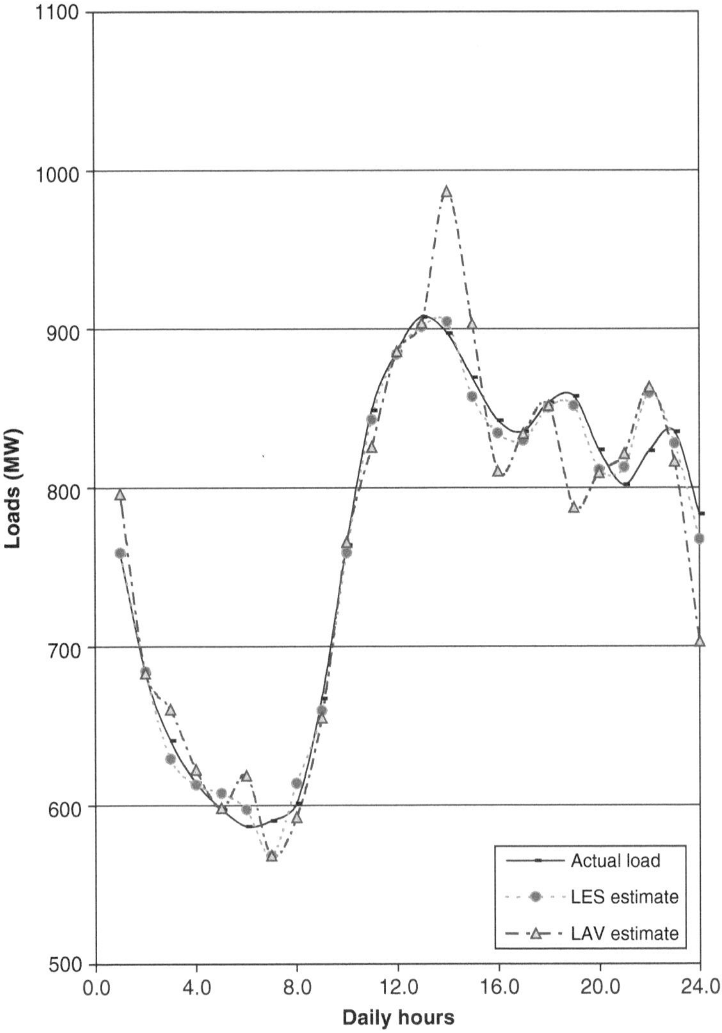

Figure 6.7 Estimated load for a summer weekend day using 24 parameter sets, model A.

Table 6.17 and Figures 6.21 and 6.22 give the predicted load for a weekend day one week ahead. Examining the table and figures reveals the following:

- The LES-predicted load error has a maximum of 20.07% (overpredicted) at hour 2 and a minimum of 0.47% (underpredicted) at hour 12.
- LAV predicts the load with a maximum error of 2.3% (underpredicted) at hour 2 and a minimum error of 0% at hours 0 and 3.
- Because all LAV-predicted load errors are less than LES-predicted load errors, LAV gives better load predictions than LES does.

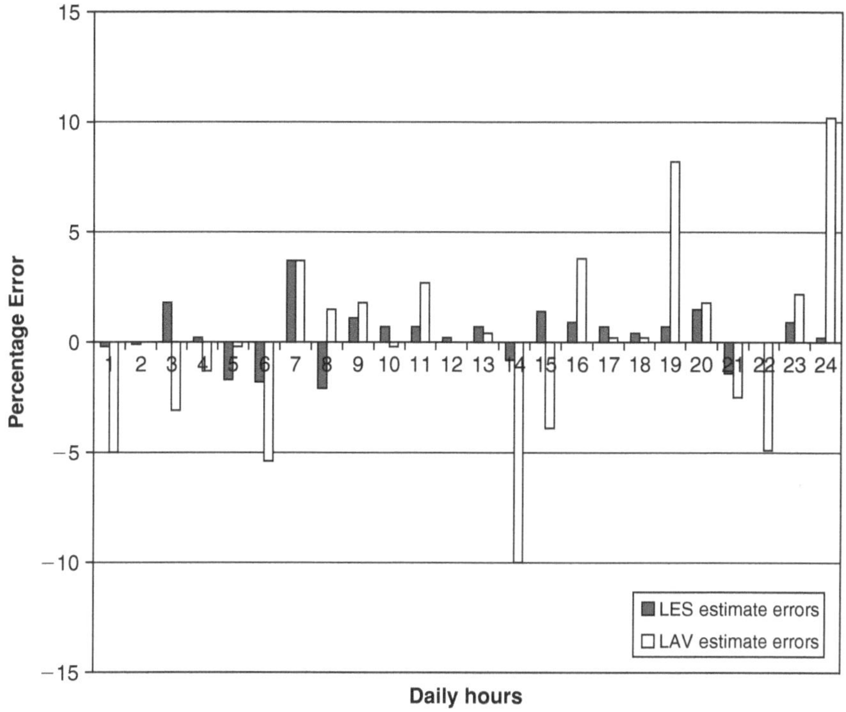

Figure 6.8 Estimated load error for a summer weekend day using 24 parameter sets, model A.

6.5.3 General Remarks for Summer Model B

The two approaches, LES and LAV, give acceptable load predictions for this model. Model B is not weather responsive. More quality data have to be used to reduce the error values. The LES and LAV tools present predicted load values for both weekday and weekend days. These algorithms will be compared to the fuzzy algorithm results in Chapter 7.

6.5.4 Winter Predictions

Appendix 6.2 gives the results for a weekday and a weekend day during the winter season. Model B is a nonsensitive weather model. This feature is reflected upon the results, which show deviations in estimated and predicted load values from the actual load. The error levels range from high to low values. The same argument applies for either weekday or weekend day.

6.6 Model C Results

Model C is a combination of a harmonic, weather-insensitive model, and a multiple linear regression model that accounts for weather parameters. In other words, it's a hybrid of models A and B. The parameters of model C are estimated for a weekday

Table 6.8 Predicted Load and Percentage Error for a Summer Weekend Day, 24 Parameter Sets, Model A

Daily Hours	Actual Load (MW)	LES Prediction	LAV Prediction	% LES Error	% LAV Error
1	716.8	736.1	761.7	−2.7	−6.27
2	637.7	655.6	643	−2.81	−0.84
3	598.5	603	618.7	−0.75	−3.38
4	573.7	588.9	581.3	−2.66	−1.32
5	558	565.6	555	−1.36	0.54
6	550.7	571.1	551.4	−3.7	−0.12
7	560.7	587.5	569	−4.78	−1.49
8	585.7	587.5	576.4	−0.37	1.58
9	659.2	645.5	649.5	2.08	1.48
10	762.4	750	748.2	1.63	1.86
11	843.9	833.5	829.4	1.23	1.72
12	875	861.6	861.8	1.53	1.51
13	881.2	867.2	867.2	1.59	1.58
14	863.2	864.8	847	−0.19	1.87
15	831.4	847.1	834.1	−1.89	−0.32
16	805.3	808.6	813.2	−0.42	−0.98
17	795.6	799.3	804.2	−4.46	−1.08
18	814	813.2	816.7	0.1	−0.33
19	808	826.2	903.6	−2.26	−11.83
20	766.4	822.6	833.5	−7.34	−8.76
21	748.2	791.5	759.2	−5.78	−1.47
22	823.3	800.2	793.2	2.81	3.66
23	801.8	807.6	840.1	−0.72	−4.78
24	744.5	753.9	888.5	−1.62	−19.34

and weekend day during the summer season and winter season. Table 6.18 shows the load parameters for a summer or winter day. Table 6.19 and Figures 6.23 and 6.24 give the estimated load and the percentage errors in the estimate during this summer weekday. Furthermore, Table 6.20 and Figures 6.25 and 6.26 give the predicted load for a summer weekday 24 hours ahead. Examining the tables and figures reveals the following:

- From Table 6.18, parameter A_0 has the largest value because it represents the basic load, while the rest of the parameters represent the variations in the load from other factors. A_0 is 1020.16 at LES estimation and 1023.68 at LAV estimation.
- The results given in Table 6.19, estimated load, indicate that the parameter estimates for model C are accurate because the errors in the estimated load power values are very small for both LES and LAV techniques. LES-estimated load error goes from the highest of 0.36% (underestimated) at hour 4 to the lowest of 0% at hour 18, whereas LAV-estimated

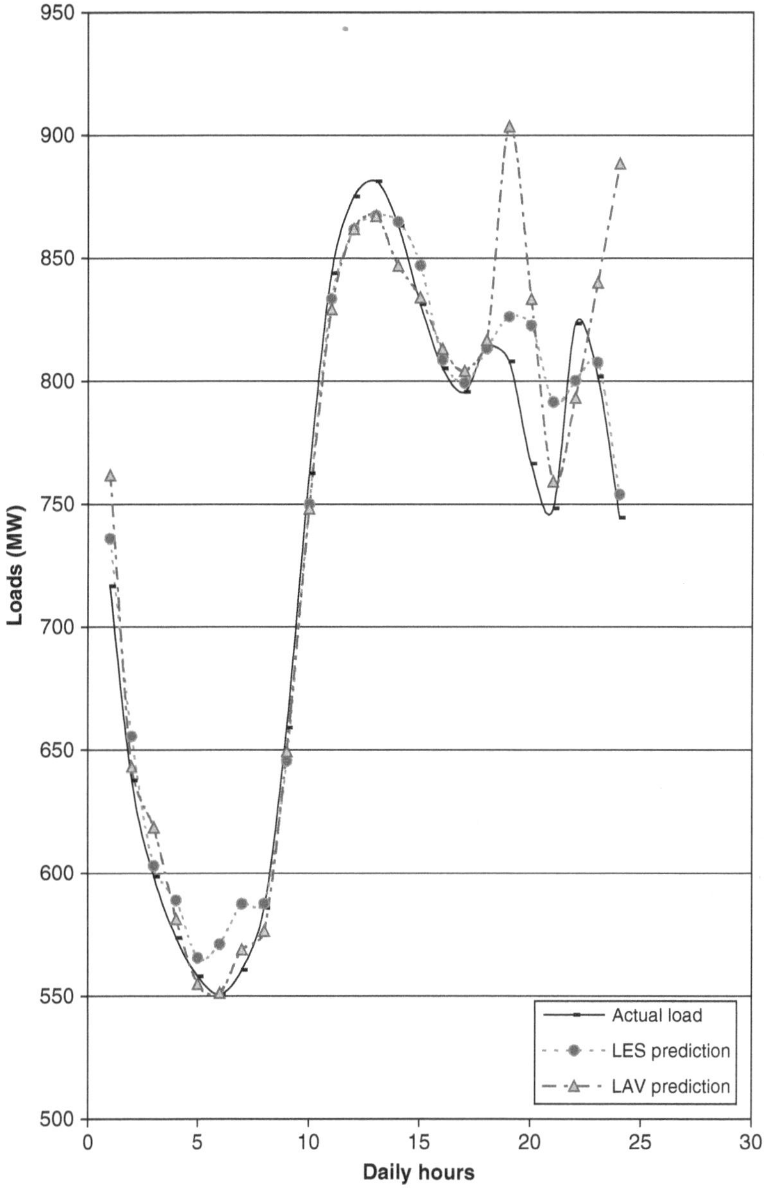

Figure 6.9 Predicted load for a summer weekend day using 24 parameter sets, model A.

load error goes from the highest of 1.86% (overestimated) at hour 22 to the lowest of 0.01% (overestimated) at hour 16.

- From Table 6.19, it is noted that the maximum error obtained at hour 22 by the LAV algorithm is 1.86% (overestimated) and the LES algorithm is 0.28% (overestimated), which is small and acceptable.

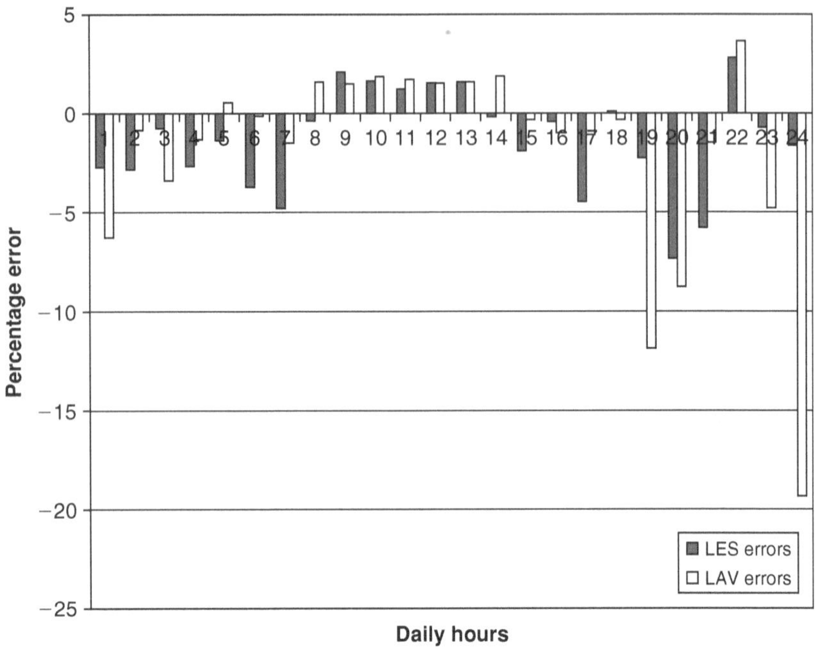

Figure 6.10 Predicted load error for a summer weekend day using 24 parameter sets, model A.

Table 6.9 Estimated Parameters for a Weekend Using the LES and LAV Algorithms

Parameter	LES Estimate	LAV Estimate
A_0	−3892.0	−3991.2
A_1	568.47	592.35
A_2	−123.38	−131.14
A_3	6.10	6.53
A_4	−8.30	−8.12
A_5	−4.97	−0.96
A_6	−0.89	−0.71
A_7	159.08	140.93
A_8	−84.80	−36.55
A_9	52.49	24.88

- Both Table 6.20, predicted load for 24 hours ahead, and Figure 6.25, giving a comparison between the predicted and actual load, show that the load is being overpredicted. LES-predicted load error has the highest value of 22.45% (overpredicted) at hour 4 and the lowest value of 0.51% (underpredicted) at hour 3. LAV-predicted load error has the highest value of 24.17% (overpredicted) at hour 4 and the lowest value of 0.7% (underpredicted) at hour 24.

Table 6.10 Estimated Load and Percentage Error for Summer Weekend Day, One Parameter Set, Model A

Daily Hours	Actual Load (MW)	LES Estimate	LAV Estimate	% LES Error	% LAV Error
1	758.1	687.1	687.1	9.4	9.4
2	683.3	672.4	677.8	1.6	0.8
3	640.8	678.1	673.8	−5.8	−5.2
4	614.2	664.1	655.6	−8.1	−6.7
5	597.3	634.3	623.9	−6.2	−4.5
6	586.8	553.4	531	5.7	−9.5
7	590.3	590.9	585.9	−0.1	0.7
8	601.3	691.6	691.6	−15	15
9	667.3	714.5	730.7	−7.1	9.5
10	764.1	753.2	759	1.4	0.7
11	848.8	842.3	854.3	0.8	−0.7
12	885.7	886.4	886.5	−0.1	−0.1
13	907.7	882.8	901	2.7	0.7
14	897.2	868.5	893.3	3.2	0.4
15	869.5	863.5	868.6	0.7	0.1
16	842.4	916.6	914.5	−8.8	−8.6
17	835.5	845.1	841	−1.1	−0.7
18	853.8	900.2	900	−5.4	5.4
19	857.7	821.5	806.6	4.2	6
20	823.9	787.5	793.6	4.4	3.7
21	801.8	809.1	807.3	−0.9	−0.7
22	823.4	794.6	791.3	3.5	3.9
23	835.3	759.7	760.3	2.1	9
24	783.1	733.9	722.3	6.3	7.8

6.6.1 General Remarks for Summer Model C

Model C considers all days of the week and does not distinguish between weekdays and weekend days. The two approaches using model C give good load predictions. Overprediction takes place more than underprediction. Therefore, the loads are overpredicted. More data have to be used to reduce error values. LES and LAV algorithms are applied as predicting tools, and in Chapter 7 these algorithms will be compared to the fuzzy algorithm in results and technique.

6.6.2 Winter Predictions

Appendix 6.3 gives the prediction results for a winter day. After examining these results, we can reach the same observations as those for the summer day. Table A6.13 shows the estimated load results to be very good. The LES-estimated load has a maximum error of

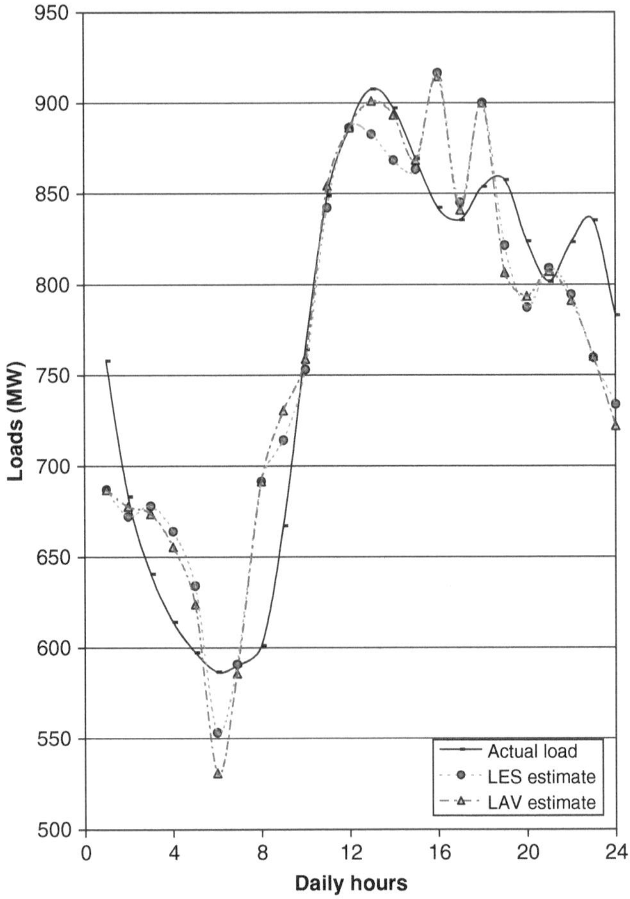

Figure 6.11 Estimated load for a summer weekend day using one parameter set, model A.

0.43% (overestimated) at hour 3 and a minimum of 0% at hour 13. The LAV-estimated load has a maximum error of 2.29% (overestimated) at hour 3 and a minimum error of 0% at hours 11, 16, 19, and 21. The level of errors is small and acceptable. Table A6.14 exhibits the predicted load results. The LES-predicted load has a maximum error of 13.01% (underpredicted) at hour 1 and a minimum of 1.12% (overpredicted) at hour 15. The LAV-predicted load has a maximum of 12.97% (underpredicted) at hour 2 and a minimum of 0.3% (overpredicted) at hour 15. In general, because model C accounts for weather and time, it exhibits better results.

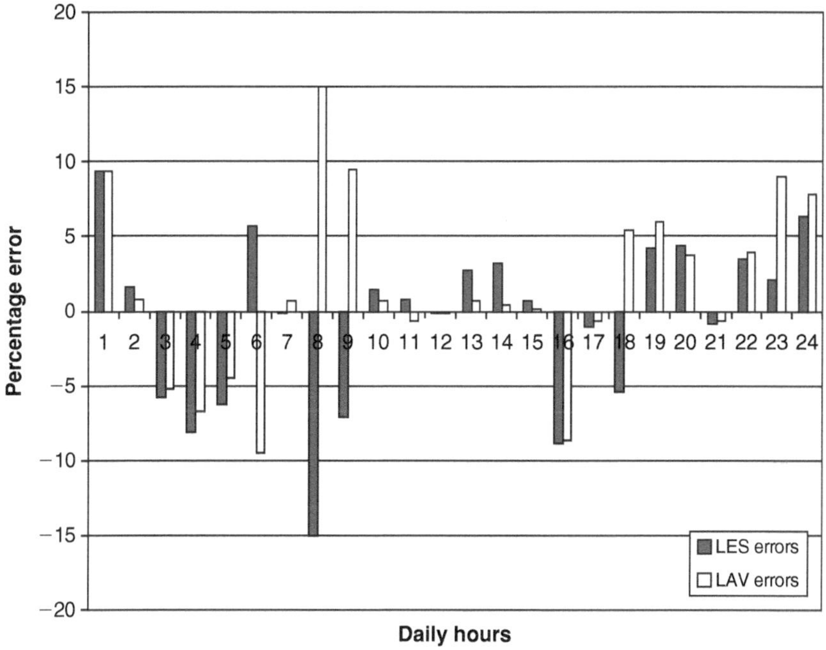

Figure 6.12 Estimated load error for a summer weekend day using one parameter set, model A.

Table 6.11 Predicted Load and Percentage Error for a Summer Weekend Day, One Parameter Set, Model A

Daily Hours	Actual Load (MW)	LES Prediction	LAV Prediction	% LES Error	% LAV Error
1	758.1	687.1	687.1	9.4	9.4
2	683.3	672.4	677.8	1.6	0.8
3	640.8	678.1	673.8	−5.8	−5.2
4	640.8	678.1	673.8	−5.8	−5.2
5	640.8	678.1	673.8	−5.8	−5.2
6	640.8	678.1	673.8	−5.8	−5.2
7	640.8	678.1	673.8	−5.8	−5.2
8	640.8	678.1	673.8	−5.8	−5.2
9	667.3	714.5	730.7	−7.1	−9.5
10	764.1	753.2	759	1.4	0.7
11	848.8	842.3	854.3	0.8	−0.7
12	885.7	886.4	886.5	−0.1	−0.1
13	907.7	882.8	901	2.7	0.7
14	897.2	868.5	893.3	3.2	0.4
15	869.5	863.5	868.6	0.7	0.1

Continued

Table 6.11 Predicted Load and Percentage Error for a Summer Weekend Day, One Parameter Set, Model A *Continued*

Daily Hours	Actual Load (MW)	LES Prediction	LAV Prediction	% LES Error	% LAV Error
16	869.5	863.5	868.6	0.7	0.1
17	869.5	863.5	868.6	0.7	0.1
18	869.5	863.5	868.6	0.7	0.1
19	869.5	863.5	868.6	0.7	0.1
20	869.5	863.5	868.6	0.7	0.1
21	801.8	809.1	807.3	–0.9	–0.7
22	823.4	794.6	791.3	3.5	3.9
23	835.3	759.7	760.3	9.1	9
24	783.1	733.9	722.3	6.3	7.8

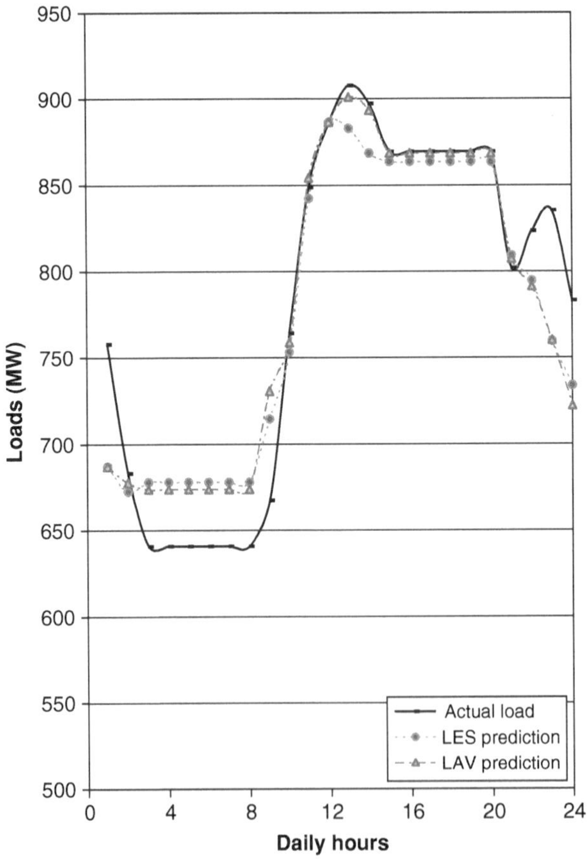

Figure 6.13 Predicted load for a summer weekend day using one parameter set, model A.

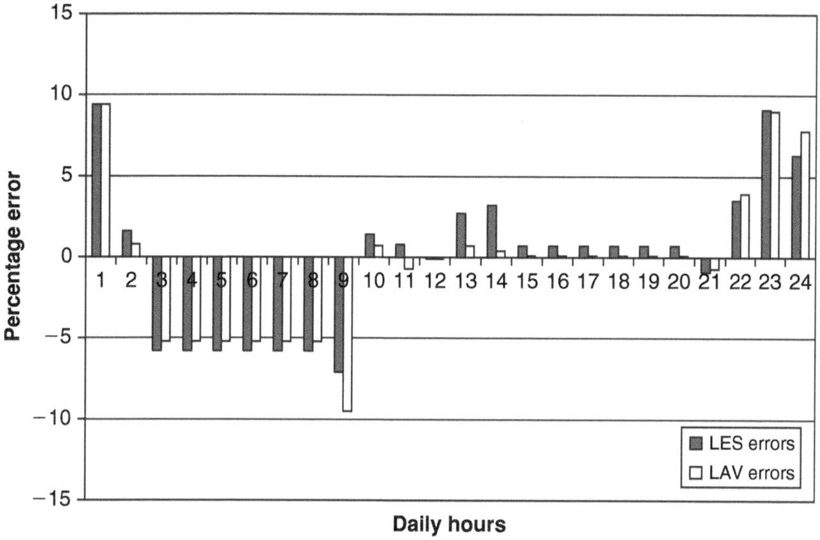

Figure 6.14 Predicted load error for a summer weekend day using one parameter set, model A.

Table 6.12a Estimated and Predicted Errors for a Winter Weekday

Algorithm	**LES %**		**LAV %**	
Type of Day	***Weekday***		***Weekday***	
Parameters Set	**24**	**1**	**24**	**1**
Estimated load maximum error	5.04	5.2	−8.64	−6.1
Estimated load minimum error	−0.1	0	−0.02	0
Predicted load maximum error	−6.48	−11.63	10.41	−17.99
Predicted load minimum error	0.07	0.41	−0.41	0.03

Table 6.12b Estimated and Predicted Errors for a Winter Weekend Day

Algorithm	LES %		LAV %	
Type of Day	*Weekend*		*Weekend*	
Parameters Set	**24**	**1**	**24**	**1**
Estimated load maximum error	−5.62	−15.22	−8.86	−11.6
Estimated load minimum error	−0.15	−0.09	0.01	0.04
Predicted load maximum error	25.81	−27.12	26.26	23.13
Predicted load minimum error	−0.41	−0.09	−1.5	0.2

Table 6.13 Load Parameters for a Summer Weekday, Model B

Parameter	LES Estimate	LAV Estimate
A_0	777.09	681.07
A_1	1.72	−168.82
B_1	12.74	186.35
A_2	−13.75	147.90
B_2	14.83	−35.02
A_3	−46.07	−84.19
B_3	27.08	−93.95
A_4	−2.18	−169.25
B_4	17.64	328.01
A_5	83.61	266.86
B_5	91.62	22.80
A_6	−7.84	−335.86
B_6	10.54	−62.62
A_7	14.35	243.44
B_7	15.06	199.83
A_8	11.67	32.27
B_8	75.36	−147.38
A_9	5.67	88.30
B_9	19.20	437.50

Table 6.14 Estimated Load and Percentage Error for a Summer Weekday, Model B

Hour	Actual Load (MW)	LES Estimate	LAV Estimate	% LES Error	% LAV Error
1	768.1	913.95	1424.26	−19	−85.4
2	678.3	855.77	795.24	−26.2	−17.2
3	636.1	774.79	1192.31	−21.8	−87.4
4	609.9	735	609.59	−20.5	0.05
5	598	702.72	599.75	−17.5	−0.29
6	595.6	712.31	594.91	−19.6	0.12
7	608.4	652.75	610.12	−7.3	−0.28
8	661	710.12	661.39	−7.4	−0.06
9	788.6	886.08	787.26	−12.4	0.17
10	908.4	1015.51	907.89	−11.8	0.06
11	982.9	1045.47	984.1	−6.4	−0.12
12	1015	1013.59	1014.47	0.14	0.05
13	1028.9	1022.18	1028.36	0.65	0.05
14	1011.1	918.6	1011.58	9.2	−0.05
15	988.6	873.21	989.87	11.7	−0.13
16	984.4	905.99	981.62	8	0.28
17	997.2	956.31	997.88	4.1	−0.07
18	1002.4	977.58	1003.96	2.5	−0.16
19	977.2	839.29	978.01	14.1	−0.08
20	929.4	791.66	929.93	14.8	−0.06
21	894.8	772.42	894.63	13.7	0.02
22	907.7	750.66	865.09	17.3	4.69
23	946.7	795.58	393.67	16	58.42
24	882.9	781.31	882.92	11.5	0

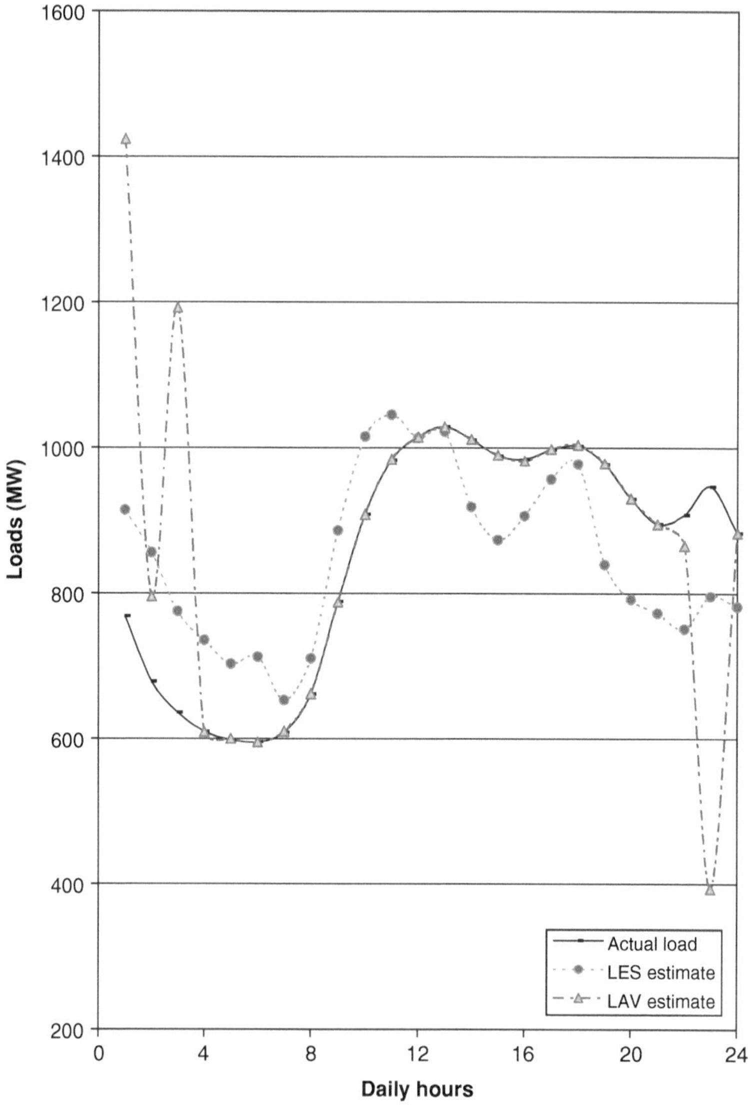

Figure 6.15 Estimated load for a summer weekday, model B.

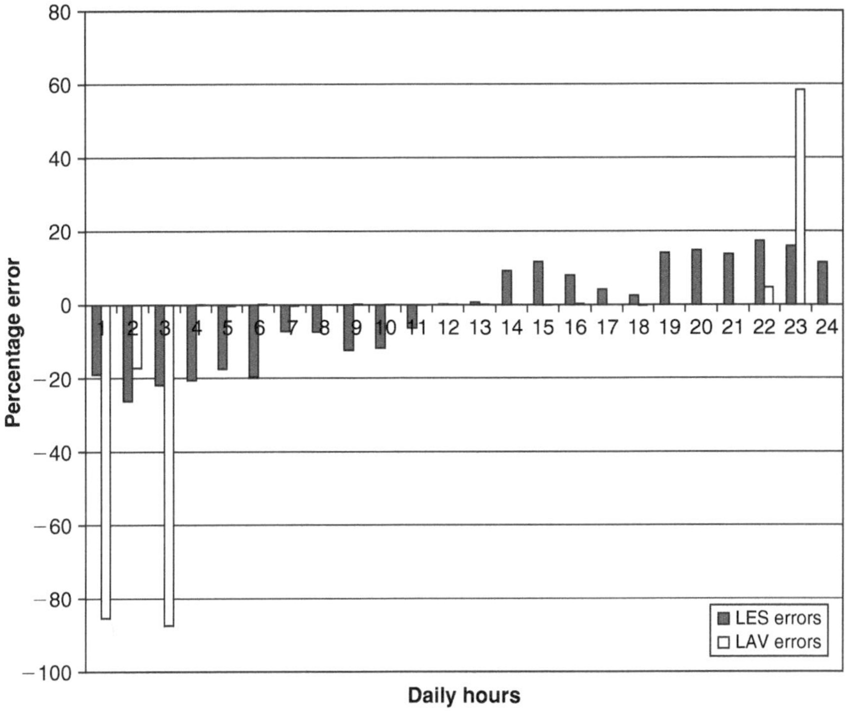

Figure 6.16 Estimated load error for a summer weekday, model B.

Table 6.15 Predicted Load and Percentage Error for a Summer Weekday, Model B

Hour	Actual Load (MW)	LES Prediction	LAV Prediction	% LES Error	% LAV Error
1	674	815.4	815.41	–20.98	–20.98
2	609.9	777.8	685.1	–27.54	–12.32
3	559.7	688	554.72	–22.92	0.89
4	537.7	657.6	538.2	–22.3	–0.09
5	536.8	637.6	535.4	–18.77	0.27
6	535.6	647.95	533.32	–20.98	0.42
7	545.6	587.13	543.41	–7.61	0.4
8	574.4	619.43	579.1	–7.84	–0.81
9	668.9	762.6	667.5	–14.01	0.22
10	787.3	889.3	789.5	–12.95	–0.28
11	875.1	933.54	872.4	–6.68	0.31
12	909.4	908.44	910.1	0.11	–0.08
13	925	919.11	926.71	0.64	–0.18
14	903	813.7	902.6	9.89	0.05

Continued

Table 6.15 Predicted Load and Percentage Error for a Summer Weekday, Model B *Continued*

Hour	Actual Load (MW)	LES Prediction	LAV Prediction	% LES Error	% LAV Error
15	876	766.6	876.7	12.49	–0.08
16	848.7	776.1	851.6	8.55	–0.35
17	848.3	808.8	852.8	4.65	–0.53
18	884.8	860.4	884.2	2.76	0.07
19	880.9	746.7	860.2	15.23	2.31
20	837.9	707.95	836.2	15.51	0.21
21	805.1	692.1	807.3	14.04	–0.27
22	824.4	673.8	921.3	18.27	–11.75
23	876.2	731.99	1035.31	16.46	–18.16
24	815.6	719.32	815.1	11.8	0.07

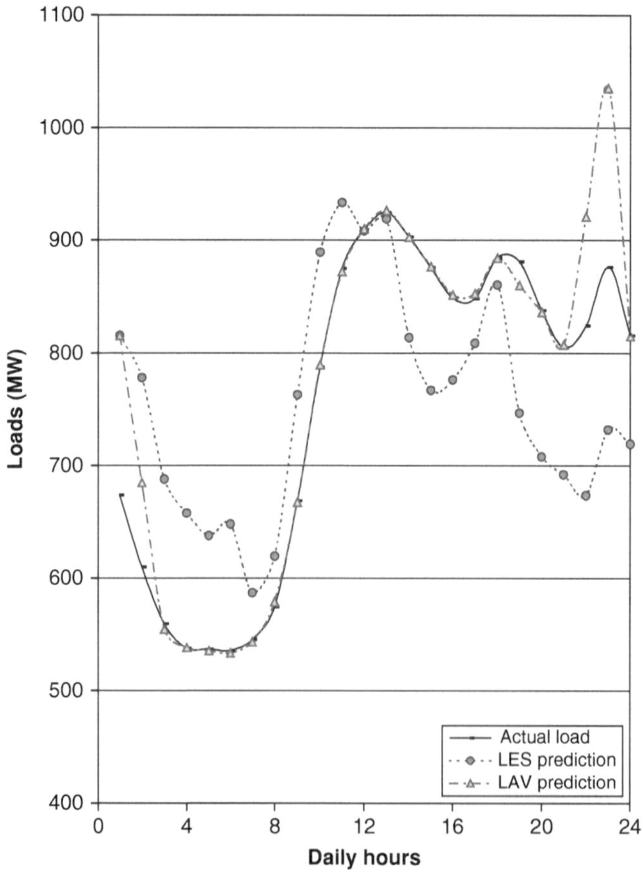

Figure 6.17 Predicted load for a summer weekday, model B.

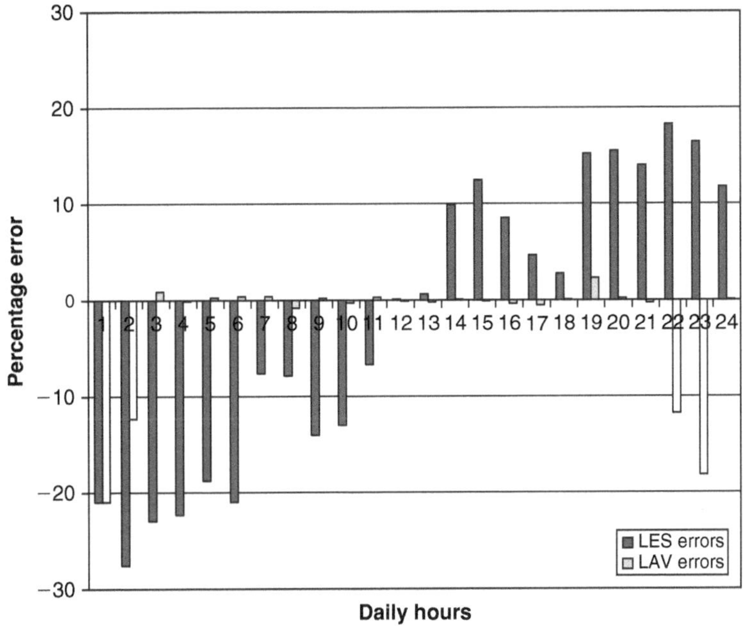

Figure 6.18 Predicted load error for a summer weekday, model B.

Table 6.16 Estimated Load and Percentage Error for a Summer Weekend Day, Model B

Hour	Actual Load (MW)	LES Estimate	LAV Estimate	% LES Error	% LAV Error
1	758.1	856.5	856.5	–12.98	–13
2	683.3	804.8	778.17	–17.78	–13.88
3	640.8	736.7	696	–14.97	8.5
4	614.2	705.6	613.83	–14.88	0.06
5	597.3	668.2	598.55	–11.92	–0.21
6	586.8	658.88	586.38	–12.28	0.07
7	590.3	619.78	591.35	–4.99	–0.18
8	601.3	635.87	601.51	–5.75	–0.04
9	667.3	737.25	666.33	–10.48	0.15
10	764.1	837.83	763.88	–9.65	0.03
11	848.8	889.14	849.58	–4.75	–0.09
12	884.7	885.69	885.42	0	0.03
13	907.7	902.45	907.21	0.58	0.05
14	897.2	832.8	897.6	7.18	–0.04
15	869.5	790.19	870.32	9.12	–0.09
16	842.4	789.04	840.56	6.33	0.22

Continued

Table 6.16 Estimated Load and Percentage Error for a Summer Weekend Day, Model B *Continued*

Hour	Actual Load (MW)	LES Estimate	LAV Estimate	% LES Error	% LAV Error
17	835.5	804.79	835.9	3.68	−0.05
18	853.8	837.54	855	1.9	−0.14
19	857.7	768.37	858.17	10.42	−0.05
20	823.9	730.4	824.31	11.35	−0.05
21	801.8	716.3	801.68	10.66	0.01
22	823.4	709.57	760.58	13.82	7.63
23	835.3	732	771.6	12.37	7.6
24	783.1	720.21	783.2	8.03	−0.01

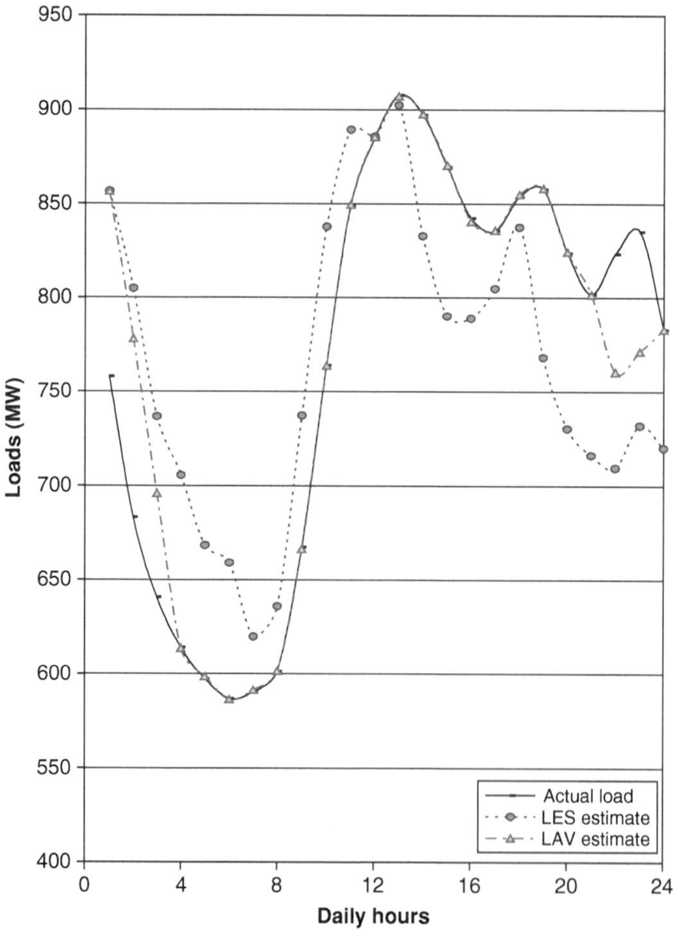

Figure 6.19 Estimated load for a summer weekend day, model B.

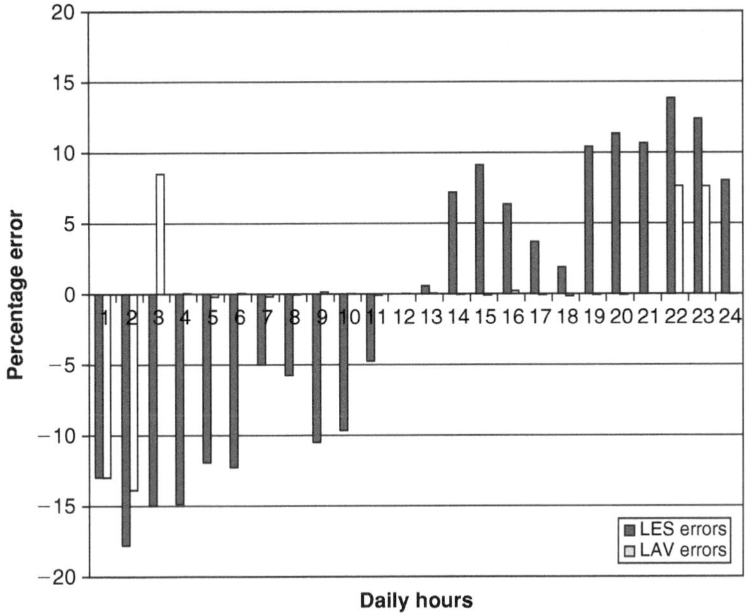

Figure 6.20 Estimated load error for a summer weekend day, model B.

Table 6.17 Predicted Load and Percentage Error for a Summer Weekend Day, Model B

Hour	Actual Load (MW)	LES Prediction	LAV Prediction	% LES Error	% LAV Error
1	950.4	1089.5	950.4	14.64	0
2	874.5	1050.04	894.7	–20.07	2.3
3	838.7	979.43	839	–16.78	0
4	814	929.21	813.16	–14.15	0.1
5	806.3	858.62	809.71	–6.49	–0.42
6	800.4	872.44	800.31	–9	0.01
7	811.1	862.16	811.44	–6.3	–0.04
8	833.2	895.18	833.49	–7.44	–0.03
9	918.6	1017.23	916.64	–10.74	0.21
10	1038.1	1121.45	1041.85	–8.03	–0.36
11	1116.9	1157.88	1120.92	–3.67	–0.36
12	1148.4	1143	1149.4	0.47	–0.09
13	1158	1140.82	1155.9	1.48	0.18
14	1135.1	1041.03	1133.6	8.29	0.13
15	1098.2	991.95	1095.1	9.68	0.28

Continued

Table 6.17 Predicted Load and Percentage Error for a Summer Weekend Day, Model B *Continued*

Hour	Actual Load (MW)	LES Prediction	LAV Prediction	% LES Error	% LAV Error
16	1074.5	1004.23	1072.3	6.54	0.21
17	1072.9	1038.91	1075.8	3.17	–0.27
18	1110.2	1110.88	1116.8	–0.06	–0.6
19	1135.1	1021.23	1132.1	10.03	0.27
20	1141.2	996.94	1148.4	12.64	–0.6
21	1159.6	1023.25	1164.8	11.76	–0.45
22	1112.8	960.62	1127.3	13.68	–1.31
23	1060.7	961.22	1063.46	9.38	–0.26
24	981.7	925	980.7	5.78	0.1

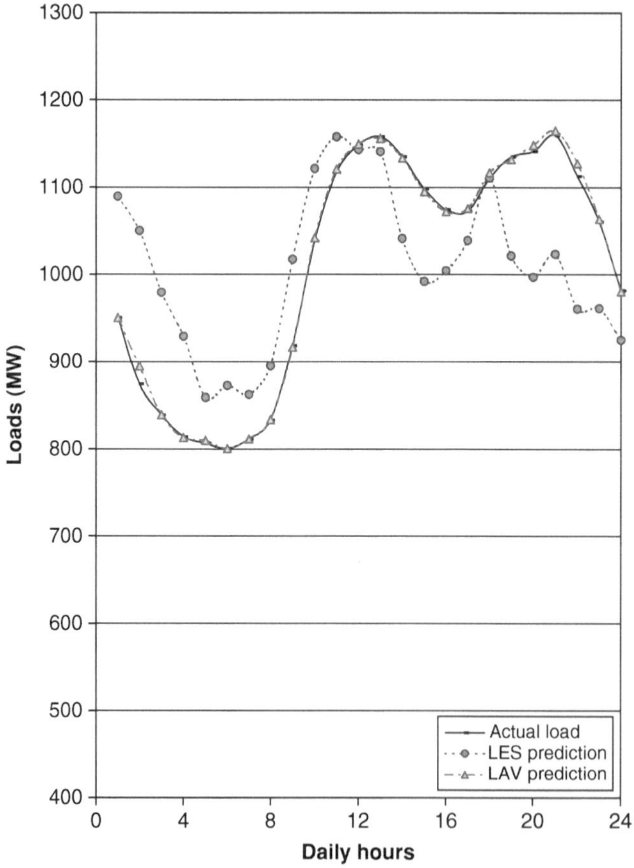

Figure 6.21 Predicted load for a summer weekend day, model B.

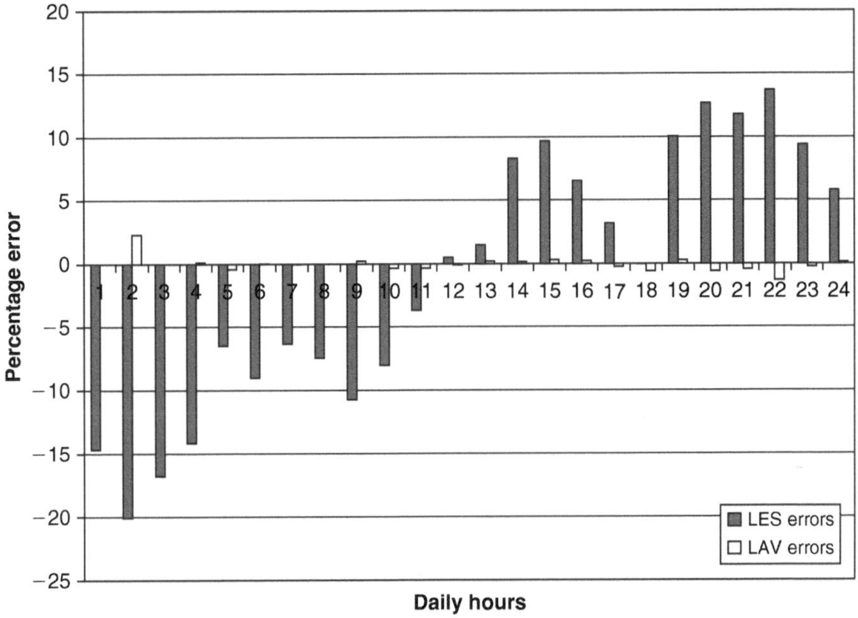

Figure 6.22 Predicted load error for a summer weekend day, model B.

Table 6.18 Load Parameters for a Summer or Winter Day, Model C

Parameter	LES Estimate	LAV Estimate
A_0	1020.16	1023.68
A_1	3.78	2.97
B_1	−2.85	−1.53
A_2	−18.56	−12.83
B_2	24.52	27.64
A_3	−24.86	−24.56
B_3	−10.3	−18.43
A_4	3.38	6.5
B_4	2.25	3.73
A_5	17.12	31.64
B_5	17.10	16.77
A_6	−5.34	−14.59
B_6	11.36	6.51
A_7	−0.44	3.43
B_7	5.5	12.72
A_8	27.35	27.26
B_8	78.71	76.95

Continued

Table 6.18 Load Parameters for a Summer or Winter Day, Model C *Continued*

Parameter	LES Estimate	LAV Estimate
A_9	−3.57	−12.48
B_9	−7.25	−16.88
C_0	−27.56	−35.72
C_1	−4.28	9.34
C_2	4.04	5.8
C_3	−7.52	−14.47

Table 6.19 Estimated Load and Percentage Error for a Summer Day, Model C

Hour	Actual Load (MW)	LES Estimate	LAV Estimate	% LES Error	% LAV Error
1	749.8	747.5	749.92	0.31	−0.02
2	666.2	667.3	665.42	−0.17	0.12
3	621.9	622.4	621.4	−0.08	0.08
4	598.5	596.4	598.85	0.36	−0.06
5	587.4	588.8	587.33	−0.23	0.01
6	586.4	586.8	586.62	−0.06	−0.04
7	603.2	604.4	603.65	−0.21	−0.07
8	656.8	656.4	656.31	0.06	0.08
9	787.2	785.8	787.56	0.18	−0.05
10	897.4	898.6	897.56	−0.13	−0.02
11	969.8	970.5	969.58	−0.07	0.02
12	1008	1006.2	1006.61	0.18	0.14
13	1019.7	1020.5	1020.4	−0.08	−0.07
14	1002.5	1003.7	1002.4	−0.12	0.05
15	990.1	988.95	989.99	0.12	0.01
16	971.7	970.98	971.78	0.07	−0.01
17	965.5	967.1	966.02	−0.17	−0.05
18	987.4	987.4	987.99	0	−0.06
19	968	969.94	968.24	−0.2	−0.03
20	923.7	922.42	924.16	0.14	−0.05
21	888.8	888.47	889.09	0.04	−0.03
22	900.1	902.66	916.85	−0.28	−1.86
23	940.5	938.93	940.7	0.17	−0.02
24	875.7	874.93	876.01	0.09	−0.04

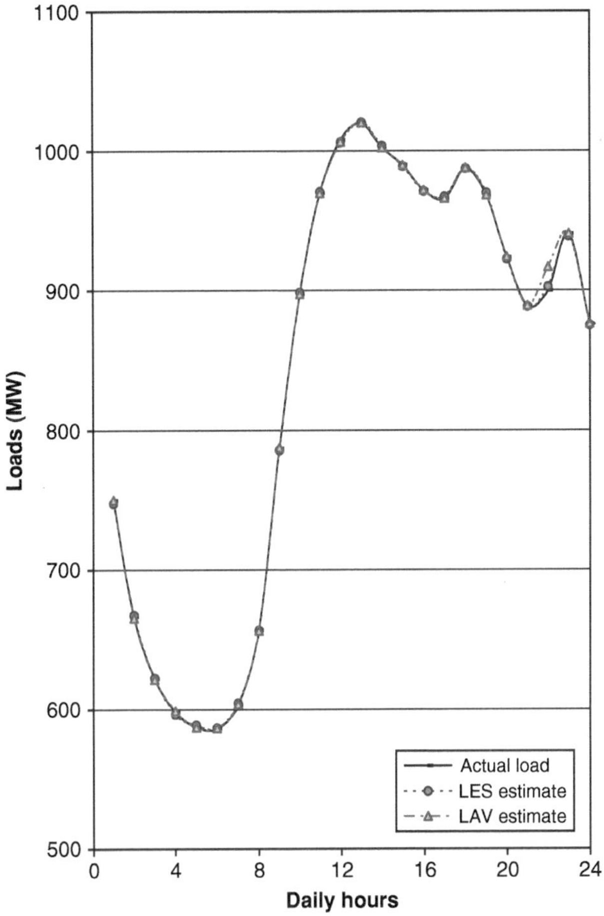

Figure 6.23 Estimated load for a summer day, model C.

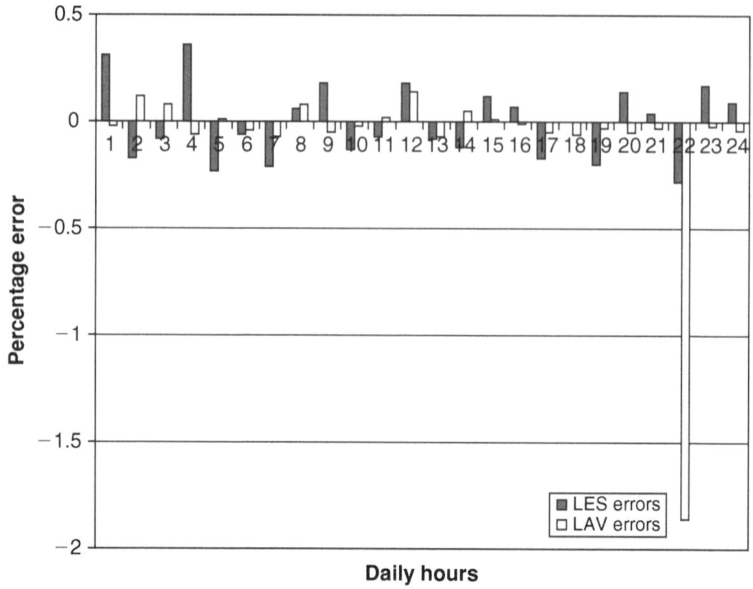

Figure 6.24 Estimated load error for a summer day, model C.

Table 6.20 Predicted Load and Percentage Error for a Summer Day, Model C

Hour	Actual Load (MW)	LES Prediction	LAV Prediction	% LES Error	% LAV Error
1	744.7	723.46	725.41	2.85	2.59
2	670.8	774.04	777.23	−15.39	−15.87
3	634.3	631.08	631.69	0.51	0.41
4	613.4	751.12	761.63	−22.45	−24.17
5	607.3	632.5	635.28	−4.15	−4.61
6	608.9	666.03	671.54	−9.38	−10.29
7	628.1	521.71	519.44	16.94	17.3
8	697.2	830.98	839.66	−19.19	−20.43
9	817.9	858.27	865.72	−4.94	−5.85
10	934.9	994.62	1000.39	−6.39	−7.01
11	997.4	1107.61	1115.68	−11.05	−11.86
12	1030.4	1155.39	1165.91	−12.13	−13.15
13	1069	1213.89	1226.49	−13.55	−14.73
14	1054.7	1121.25	1129.7	−6.31	−7.11
15	1043.3	1250.67	1268.26	−19.88	−21.56
16	1028.5	1112.96	1126.19	−8.21	−9.5
17	1033.2	1119.93	1131.55	−8.39	−9.52
18	1058.1	906.4	908.96	14.34	14.09

Continued

Table 6.20 Predicted Load and Percentage Error for a Summer Day, Model C *Continued*

Hour	Actual Load (MW)	LES Prediction	LAV Prediction	% LES Error	% LAV Error
19	1036.2	1158.78	1169.03	−11.83	−12.82
20	970.1	1027.05	1037.87	−5.87	−6.99
21	930.2	1013.43	1023.91	−8.95	−10.07
22	962.9	941.63	961.87	2.21	0.11
23	996.4	1014.58	1023.29	−1.82	−2.7
24	925.4	912.85	918.96	1.36	0.7

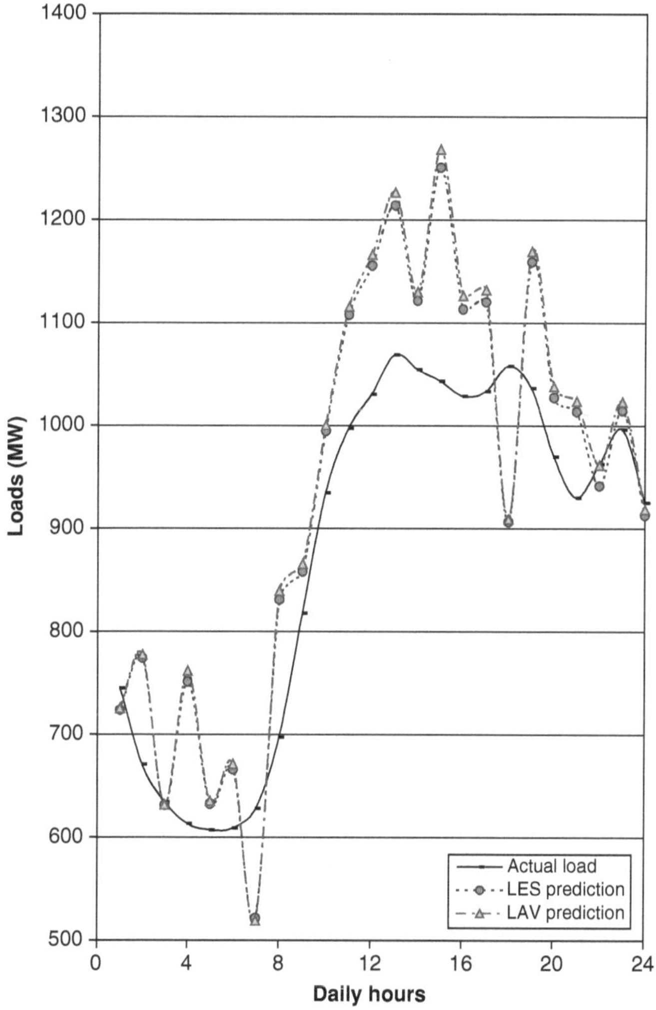

Figure 6.25 Predicted load for a summer day, model C.

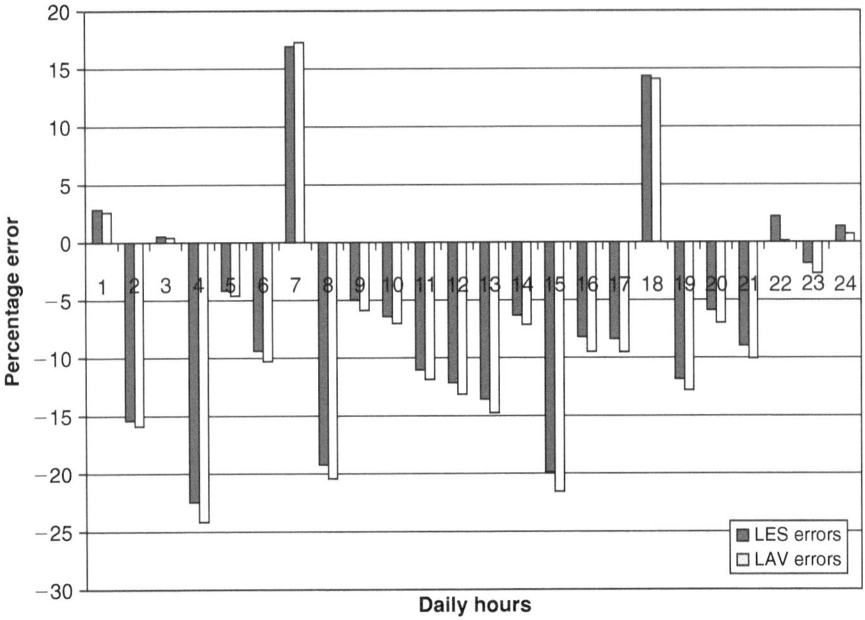

Figure 6.26 Predicted load error for a summer day, model C.

6.7 Conclusion

In this chapter the LES and LAV parameter estimation algorithms were used for static estimation of the parameters of different load models. Three models were used—namely, models A, B, and C. These models were used to predict load power for the next 24 hours in a weekday and a weekend ahead for a weekend day. It was found that model A gives acceptable load predictions.

Model A possesses the advantage of being weather sensitive but suffers from the following: (1) It needs 24 separate parameter sets to predict the load 24 hours ahead as accurately as possible, and this needs more computing time; it was also found that one set of parameters gives acceptable results. (2) Separate models must be used for weekdays and weekend days both with summer and winter formulations. Model B does not account for weather effects but is a function of the hour (time) considered; it produces acceptable results and takes less computing time. It can be used only for a case in which the weather variations are small during the day.

Model C is the most suitable model because it takes into account both time and weather during summer and winter seasons. It eliminates the use of separate models for both weekdays and weekend days.

Appendix 6.1 Winter Static Load Results for Model A

Table A6.1 Estimated Load and Percentage Error for a Winter Weekday Using 24 Parameter Sets, Model A

Daily Hours	Actual Load (MW)	LES Estimation	LAV Estimation	% LES Error	% LAV Error
1	943.4	896.3	943.5	5	−0.02
2	850.5	834	863.1	1.94	−1.48
3	811.2	797.3	807.2	1.72	0.5
4	793.6	769.5	792.4	3.03	0.16
5	794.6	754.5	863.3	5.04	−8.64
6	810.3	816.2	816.6	−0.72	−0.78
7	855.2	874.1	845.9	−2.21	1.09
8	991	996.6	994.6	−0.57	−0.36
9	1198.5	1198.9	1193.4	−0.03	0.43
10	1302.3	1304.5	1305.8	−0.17	−0.27
11	1331.8	1331.7	1332.5	0.01	−0.05
12	1344.7	1345.2	1343.4	−0.04	0.1
13	1366.1	1365.5	1364.9	0.04	0.09
14	1346.2	1345.8	1345.1	0.03	0.08
15	1332.7	1333.7	1334.1	−0.07	−0.11
16	1320.6	1321.7	1321.3	−0.08	−0.05
17	1341.4	1342.8	1342.7	−0.1	−0.1
18	1405.2	1407.6	1408.6	−0.17	−0.24
19	1403.3	1405.3	1423.5	−0.14	−1.44
20	1380.9	1384.3	1385.2	−0.25	−0.31
21	1406.5	1407	1410	−0.03	−0.25
22	1440.9	1443.9	1443.7	−0.21	−0.19
23	1390.9	1387.5	1368.7	0.25	0.3
24	1281.1	1282.5	1283.4	−0.11	−0.18

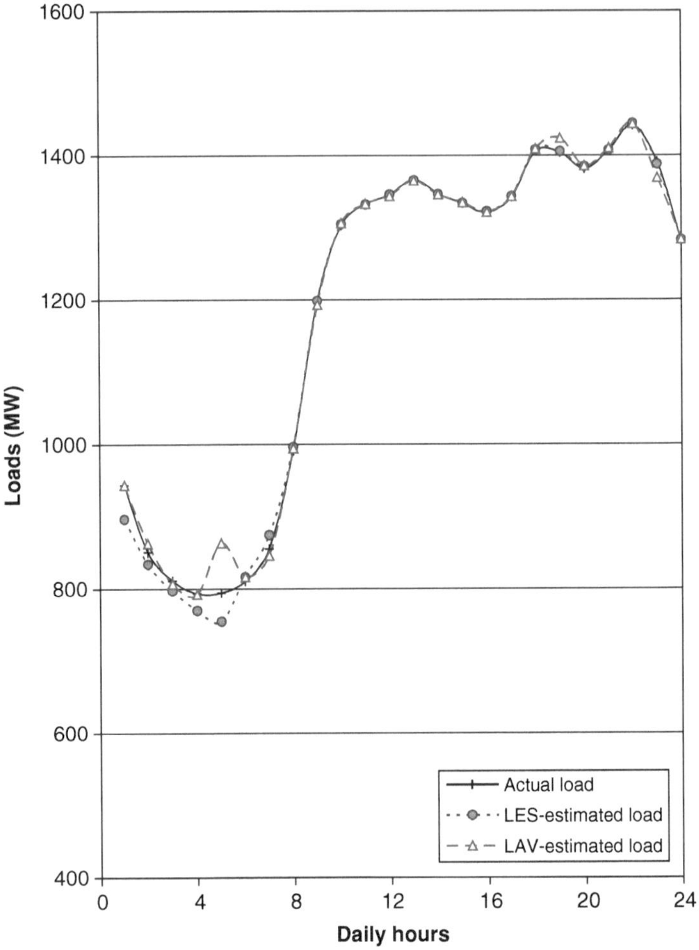

Figure A6.1 Estimated load for a winter weekday using 24 parameter sets, model A.

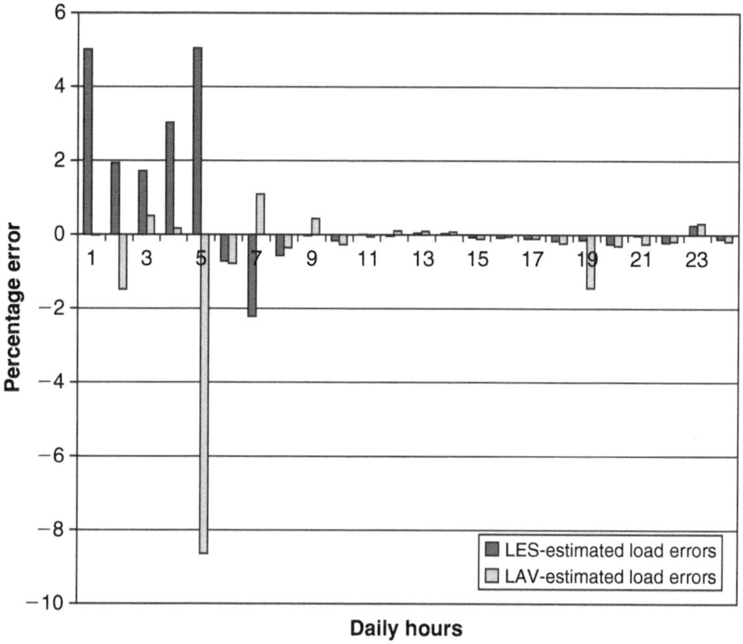

Figure A6.2 Estimated Load Error for a Winter Weekday Using 24 Parameter Sets, model A.

Table A6.2 Estimated Load and Percentage Error for a Winter Weekday Using One Parameter Set, Model A

Daily Hours	Actual Load (MW)	LES Estimation	LAV Estimation	% LES Error	% LAV Error
1	1331.8	1330.2	1331.8	0.1	0
2	1344.7	1381.1	1360.9	−2.7	−1.2
3	1366.1	1340.3	1366	1.9	0
4	1346.2	1348.3	1346.2	−0.2	0
5	1332.7	1366.2	1360.1	−2.5	−2.1
6	1320.6	1288.9	1309.4	2.4	0.9
7	1341.4	1342.3	1341.4	−0.1	0
8	991	1034.3	1041.6	−4.4	−5.1
9	1198.5	1136	1134	5.2	5.4
10	1302.3	1261.7	1274.5	3.1	2.1
11	1331.8	1330.2	1331.8	0.1	0
12	1344.7	1381.1	1360.9	−2.7	−1.2
13	1366.1	1340.3	1366	1.9	0
14	1346.2	1348.3	1346.2	−0.2	0
15	1332.7	1366.2	1360.1	−2.5	−2.1

Continued

Table A6.2 Estimated Load and Percentage Error for a Winter Weekday Using One Parameter Set, Model A *Continued*

Daily Hours	Actual Load (MW)	LES Estimation	LAV Estimation	% LES Error	% LAV Error
16	1320.6	1288.9	1309.4	2.4	0.9
17	1341.4	1342.3	1341.4	−0.1	0
18	1405.2	1416	1405.3	−0.8	0
19	1403.3	1432.7	1425.3	−2.1	−1.6
20	1380.9	1391.8	1380.9	−0.8	0
21	1406.5	1373.8	1397.1	2.3	0.7
22	1440.9	1413.8	1413.3	1.9	1.9
23	1390.9	1390.7	1390.8	0	0
24	1281.1	1317.4	1359.6	−2.8	−6.1

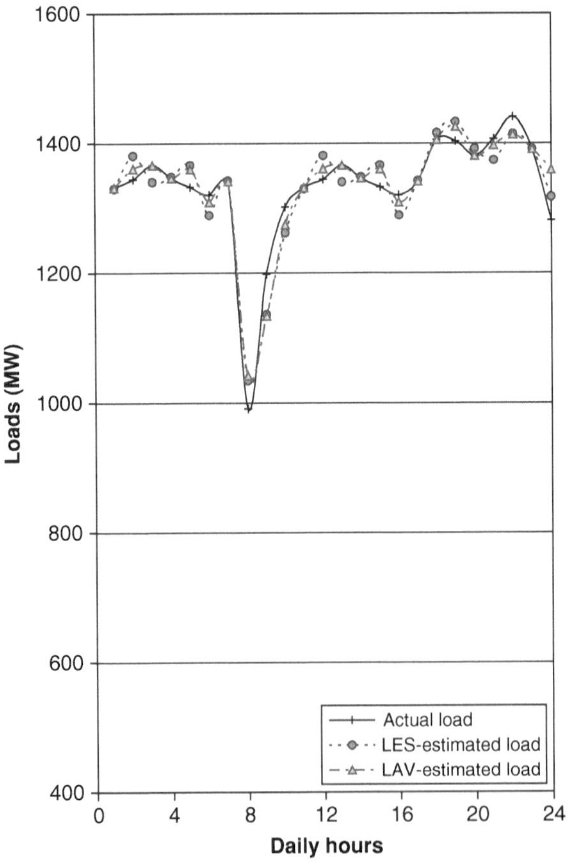

Figure A6.3 Estimated load for a winter weekday using one parameter set, model A.

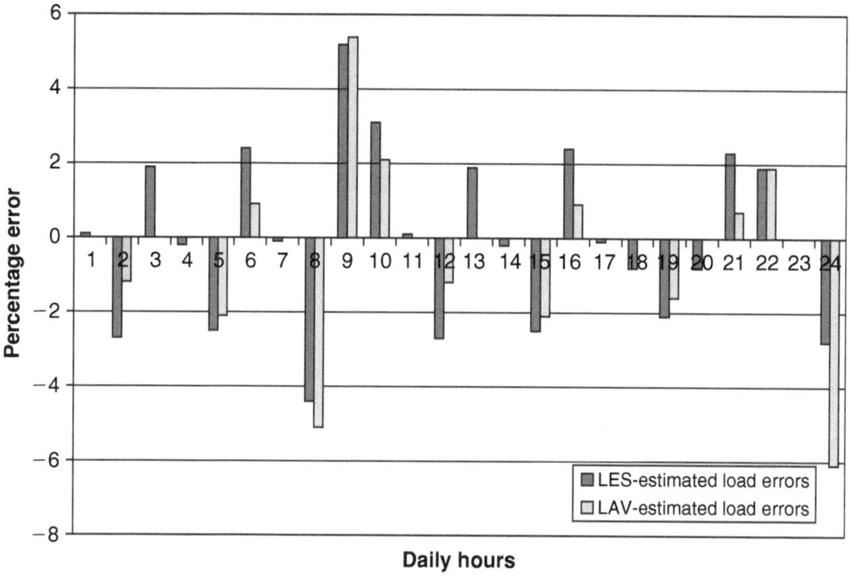

Figure A6.4 Estimated load error for a winter weekday using one parameter set, model A.

Table A6.3 Predicted Load and Percentage Error for a Winter Weekday Using 24 Parameter Sets, Model A

Daily Hours	Actual Load (MW)	LES Prediction	LAV Prediction	% LES Error	% LAV Error
1	779.2	795.6	835.7	−2.11	−7.25
2	698.6	724.8	737.7	−3.75	−5.6
3	665.8	674.9	703.4	−1.36	−5.65
4	658.5	669.7	700.5	−1.71	−6.38
5	660.1	697.6	707.4	−5.68	−7.17
6	674.7	715.4	706.3	−6.03	−4.68
7	714.3	745	767.3	−4.3	−7.43
8	827.4	808.7	741.3	2.26	10.41
9	1003.5	1033.4	1031.4	−2.98	−2.78
10	1065.1	1089.7	1098.9	−2.31	−3.17
11	1062.1	1063.7	1073.2	−0.15	−1.05
12	1044.2	1043.4	1060.4	0.07	−1.55
13	1030.9	1033.3	1043.9	−0.23	−1.26
14	996.2	1006	1005	−0.99	−0.89
15	971.3	977.6	975.3	−0.65	−0.41
16	946.9	954.8	951.3	−0.84	−0.46
17	946.9	959.3	958.8	−1.31	−1.26

Continued

Table A6.3 Predicted Load and Percentage Error for a Winter Weekday Using 24 Parameter Sets, Model A *Continued*

Daily Hours	Actual Load (MW)	LES Prediction	LAV Prediction	% LES Error	% LAV Error
18	975.9	1001.6	996.9	−2.63	−2.16
19	965.8	993	1016.8	−2.82	−5.28
20	917.2	976.6	977.4	−6.48	−6.57
21	902.1	941.5	879.2	−4.37	2.53
22	971.3	996.6	976.3	−2.61	−0.51
23	981.8	987.5	995.1	−0.58	−1.35
24	898.1	907.1	908.5	−1	−1.16

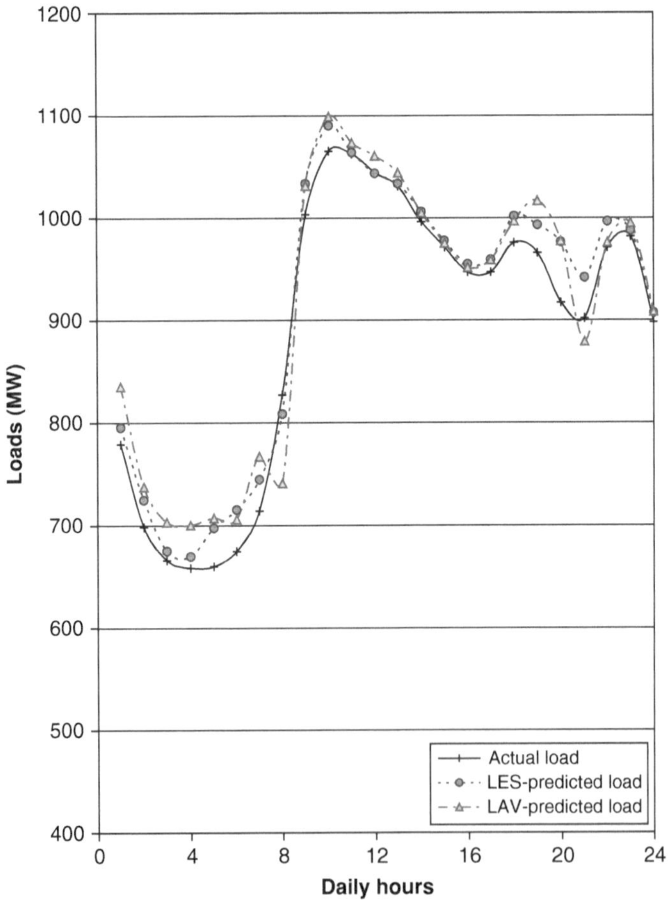

Figure A6.5 Predicted load for a winter weekday using 24 parameter sets, model A.

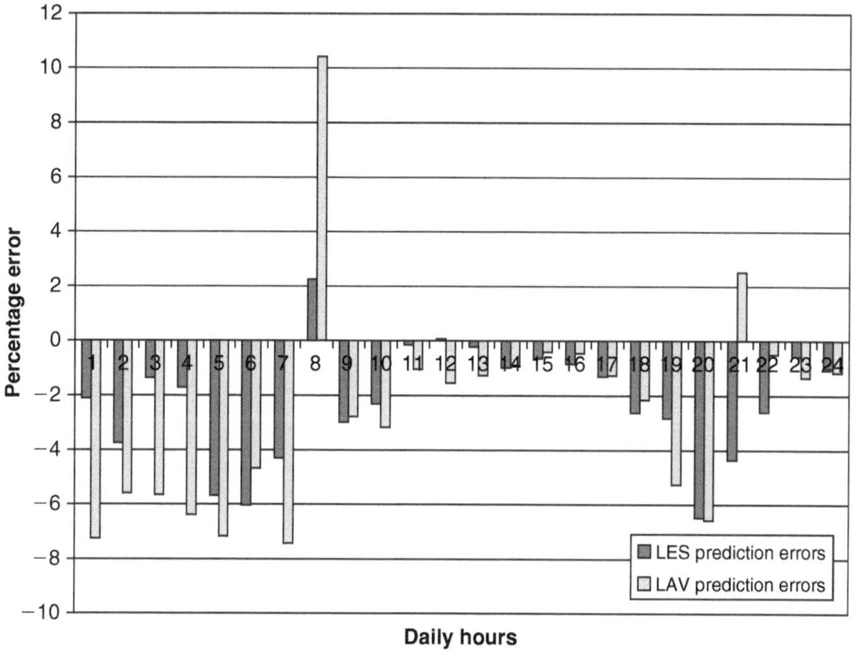

Figure A6.6 Predicted load error for a winter weekday using 24 parameter sets, model A.

Table A6.4 Predicted Load and Percentage Error for a Winter Weekday Using One Parameter Set, Model A

Daily Hours	Actual Load (MW)	LES Prediction	LAV Prediction	% LES Error	% LAV Error
1	883.7	913.6	886.2	−3.38	−0.29
2	806.5	900.3	887.7	−11.63	−9.15
3	779	933.3	949.9	−19.8	−17.99
4	772.4	824.8	850.5	−6.79	−9.18
5	772.4	824.8	850.5	−6.79	−9.18
6	772.4	824.8	850.5	−6.79	−9.18
7	772.4	824.8	850.5	−6.79	−9.18
8	966.8	949.5	968.5	1.79	−0.17
9	1145.8	1175	1144.8	−2.55	0.09
10	1225.8	1213.3	1223.7	1.02	0.17
11	1220.9	1209.2	1220.3	0.96	0.05
12	1188.1	1144.9	1169.4	3.64	1.6
13	1174.1	1085.3	1091.5	7.57	7.57
14	1130.2	1037.4	1053.3	8.21	7.3
15	1108.7	1047.5	1069.9	5.52	3.63

Continued

Table A6.4 Predicted Load and Percentage Error for a Winter Weekday Using One Parameter Set, Model A *Continued*

Daily Hours	Actual Load (MW)	LES Prediction	LAV Prediction	% LES Error	% LAV Error
16	1082.2	1069.2	1080	1.2	0.2
17	1082.2	1069.2	1080	1.2	0.2
18	1082.2	1069.2	1080	1.2	0.2
19	1082.2	1069.2	1080	1.2	0.2
20	1120.2	1065.5	1056	4.88	6.08
21	1128.4	1138.3	1128.1	–0.88	0.03
22	1164.8	1160	1162.2	0.41	0.23
23	1126.5	1205.4	1220.7	–7	–7.72
24	1026.5	1225.7	1225.7	–19.4	–16.25

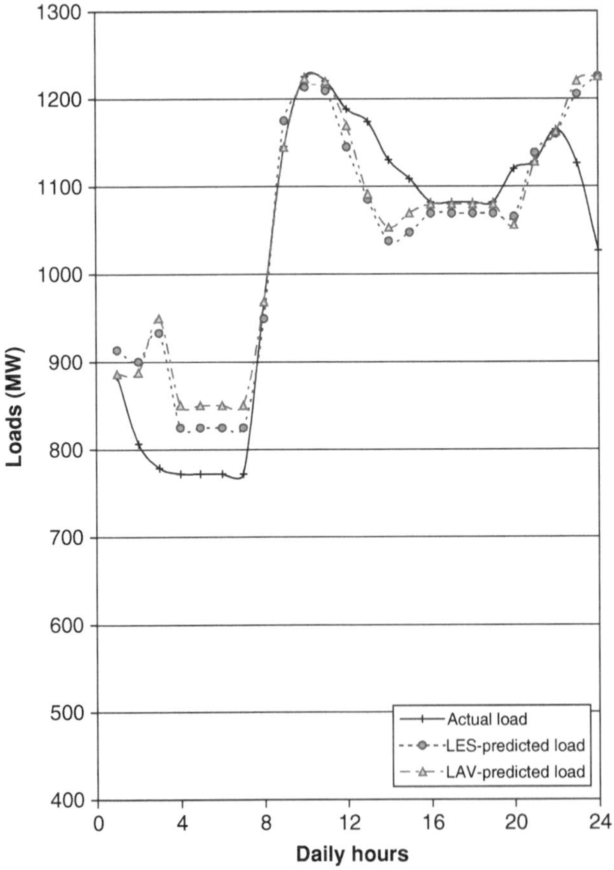

Figure A6.7 Predicted load for a winter weekday using one parameter set, model A.

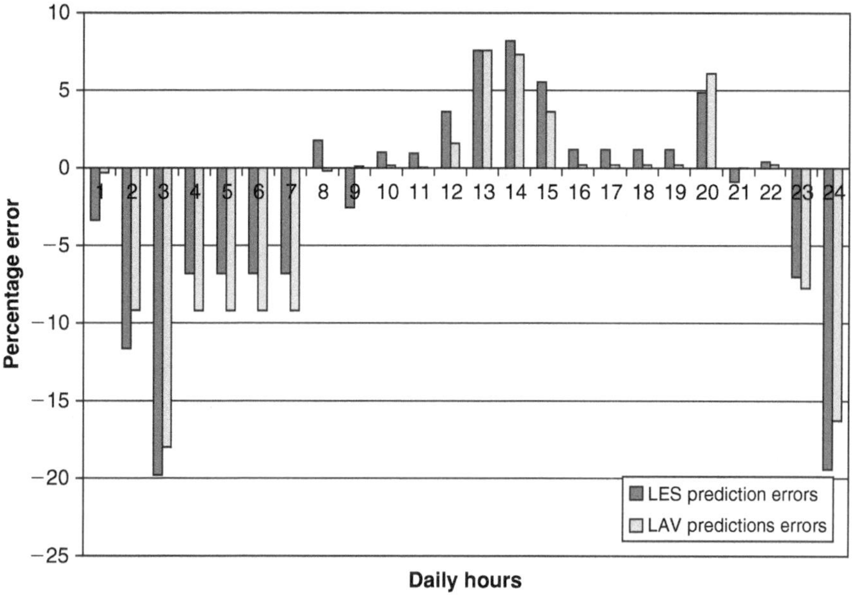

Figure A6.8 Predicted load error for a winter weekday using one parameter set, model A.

Table A6.5 Estimated Load and Percentage Error for a Winter Weekend Day Using 24 Parameter Sets, Model A

Daily Hours	Actual Load (MW)	LES Estimation	LAV Estimation	% LES Error	% LAV Error
1	776.8	791.8	793.5	−1.93	−2.15
2	710	721.5	709.1	−1.62	0.12
3	667.1	697.8	717.8	−4.6	−7.6
4	647.2	672.5	609.1	−3.9	5.89
5	639.3	675.2	681.7	−5.62	−6.63
6	642.8	652.2	642.7	−1.46	0.01
7	657.2	662.9	679.6	−0.87	−3.4
8	689.3	694.7	716.9	−0.78	−4.01
9	767.5	774.1	767.8	−0.86	−0.03
10	898	925.2	898.4	−3.03	−0.05
11	995.1	1054.6	1056.6	−5.98	−6.18
12	1016.2	1042.2	1015.8	−2.56	0.04
13	1008.1	1025.9	1008.3	−1.77	−0.02
14	977.9	1014.8	984	−3.77	−0.62
15	940.1	964	960.5	−2.54	−2.17
16	905.1	919.1	969.2	−1.55	−7.09

Continued

Table A6.5 Estimated Load and Percentage Error for a Winter Weekend Day Using 24 Parameter Sets, Model A *Continued*

Daily Hours	Actual Load (MW)	LES Estimation	LAV Estimation	% LES Error	% LAV Error
17	892.8	885.1	894.8	0.86	–0.22
18	915.4	946.3	996.5	–3.37	–8.86
19	915.1	940.7	957.7	–2.8	–4.6
20	887	902.1	918.8	–1.7	–3.58
21	900.2	891.2	897.4	1	0.31
22	961.4	930.4	892.3	3.22	7.19
23	953.1	954.5	958.4	–0.15	–0.56
24	903.7	889.2	899.3	1.6	0.49

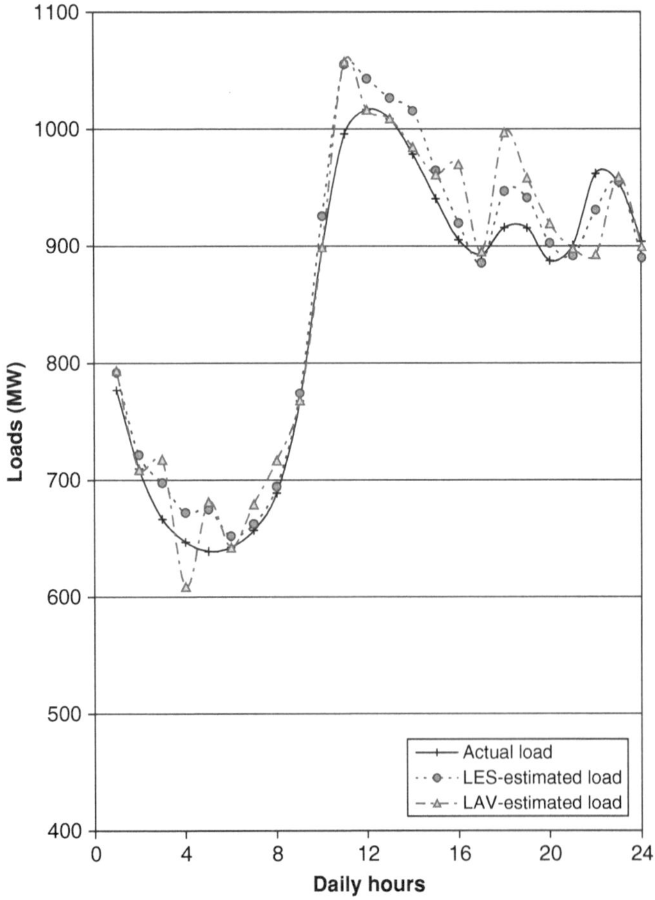

Figure A6.9 Estimated load for a winter weekend day using 24 parameter sets, model A.

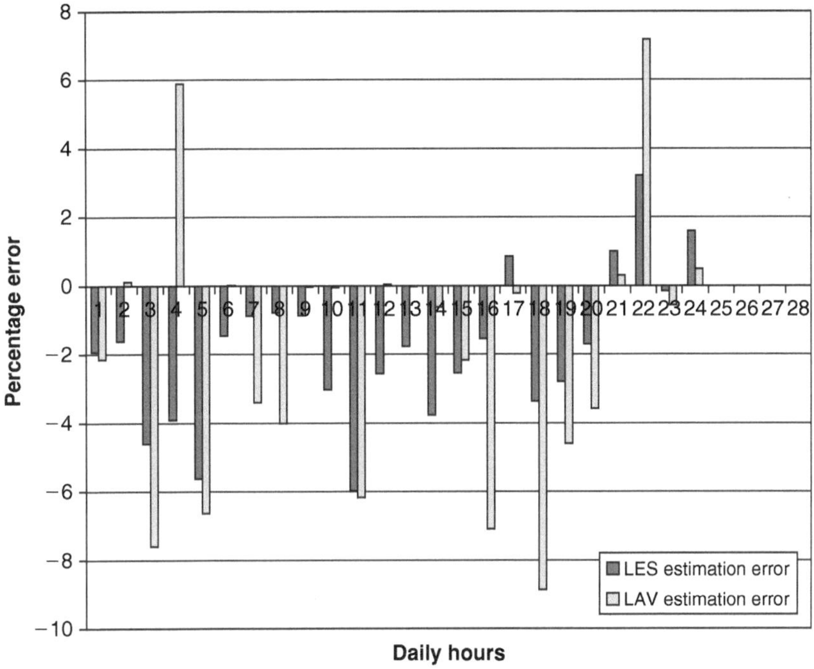

Figure A6.10 Estimated load error for a winter weekend day using 24 parameter sets, model A.

Table A6.6 Estimated Load and Percentage Error for a Winter Weekend Day Using One Parameter Set, Model A

Daily Hours	Actual Load (MW)	LES Estimation	LAV Estimation	% LES Error	% LAV Error
1	776.8	797.7	780.9	–2.69	–0.53
2	710	764.6	785.3	–7.68	–10.6
3	667.1	755.1	745.1	–14	–11.6
4	647.2	745.7	705	–15.22	–8.94
5	639.3	648.6	634.2	–1.45	0.8
6	642.8	630.6	640.5	1.89	0.35
7	657.2	662.1	660.7	–0.75	–0.53
8	689.3	731.2	717.6	–6.07	–4.1
9	767.5	811.5	799.9	–5.73	–4.22
10	898	865.4	894	3.63	0.45
11	995.1	879.6	928.2	11.61	6.7
12	1016.2	1053.7	962.5	–3.69	5.28
13	1008.1	1030.5	1006.1	–2.22	0.2
14	977.9	935.6	973.7	4.33	0.43

Continued

Table A6.6 Estimated Load and Percentage Error for a Winter Weekend Day Using One Parameter Set, Model A *Continued*

Daily Hours	Actual Load (MW)	LES Estimation	LAV Estimation	% LES Error	% LAV Error
15	940.1	871.3	900.8	7.32	4.18
16	905.1	935.6	955.6	–3.37	–5.58
17	892.8	814.3	892.4	8.8	0.04
18	915.4	834.5	829.1	7.7	9.4
19	915.1	854.8	859.4	6.59	6.1
20	887	900.7	890.7	–1.54	–0.41
21	900.2	931.5	905.1	–3.48	–0.54
22	961.4	933.3	965.9	2.93	–0.47
23	953.1	974.8	956.6	–2.28	–0.37
24	903.7	904.5	898.7	–0.09	0.55

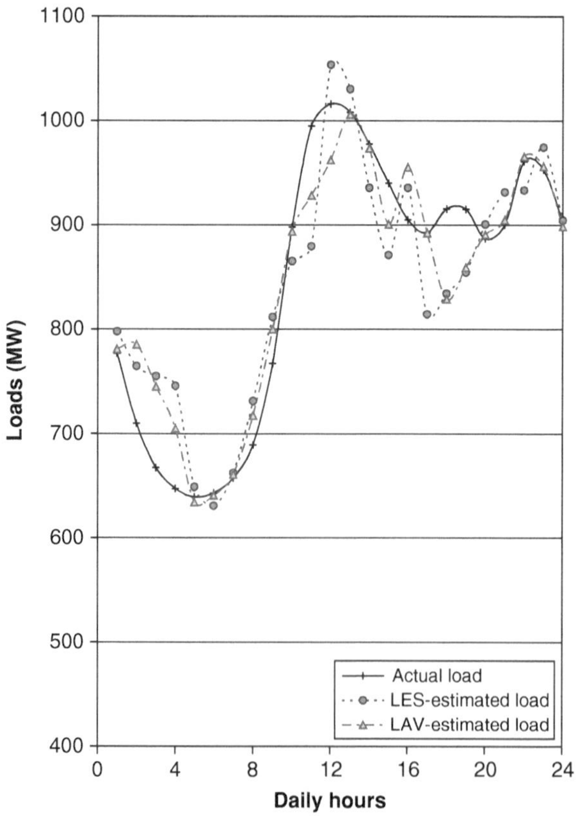

Figure A6.11 Estimated load for a winter weekend day using one parameter set, model A.

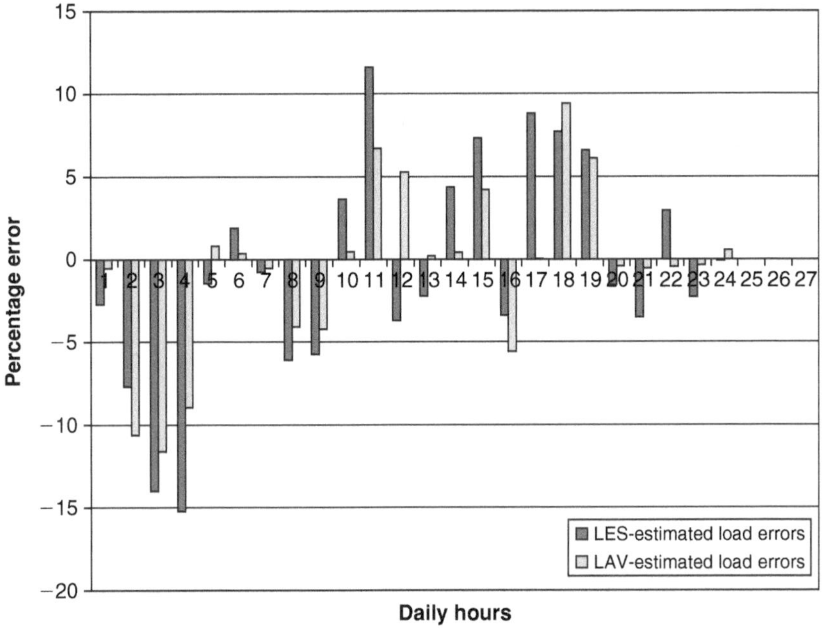

Figure A6.12 Estimated load error for a winter weekend day using one parameter set, model A.

Table A6.7 Predicted Load and Percentage Error for a Winter Weekend Day Using 24 Parameter Sets, Model A

Daily Hours	Actual Load (MW)	LES Prediction	LAV Prediction	% LES Error	% LAV Error
1	814.4	790.8	788.4	2.9	3.2
2	738.5	716.3	708.6	3.01	4.05
3	712	674.9	678.2	5.21	4.74
4	699.9	638	659.2	8.85	5.81
5	705.5	634.7	647.6	10.03	8.21
6	717.1	619.8	614.7	13.57	14.28
7	756.4	658.6	664.8	12.93	12.11
8	866.6	677	679.4	21.88	21.6
9	1035.1	767.9	763.3	25.81	26.26
10	1095.8	850.7	824.7	22.37	24.74
11	1085.7	980.7	965.3	9.67	11.09
12	1057.4	979.3	1004.3	7.38	5.02
13	1042.7	1033.3	1023	0.9	1.89
14	1003.3	1024.6	1038.9	−2.12	−3.55
15	976.7	983	992.6	−0.64	−1.63

Continued

Table A6.7 Predicted Load and Percentage Error for a Winter Weekend Day Using 24 Parameter Sets, Model A *Continued*

Daily Hours	Actual Load (MW)	LES Prediction	LAV Prediction	% LES Error	% LAV Error
16	951.6	977.1	974.3	−2.68	−2.38
17	959.5	978.5	973.9	−1.98	−1.5
18	1000.5	989.2	1051.6	1.13	−5.1
19	1006.2	1025.3	1062.6	−1.9	−5.6
20	973.4	1027.6	1035.5	−5.57	−6.38
21	972.2	1013.3	1018	−4.23	−4.71
22	1021	1060.8	1048.2	−3.9	−2.67
23	997.3	1001.4	1013.8	−0.41	−1.65
24	909.5	938.3	939.7	−3.16	−3.32

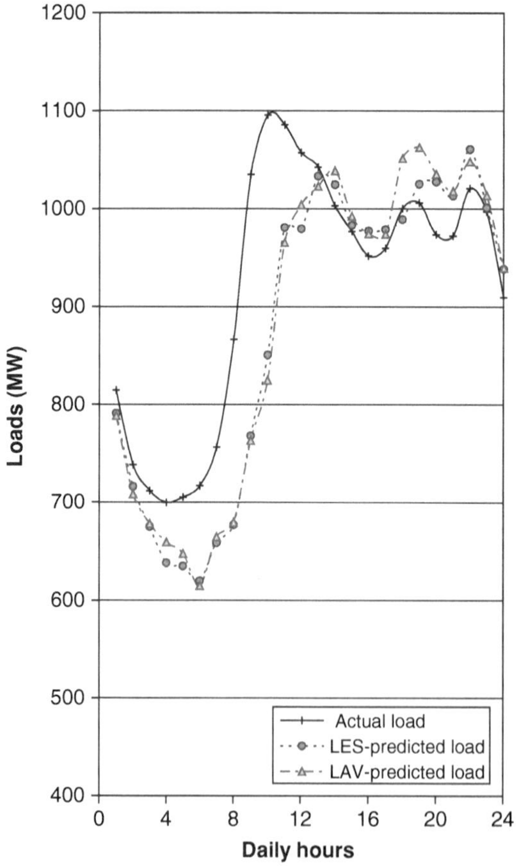

Figure A6.13 Predicted load for a winter weekend day using 24 parameter sets, model A.

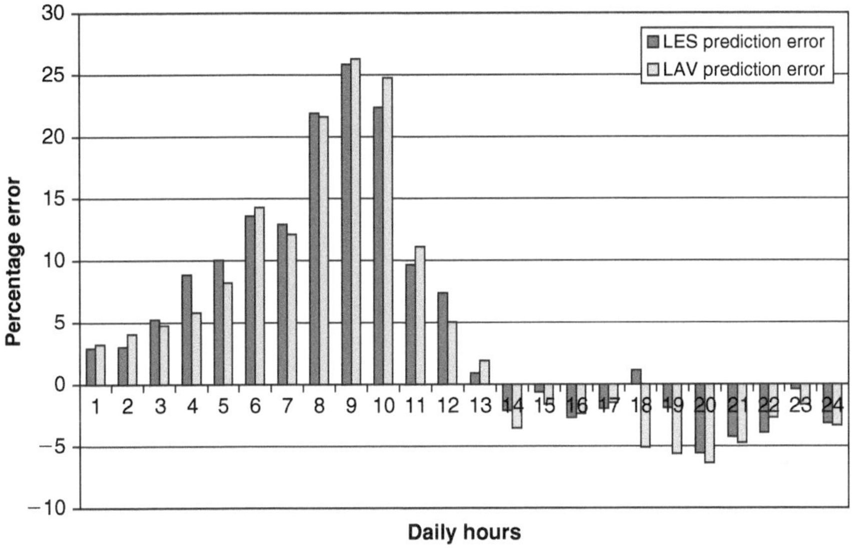

Figure A6.14 Predicted load error for a winter weekend day using 24 parameter sets, model A.

Table A6.8 Predicted Load and Percentage Error for a Winter Weekend Day Using One Parameter Set, Model A

Daily Hours	Actual Load (MW)	LES Prediction	LAV Prediction	% LES Error	% LAV Error
1	776.8	797.7	780.9	−2.69	−0.53
2	710	764.6	785.3	−7.68	−10.6
3	667.1	848	813.2	−27.12	−21.9
4	647.2	848	813.2	−27.12	−21.9
5	639.3	848	813.2	−27.12	−21.9
6	642.8	630.6	640.5	1.89	0.35
7	657.2	662.1	660.7	−0.75	−0.53
8	689.3	731.2	717.6	−6.07	−4.1
9	767.5	811.5	799.9	−5.73	−4.22
10	898	865.4	894	3.63	0.45
11	995.1	879.6	843.4	11.61	15.25
12	1016.2	1053.7	962.5	−3.69	5.28
13	1008.1	1030.5	1006.1	−2.22	0.2
14	977.9	935.6	973.7	4.33	0.43
15	940.1	871.3	900.8	7.32	4.18
16	905.1	871.3	900.8	7.32	4.18
17	892.8	871.3	900.8	7.32	4.18

Continued

Table A6.8 Predicted Load and Percentage Error for a Winter Weekend Day Using One Parameter Set, Model A *Continued*

Daily Hours	Actual Load (MW)	LES Prediction	LAV Prediction	% LES Error	% LAV Error
18	915.4	761.6	703.7	16.8	23.13
19	915.1	854.8	798.7	6.59	12.72
20	887	900.7	890.7	–1.54	–0.41
21	900.2	931.5	905.1	–3.48	–0.54
22	961.4	933.3	965.9	2.93	–0.47
23	953.1	974.8	956.6	–2.28	–0.37
24	903.7	904.5	898.7	–0.09	0.55

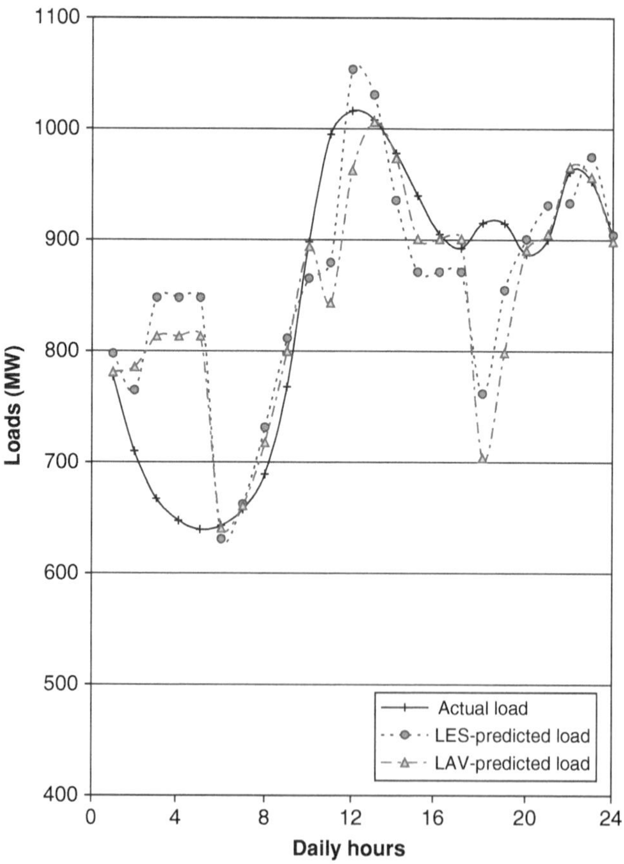

Figure A6.15 Predicted load for a winter weekend day using one parameter set, model A.

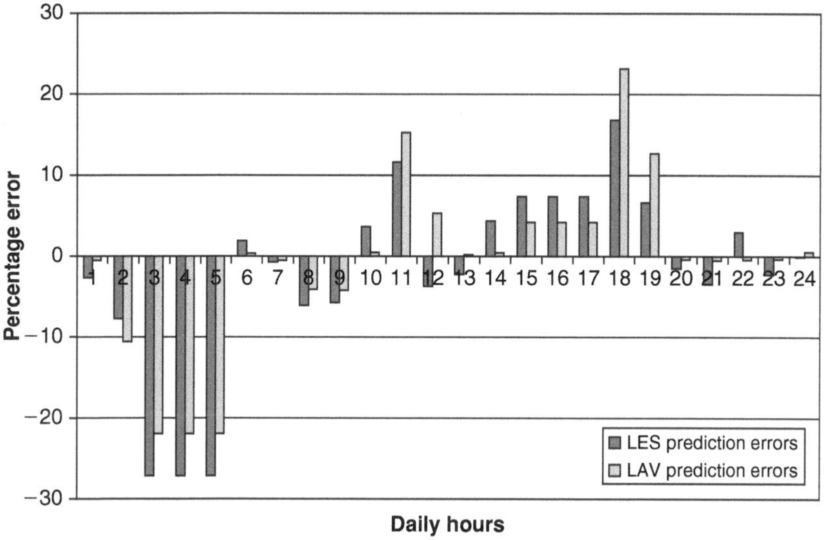

Figure A6.16 Predicted load error for a winter weekend day using one parameter set, model A.

Appendix 6.2 Winter Static Load Results for Model B

Table A6.9 Estimated Load and Percentage Error for a Winter Weekday, Model B

Daily Hours	Actual Load (MW)	LES Estimation	LAV Estimation	% LES Error	% LAV Error
1	943.4	1199.08	780	–27.1	17.3
2	850.5	1157.67	715	–36.1	15.9
3	811.2	1072.45	650	–32.2	19.9
4	793.6	1023.73	793.24	–29	0
5	794.6	969.21	797.75	–22	–0.4
6	810.3	1014.77	809.27	–25.2	0.1
7	855.2	929.75	857.75	–8.7	–0.3
8	991	1085.1	991.43	–9.5	0
9	1198.5	1375.46	1196.94	–14.8	0.1
10	1302.3	1486.53	1301.21	–14.1	0.1
11	1331.8	1446.21	1333.82	–8.6	–0.2
12	1344.7	1344.73	1344.4	0	0
13	1366.1	1350.99	1365.11	1.1	0.1
14	1346.2	1182.07	1347.21	12.2	–0.1
15	1332.7	1132.32	1334.5	15	–0.1

Continued

Table A6.9 Estimated Load and Percentage Error for a Winter Weekday, Model B *Continued*

Daily Hours	Actual Load (MW)	LES Estimation	LAV Estimation	% LES Error	% LAV Error
16	1320.6	1174.25	1315.78	11.1	0.4
17	1341.4	1264.18	1342.01	5.8	0
18	1405.2	1366.85	1407.81	2.7	–0.2
19	1403.3	1162.88	1404.19	17.1	–0.1
20	1380.9	1144.36	1382.12	17.1	–0.1
21	1406.5	1174.45	1406.14	16.5	0
22	1440.9	1154.32	1339.92	19.9	7
23	1390.9	1134.51	887	18.4	36.2
24	1281.1	1099.54	1281.07	14.2	0

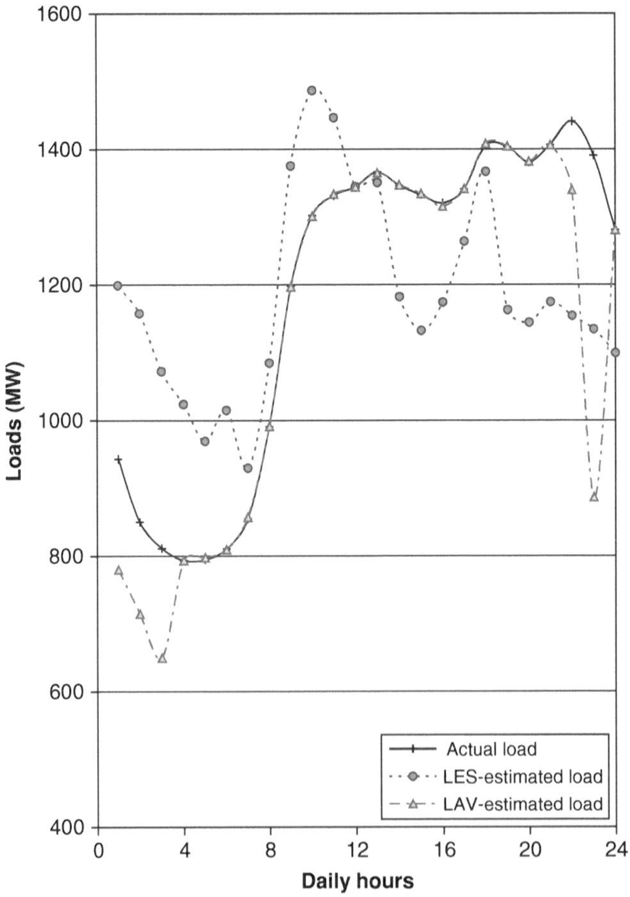

Figure A6.17 Estimated load for a winter weekday, model B.

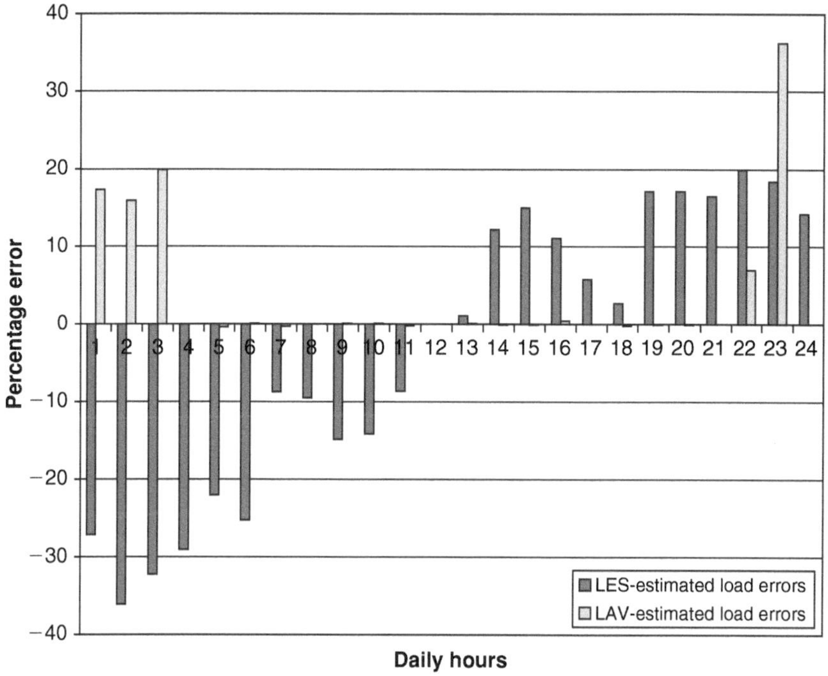

Figure A6.18 Estimated load error for a winter weekday, model B.

Table A6.10 Predicted Load and Percentage Error for a Winter Weekday, Model B

Daily Hours	Actual Load (MW)	LES Prediction	LAV Prediction	% LES Error	% LAV Error
1	1096.8	1217.08	1304.44	−10.97	−18.93
2	1014.4	1172.24	1144.71	−15.56	−12.8
3	988.9	1086.51	984.98	−9.87	0.4
4	983.3	1038.37	983.15	−5.6	0.02
5	991.3	1012.94	992.67	−2.18	−0.14
6	1003.1	1063.71	1002.5	−6.04	0.06
7	1050.2	1107.28	1051.01	−5.43	−0.08
8	1187.1	1236.25	1187.86	−4.14	−0.06
9	1384	1450.19	1382.71	−4.78	0.09
10	1433	1491.2	1434.98	−4.06	−0.14
11	1404.4	1428.25	1406.94	−1.7	−0.18
12	1354.1	1340.07	1354.2	1.04	−0.01
13	1319.2	1303.76	1319	1.17	0.02
14	1261.3	1186.46	1259.87	5.93	0.11

Continued

Table A6.10 Predicted Load and Percentage Error for a Winter Weekday, Model B *Continued*

Daily Hours	Actual Load (MW)	LES Prediction	LAV Prediction	% LES Error	% LAV Error
15	1221.7	1133.64	1219.76	7.21	0.16
16	1184.5	1139.18	1183.88	3.83	0.05
17	1187.9	1180.8	1190.33	0.6	–0.2
18	1240.1	1249.37	1243.48	–0.75	–0.27
19	1307.1	1204.12	1200.94	7.88	8.1
20	1361.9	1221.47	1200.94	10.31	11.8
21	1349.6	1249.4	1199.1	7.42	11.2
22	1313.9	1225.59	1313.89	6.72	0
23	1262.5	1202.34	1264.48	4.76	–0.16
24	1166.7	1128.54	1166.23	3.27	0.04

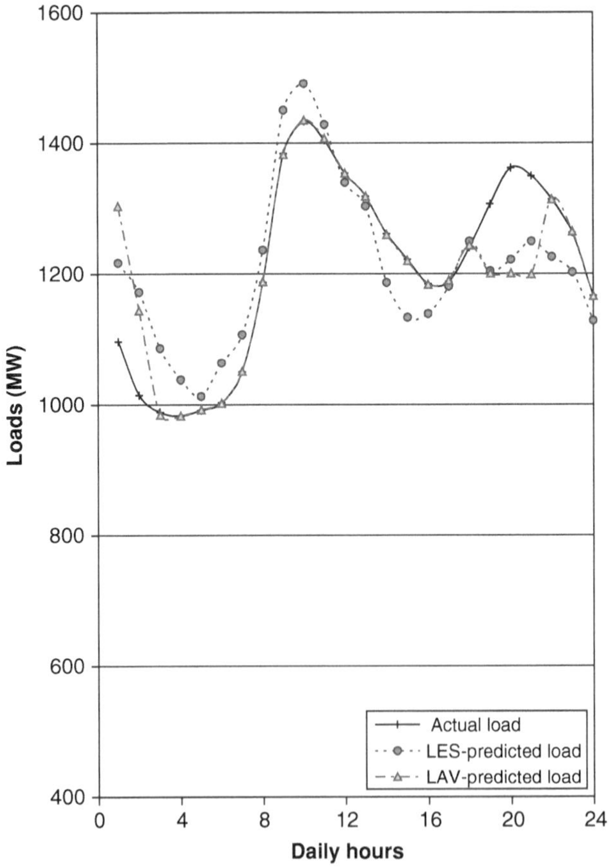

Figure A6.19 Predicted load for a winter weekday, model B.

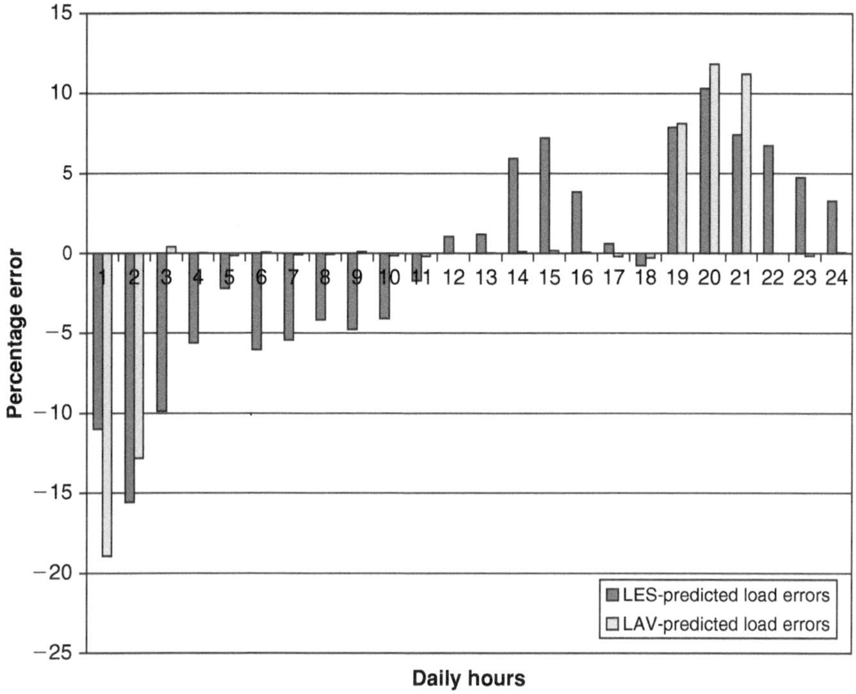

Figure A6.20 Predicted load error for a winter weekday, model B.

Table A6.11 Estimated Load and Percentage Error for a Winter Weekend Day, Model B

Daily Hours	Actual Load (MW)	LES Estimation	LAV Estimation	% LES Error	% LAV Error
1	916.8	899.66	780	1.9	14.9
2	846.7	846.56	836.88	0	15.5
3	810.4	769.84	650	5	19.8
4	801.4	748.77	652.21	6.6	18.6
5	796.9	699.11	647.98	12.3	18.7
6	808.4	706.74	645.76	12.6	20.1
7	826	696.44	659.44	15.7	20.2
8	869.7	729.43	687.89	16.1	20.9
9	939.8	847.3	765.89	9.8	18.5
10	1031.8	959.46	892.76	7	13.5
11	1092.4	999.63	972.25	8.5	11
12	1093.8	986.42	990.06	9.8	9.5
13	1072.2	973.71	981.79	9.2	8.4

Continued

Table A6.11 Estimated Load and Percentage Error for a Winter Weekend Day, Model B *Continued*

Daily Hours	Actual Load (MW)	LES Estimation	LAV Estimation	% LES Error	% LAV Error
14	1030.2	876.4	950.87	14.9	7.7
15	981.6	827.58	908.69	15.7	7.4
16	937.4	823.16	871.66	12.2	7
17	912	830	862.19	9	5.5
18	928.4	881.36	893.76	5.1	3.7
19	938.2	816.56	908.56	13	3.2
20	915.2	790.67	901.3	13.6	1.5
21	930.8	795.04	894.04	14.6	3.9
22	1004.6	798.32	954.98	20.5	4.9
23	994.3	810.64	887	18.5	10.8
24	942.9	789.79	835	16.2	11.4

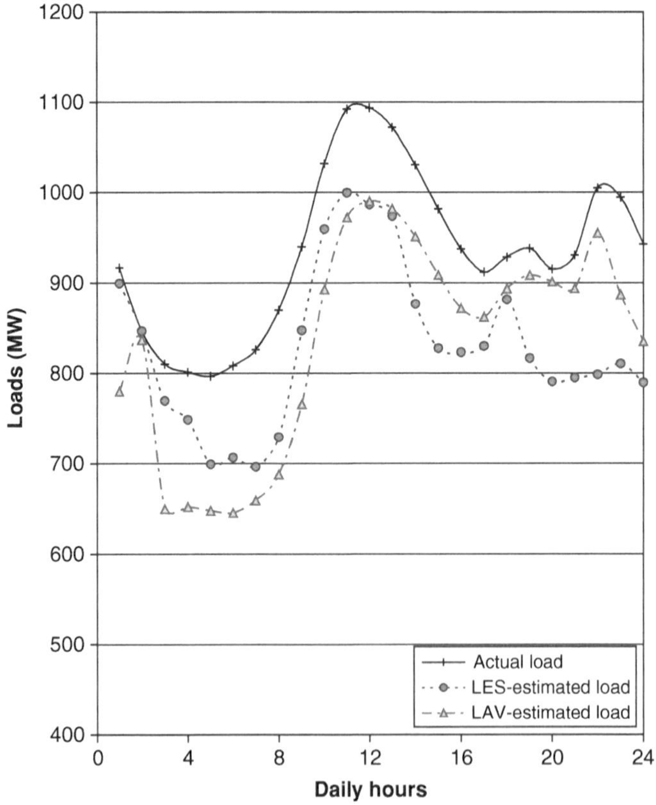

Figure A6.21 Estimated load for a winter weekend day, model B.

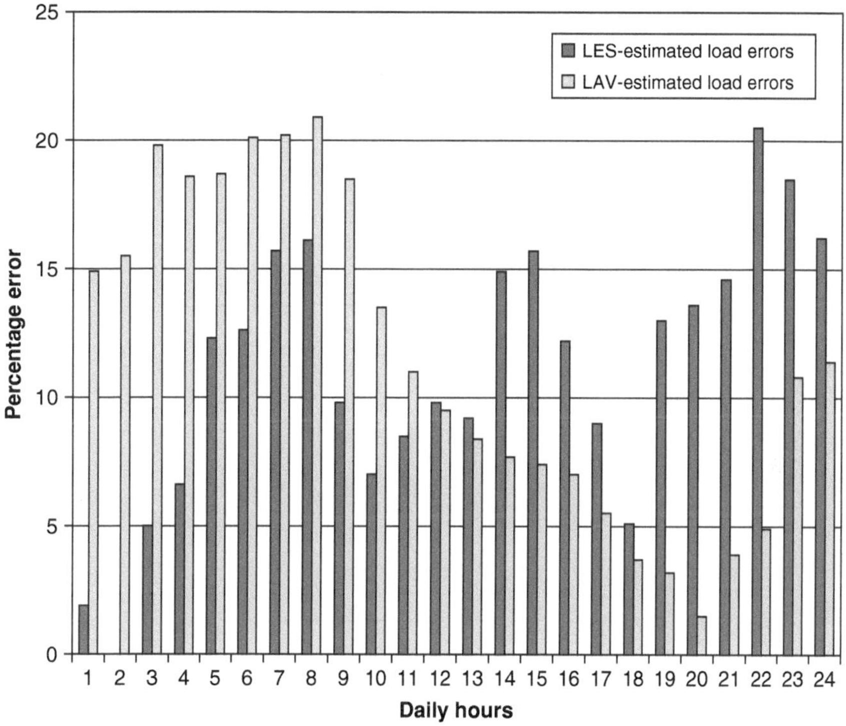

Figure A6.22 Estimated load error for a winter weekend day, model B.

Table A6.12 Predicted Load and Percentage Error for a Winter Weekend Day, Model B

Daily Hours	Actual Load (MW)	LES Prediction	LAV Prediction	% LES Error	% LAV Error
1	950.4	1089.5	938.83	−14.64	1.21
2	874.5	1050.04	896.94	−20.07	−2.56
3	838.7	979.43	805.05	−16.78	4.01
4	814	929.21	813.16	−14.15	0.1
5	806.3	858.62	809.71	−6.49	−0.42
6	800.4	872.44	800.31	−9	0.01
7	811.1	862.16	811.44	−6.3	−0.04
8	833.2	895.18	833.49	−7.44	−0.03
9	918.6	1017.23	916.64	−10.74	0.21
10	1038.1	1121.45	1041.85	−8.03	−0.36
11	1116.9	1157.88	1120.92	−3.67	−0.36
12	1148.4	1143	1149.4	0.47	−0.09

Continued

Table A6.12 Predicted Load and Percentage Error for a Winter Weekend Day, Model B *Continued*

Daily Hours	Actual Load (MW)	LES Prediction	LAV Prediction	% LES Error	% LAV Error
13	1158	1140.82	1155.89	1.48	0.18
14	1135.1	1041.03	1133.63	8.29	0.13
15	1098.2	991.95	1095.1	9.68	0.28
16	1074.5	1004.23	1072.26	6.54	0.21
17	1072.9	1038.91	1075.77	3.17	−0.27
18	1110.2	1110.88	1116.84	−0.06	−0.6
19	1135.1	1021.23	1132.05	10.03	0.27
20	1141.2	996.94	1148.43	12.64	−0.63
21	1159.6	1023.25	1164.81	11.76	−0.45
22	1112.8	960.62	1127.33	13.68	−1.31
23	1060.7	961.22	1063.46	9.38	−0.26
24	981.7	925	980.73	5.78	0.1

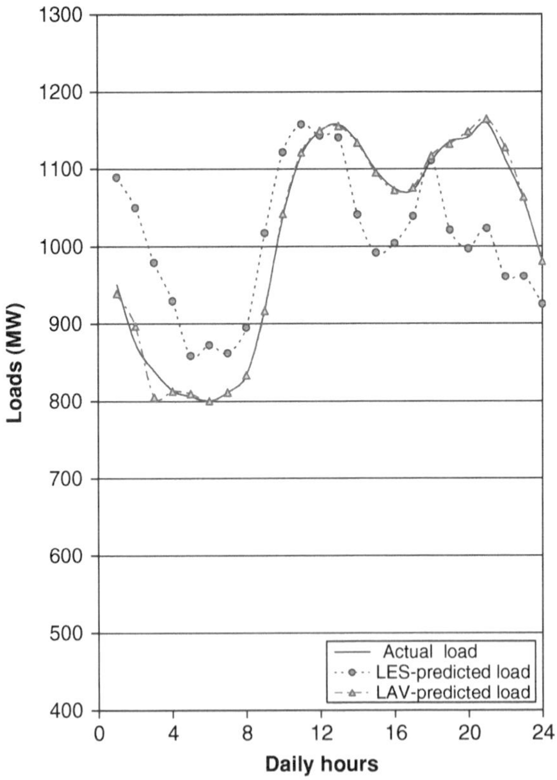

Figure A6.23 Predicted load for a winter weekend day, model B.

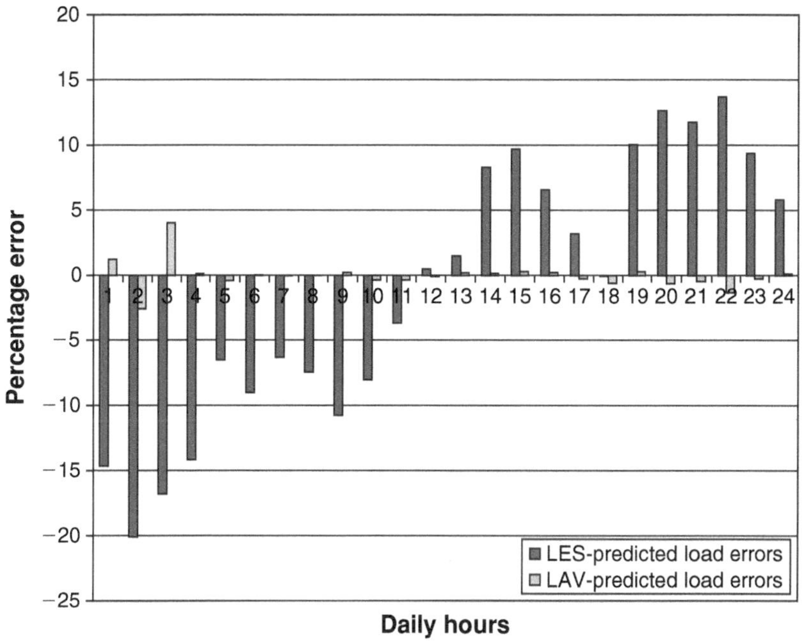

Figure A6.24 Predicted load error for a winter weekend day, model B.

Appendix 6.3 Winter Static Load Results for Model C

Table A6.13 Estimated Load and Percentage Error for a Winter Day, Model C

Daily Hours	Actual Load (MW)	LES Estimation	LAV Estimation	% LES Error	% LAV Error
1	943.4	942.15	943.82	0.13	−0.04
2	850.5	850.56	849.79	−0.01	0.08
3	811.2	814.68	829.75	−0.43	−2.29
4	793.6	791.68	793.89	0.24	−0.04
5	794.6	792.57	794.7	0.26	−0.01
6	810.3	810.78	810.07	−0.06	0.03
7	855.2	856.41	855.73	−0.14	−0.06
8	991	992.96	991.31	−0.2	−0.03
9	1198.5	1197.02	1198.36	0.12	0.01
10	1302.3	1301.02	1301.63	0.1	0.05
11	1331.8	1333.28	1331.78	−0.11	0
12	1344.7	1343.98	1344.41	0.05	0.02
13	1366.1	1366.08	1366.82	0	−0.05

Continued

Table A6.13 Estimated Load and Percentage Error for a Winter Day, Model C *Continued*

Daily Hours	Actual Load (MW)	LES Estimation	LAV Estimation	% LES Error	% LAV Error
14	1346.2	1347.41	1346.24	–0.09	0
15	1332.7	1332.93	1333	–0.02	–0.02
16	1320.6	1318.58	1320.56	0.15	0
17	1341.4	1342.45	1341.48	–0.08	–0.01
18	1405.2	1406.36	1405.07	–0.08	0.01
19	1403.3	1404.3	1403.34	–0.07	0
20	1380.9	1380.14	1381.19	0.06	–0.02
21	1406.5	1402.9	1406.43	0.26	0
22	1440.9	1443.07	1440.71	–0.15	0.01
23	1390.9	1392.63	1390.79	–0.12	0.01
24	1281.1	1280.05	1281.77	0.08	–0.05

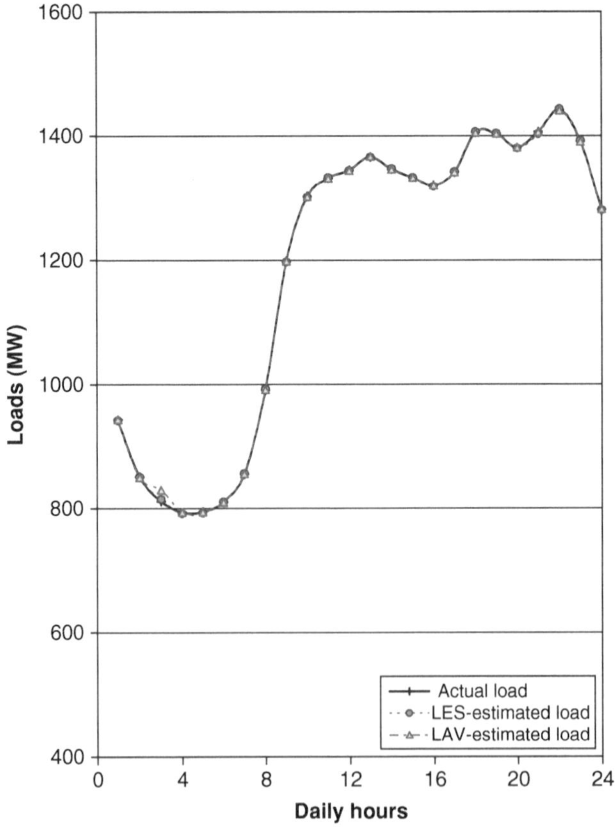

Figure A6.25 Estimated load for a winter day, model C.

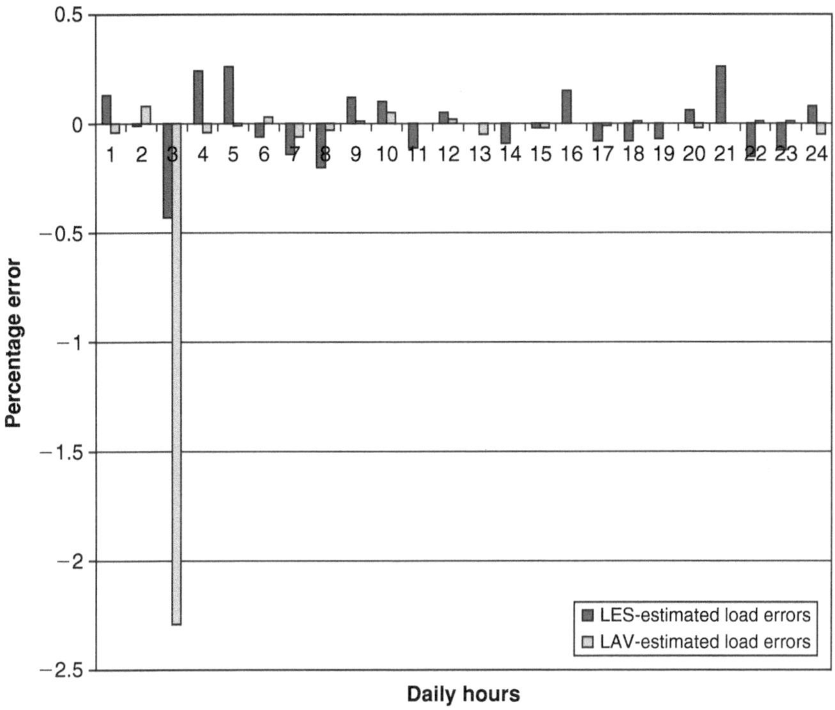

Figure A6.26 Estimated load error for a winter day, model C.

Table A6.14 Predicted Load and Percentage Error for a Winter Day, Model C

Daily Hours	Actual Load (MW)	LES Prediction	LAV Prediction	% LES Error	% LAV Error
1	1117.6	972.23	974.73	13.01	12.78
2	1006.4	875.64	875.91	12.99	12.97
3	943.6	904.36	922.82	4.16	2.2
4	871.1	905.33	912.45	–3.93	–4.75
5	813	861.62	867.75	–5.98	–6.73
6	869.7	848.86	851	2.4	2.15
7	914.8	847.76	847.99	7.33	7.3
8	978.7	922.48	919.01	5.74	6.1
9	1157.3	1053.34	1049.43	8.98	9.32
10	1223	1103.54	1095.86	9.77	10.4
11	1216.8	1124.58	1113.24	7.58	8.51
12	1284.3	1166.26	1157.16	9.19	9.9
13	1258.8	1162.27	1152.49	7.67	8.45

Continued

Table A6.14 Predicted Load and Percentage Error for a Winter Day, Model C *Continued*

Daily Hours	Actual Load (MW)	LES Prediction	LAV Prediction	% LES Error	% LAV Error
14	1207.8	1158.97	1147.51	4.04	4.99
15	1155.4	1168.28	1158.85	−1.12	−0.3
16	1110.6	1114.91	1106.27	−0.39	0.39
17	1094.1	1138.9	1127.06	−4.09	−3.01
18	1113.3	1239.31	1228.27	−11.32	−10.33
19	1186.4	1222.02	1211.05	−3	−2.08
20	1139	1205.36	1196.66	−5.83	−5.06
21	1152.3	1217.28	1210.77	−5.64	−5.07
22	1226.5	1253.05	1240.43	−2.16	−1.14
23	1198.4	1220.65	1209.09	−1.86	−0.89
24	1111.8	1066.59	1057.3	4.07	4.9

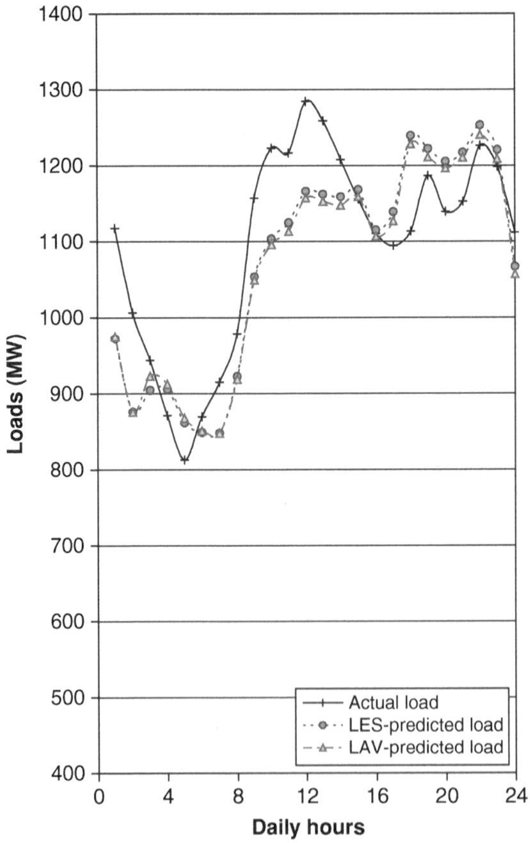

Figure A6.27 Predicted load for a winter day, model C.

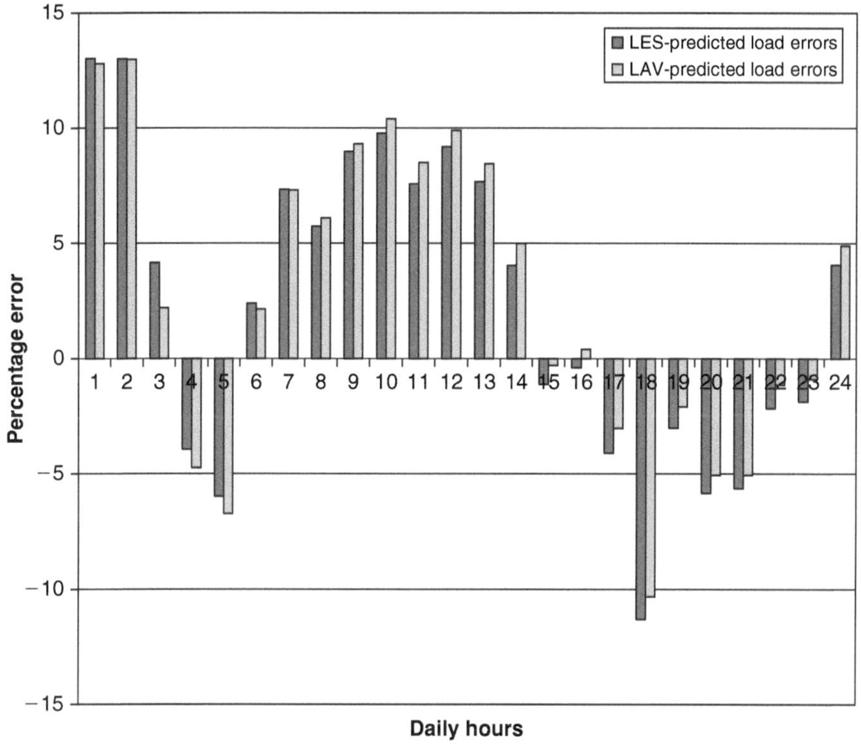

Figure A6.28 Predicted load error for a winter day, model C.

7 Load-Forecasting Results Using Fuzzy Systems

7.1 Objectives

Chapter 6 described the short-term load-forecasting problem and the least error squares (LES) and least absolute value (LAV) parameter estimation algorithms used to estimate the load model parameters. The error in the estimates is calculated for both techniques. The three models proposed in Chapter 3 are used in that chapter to forecast the load in different days for different seasons. In this chapter, the objectives are

- Testing the fuzzy load models developed in Chapter 4.
- Estimating the fuzzy parameters of these models using the past history data for summer weekday and weekend days as well as for winter weekday and weekend days.
- Using these models to predict the fuzzy load power for 24 hours ahead, in both summer and winter seasons.

The results are given in the form of tables and figures for the estimated and predicted loads.

7.2 Fuzzy Load Model A

The fuzzy load model A for summer developed in Chapter 4 is tested in this section. First, the load power data are assumed to be crisp values, and the load parameters are fuzzy. Then both the load power data and load parameters are assumed to be fuzzy. We will find that nine fuzzy parameters are enough to model this type of load.

7.2.1 Load Parameters for a Summer Weekday

Table 7.1 gives the estimated fuzzy parameters for three cases. In the first case, the load power has crisp values. In the other two cases, the load power is fuzzy data, and it is assumed that load power has deviated by 5% and 20% from the original case to simulate the fuzziness in these values. Examining this table reveals the following:

- The only fuzzy parameter is A_0, which conforms with the assumption that the load power has a crisp value, and the spreads of the parameters are to be minimized.
- We can note that three parameters are adequate to represent the load for the crisp case, six parameters for the 5% load deviation case, and five parameters for the 20% load deviation case because the output of the linear optimization problem produces only these parameters.

DOI: 10.1016/B978-0-12-381543-9.00007-5

Table 7.1 Fuzzy Parameter for a Summer Weekday, Model A

Parameters	Crisp Load		5% Load Deviation		20% Load Deviation	
	Middle	Spread	Middle	Spread	Middle	Spread
$\underline{A}_0$	0.0	335.69	0.0	288.4	0.0	391.91
$\underline{A}_1$	0.0	0.0	8.5	0.0	0.0	0.0
$\underline{A}_2$	0.0	0.0	0.0	0.0	0.0	0.0
$\underline{A}_3$	0.0	0.0	0.0	0.0	0.00807	0.0
$\underline{A}_4$	0.0	0.0	3.75	0.0	3.229	0.0
$\underline{A}_5$	0.0	0.0	0.579	0.0	0.612	0.0
$\underline{A}_6$	0.0	0.0	0.0	0.0	0.0	0.0
$\underline{A}_7$	22.94	0.0	1.919	0.0	22.92	0.0
$\underline{A}_8$	0.0	0.0	0.0	0.0	0.0	0.0
$\underline{A}_9$	1.03	0.0	21.82	0.0	0.0	0.0

- The middle of some parameters ($\underline{A}_1$) *is not* zeros at 5% load deviation and *is* zeros for 20% load deviation and vice versa. ($\underline{A}_3$ has a middle value at 20% load deviation, but the middle is zero at 5% load deviation.)
- Among the values of the parameters, the $\underline{A}_0$ parameter is the largest one. This parameter represents the base load. The extra power component that comes from other parameters represents the variation in the load power due to the variation in weather conditions.

7.2.2 Load Estimation for a Summer Weekday

Using the estimated fuzzy parameters in Table 7.1, Figures 7.1 to 7.3 give the actual and the estimated load during the same period of time for the three load power conditions. Examining these figures reveals the following:

- The estimated fuzzy load contains the given load values within the allowable range specified by the spread in the parameters.
- The estimation results are good because the given load has never gone outside the range given by the spreads of the fuzzy parameters.
- The problem involving crisp values for load power at any hour, mentioned in Chapter 6, is now solved by transforming the load at the hour into a soft load, and a range of lower load to upper load is allowed.
- As the load deviation percentage increases, the spread between the upper load and lower load increases.

7.2.3 Load Prediction for a Summer Weekday

The estimated fuzzy parameters are used to predict the load 24 hours ahead for a summer workday. Figures 7.4 to 7.6 give the results obtained for the three fuzzy ranges for this day. Examining these figures reveals the following:

- The estimated parameters produce good predictions for the load at every hour in question.
- The given load is within the range produced by the estimated parameter spreads.

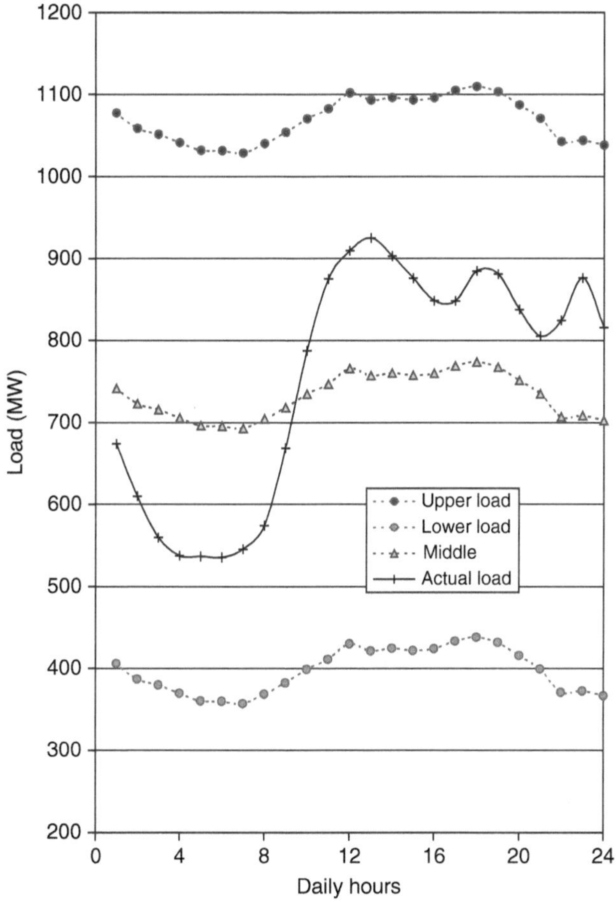

Figure 7.1 Estimated load for a summer weekday, crisp load, model A.

- The actual load deviates a very small amount from these ranges. These deviations can be neglected for such types of forecasting.
- At a given hour, the upper and lower values can be considered as constraints on the load at this hour.

7.2.4 Load Parameters for a Summer Weekend Day

The proposed fuzzy model also is used to predict the load on a summer weekend day. The fuzzy parameters are estimated first. Table 7.2 gives the estimated fuzzy parameters, and Figures 7.7 to 7.9 depict the results for the load deviation ranges. Examining the table and figures reveals the following:

- Among the parameters, $\underline{A}_0$ is the only parameter showing fuzziness for the crisp and other two cases because it has spread values. The objective is to minimize the spread of each fuzzy parameter.

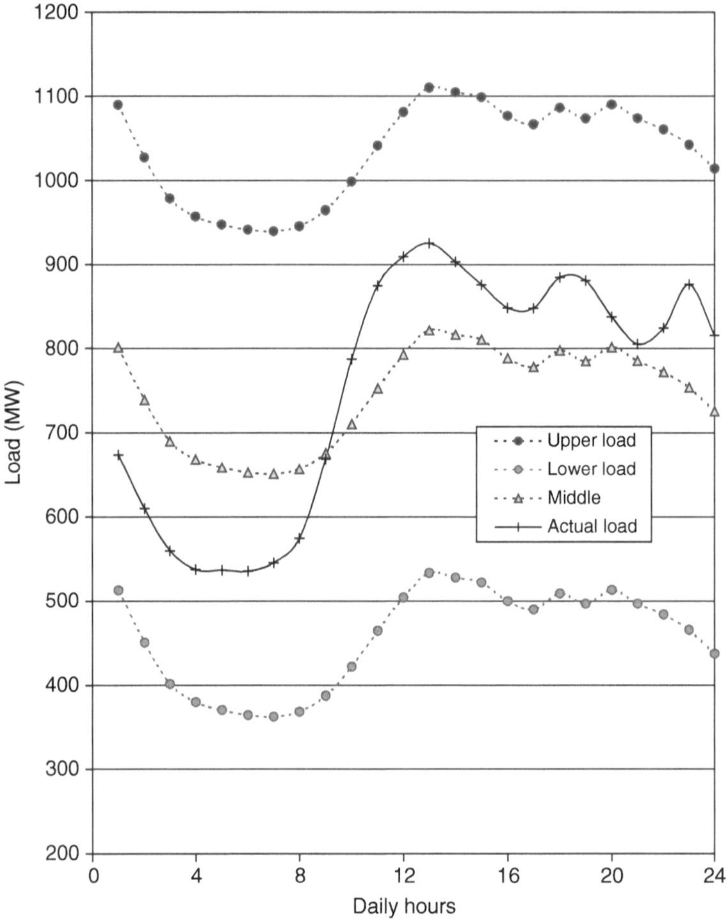

Figure 7.2 Estimated load for a summer weekday (5% load deviation), model A.

- Five fuzzy parameters are adequate to model this type of load for this specific day and season.
- The actual load is in the range given by the estimated spread and does not cross the border of the estimated load.
- The actual load lies between the upper and lower fuzzy ranges of the loads.

7.2.5 Load Prediction for a Summer Weekend Day

The estimated fuzzy parameters are used to predict the load ahead in a weekend day. The results obtained are given in Figures 7.10 to 7.12. Examining these figures reveals the following:

- A good load prediction is obtained for a specified weekend day.
- A range is allowed for the load power to vary at every specified hour, and this range increases as the load deviation increases.

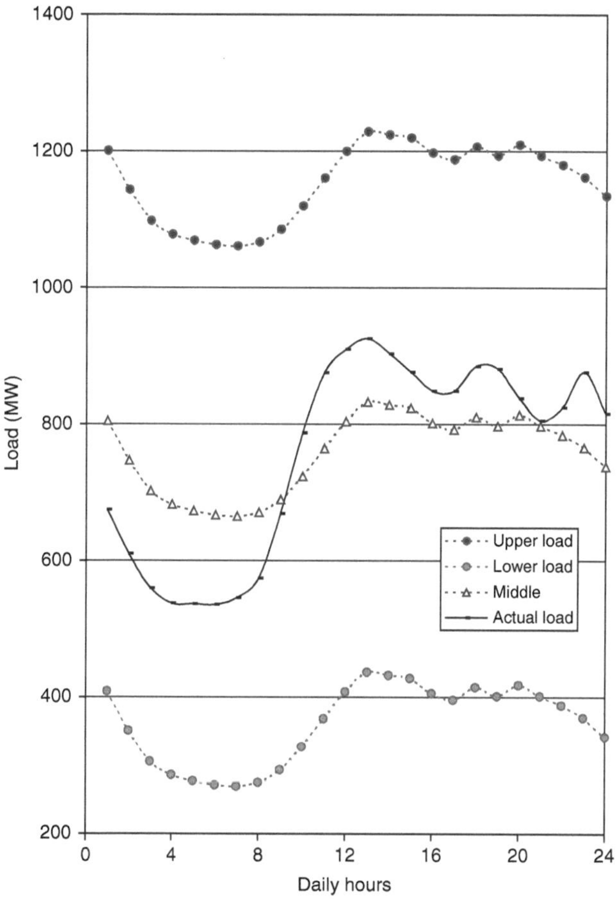

Figure 7.3 Estimated load for a summer weekday (20% load deviation), model A.

- The actual load never crosses the limits determined by the spreads of the load parameters. These limits are an upper load and a lower load.
- At a given hour, the upper and lower load powers can be considered as constraints on the actual load at this hour.
- The actual powers, 24 hours ahead, in all curves do not violate the upper and lower constraints' power load.

In conclusion, fuzzy load model A is adequate to present the load for the summer weekday and weekend days.

7.2.6 Load Estimation and Prediction for a Winter Weekday and a Winter Weekend Day

The results for a winter weekday and a winter weekend day are given in Appendix 7.1. The same concluding remarks can be reached for the data listed.

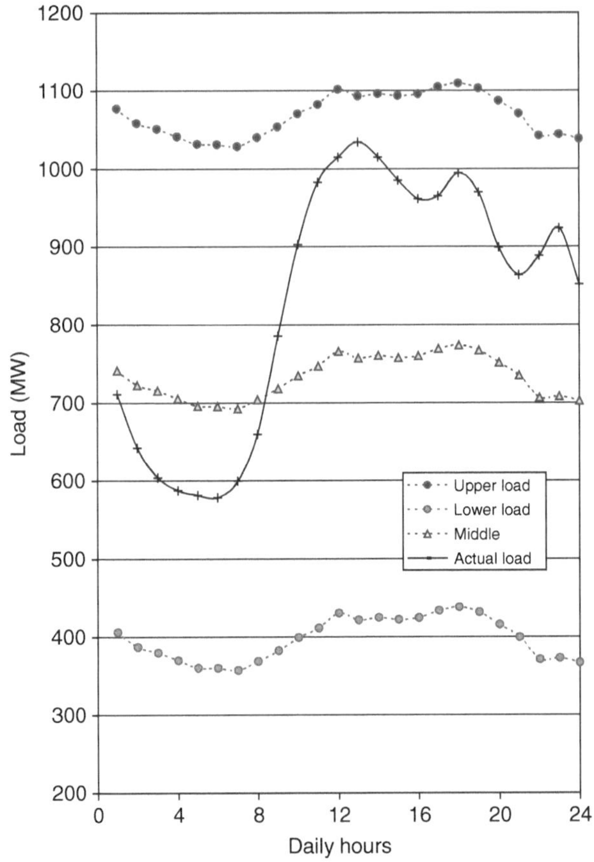

Figure 7.4 Predicted load for a summer weekday, crisp load, model A.

7.3 Fuzzy Load Model B

Fuzzy load model B is a harmonic model and is not sensitive to the weather parameters (temperature, wind speed, humidity, etc.). Nine parameters are chosen for the sine term and nine for the cosine term in addition to the base load parameter. The load deviation for this load model takes values of 0% (crisp load power), 5%, 10%, and 20% to simulate the fuzziness of the load power.

7.3.1 Load Parameters for Model B

Table 7.3 gives the variation of the fuzzy parameters for load model B at percentage load deviations. Examining this table reveals the following:

- Among the load parameters, only parameters A_0, the base load parameter, and A_5 are fuzzy.

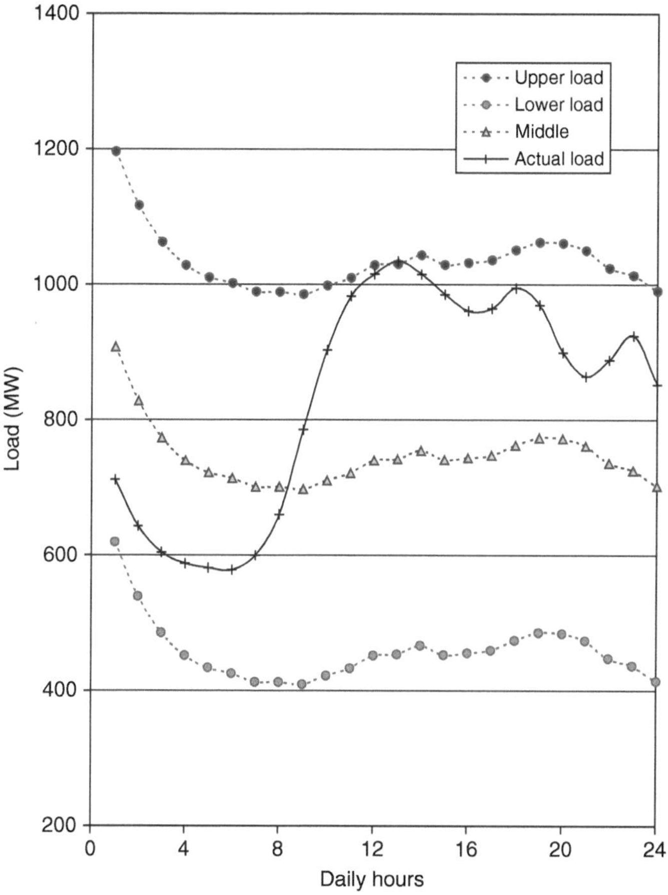

Figure 7.5 Predicted load for a summer weekday (5% load deviation), model A.

- Parameter A_5 has a zero value at the middle and a different spread value in all cases considered. For example, in the 20% load deviation case:
  - The upper parameter value = middle + spread = 0 + 20.196 = 20.196.
  - The lower parameter value = middle − spread = 0 − 20.196 = −20.196.
  - The membership for A_5 is a line on the x-axis centered at the origin with a zero middle value and a spread of 20.196.

  The spread increases with the increase of the degree of fuzziness:
  - For 0% load deviation, the spread is 14.43.
  - For 5% load deviation, the spread is 18.5504.
  - For 10% load deviation, the spread is 19.411.

  This indicates the fuzziness effect in the load's nature, where for increasing the degree of fuzziness, the spread increases and then the range between upper and lower limits increases.
- For A_0 in 20% load deviation:
  - The upper parameter value = 886.038 + 424.722 = 1310.76 MW.
  - The lower parameter value = 886.038 − 424.722 = 461.316 MW.

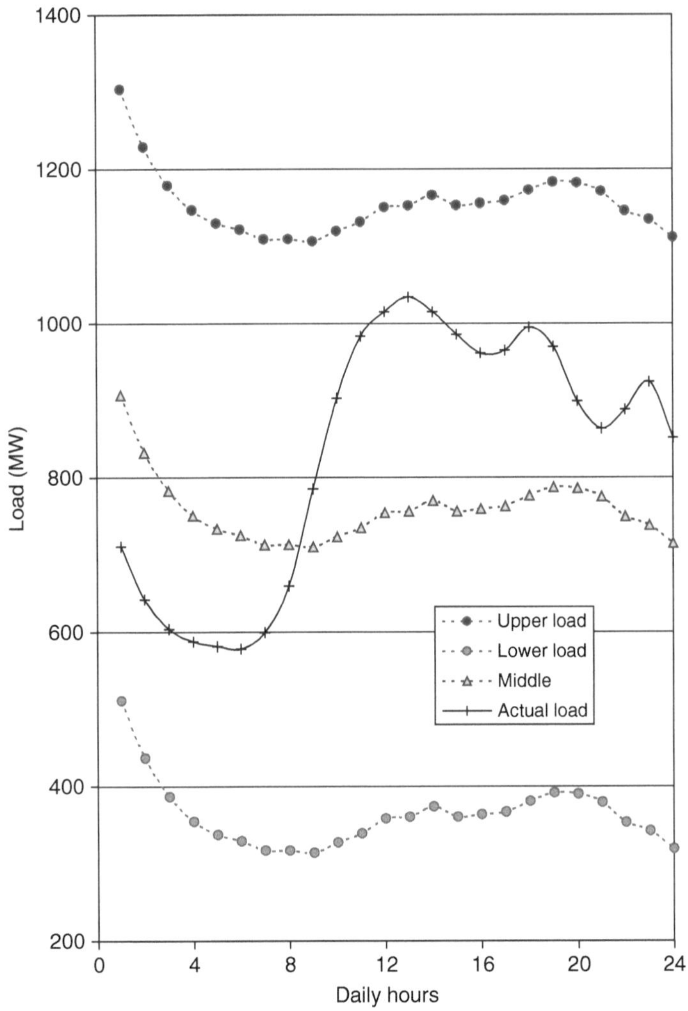

Figure 7.6 Predicted load for a summer weekday (20% load deviation), model A.

- The predicted base load will fall in the range between 1310.76 and 461.316. Both spreads from A_0 and A_5 contribute to the total spread between upper and lower load values.
- The total spread = 424.722 + 20.196 [sin (5ω*t*)].
- The effect of the large middle and spread values of A_0 shows the fuzziness in the large range where predicted load should lie in it.

- Parameter A_0 has a large middle and spread because A_0 represents the base load, whereas the other parameters (either fuzzy or not) are contributing to the excess power variations due to other load factors.
- Both the middle and spread of the base load parameter increase due to the increase in load deviation.
- All load parameters follow the same pattern of variation at each load deviation.

Table 7.2 Fuzzy Parameters for a Summer Weekend Day, Model A

Parameters	Crisp Load		5% Load Deviation		20% Load Deviation	
	Middle	Spread	Middle	Spread	Middle	Spread
$\underline{A}_0$	0.0	247.328	0.0	283.0	0.0	391.9
$\underline{A}_1$	0.0	0.0	0.0	0.0	0.0	0.0
$\underline{A}_2$	0.0	0.0	0.0	0.0	0.0	0.0
$\underline{A}_3$	0.0069	0.0	0.0072	0.0	0.0081	0.0
$\underline{A}_4$	2.779	0.0	2.891	0.0	3.23	0.0
$\underline{A}_5$	0.307	0.0	0.383	0.0	0.612	0.0
$\underline{A}_6$	0.0	0.0	0.0	0.0	0.0	0.0
$\underline{A}_7$	22.576	0.0	22.662	0.0	22.92	0.0
$\underline{A}_8$	0.0	0.0	0.0	0.0	0.0	0.0
$\underline{A}_9$	0.0	0.0	0.0	0.0	0.0	0.0

7.3.2 Load Estimation and Prediction

The estimated and predicted loads for a summer day, either weekday or weekend day, are given in Figures 7.13 to 7.20 for the ranges of load deviation. Examining these figures reveals the following:

- Load model B estimates and predicts the load power at any weekday in any season given that the actual load does not violate the upper and lower load values.
- As the load deviation increases, the range between the upper and lower loads increases due to increases in the spread of the fuzzy parameters.

In conclusion, model B is as good as model A. Despite the fact that model B does not account for the weather variation, the predicted load does not violate the upper or lower load limits.

7.4 Fuzzy Load Model C

In this section the fuzzy load model C developed in Chapter 4 is tested for summer weekday or weekend days. This model is a hybrid combination of models A and B. Summer weekday load data are used to estimate the fuzzy parameters of the model. These parameters are then used to predict the load power one day ahead. The load deviation that creates fuzziness is changed from 0%, 5%, to 20%, with a degree of fuzziness of 50%.

7.4.1 Load Parameters for Model C

Table 7.4 gives the estimated 24 fuzzy parameters—23 parameters and a base load parameter—at different load deviations. Examining this table reveals the following:

- Most of the load parameters are crisp because the spreads are zeros. There are three fuzzy parameters, and they are the same parameters in the three cases of the load deviation (A_0, A_4, and B_8).

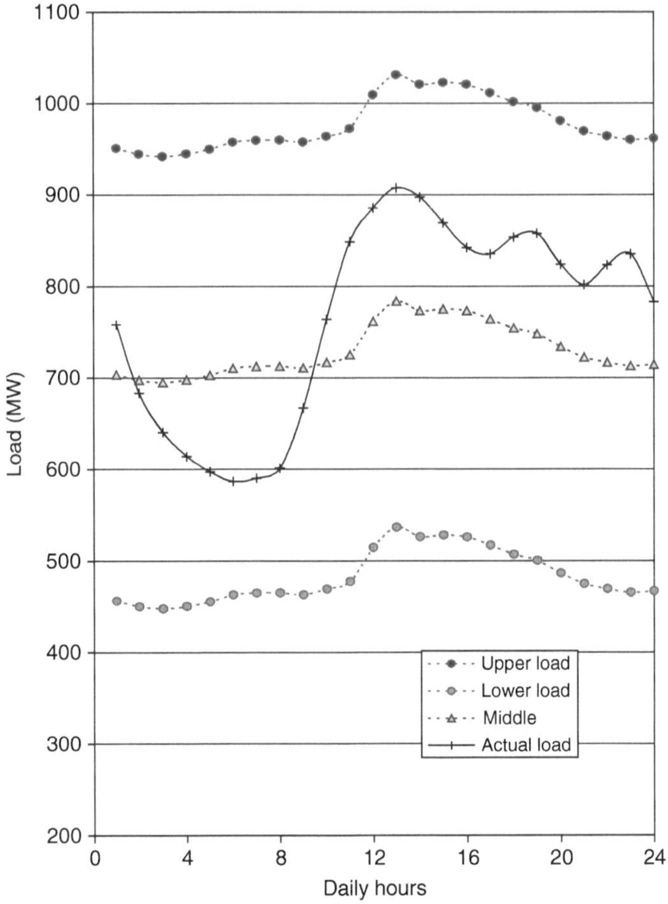

Figure 7.7 Predicted load for a summer weekend day, crisp load, model A.

- As the load deviation increases, the spreads of these parameters increase to include the parameter memberships in the solution.
- There are large middle and spread values for A_0 in the three cases because A_0 represents the base load.

7.4.2 Load Estimation and Prediction for a Summer Day

The estimated parameters for fuzzy load model C are used to predict the load power 24 hours ahead, for either a weekday or weekend day. Figures 7.21 to 7.28 give the estimated and predicted loads at the given load deviations. Examining these figures reveals the following:

- As load deviation equals 5%, the actual load is greater than the upper limit for two hours only by about 1.7% and 4.1%, which is still an acceptable amount. However, if the load deviation is increased to 10%, the actual load does not violate the upper load.

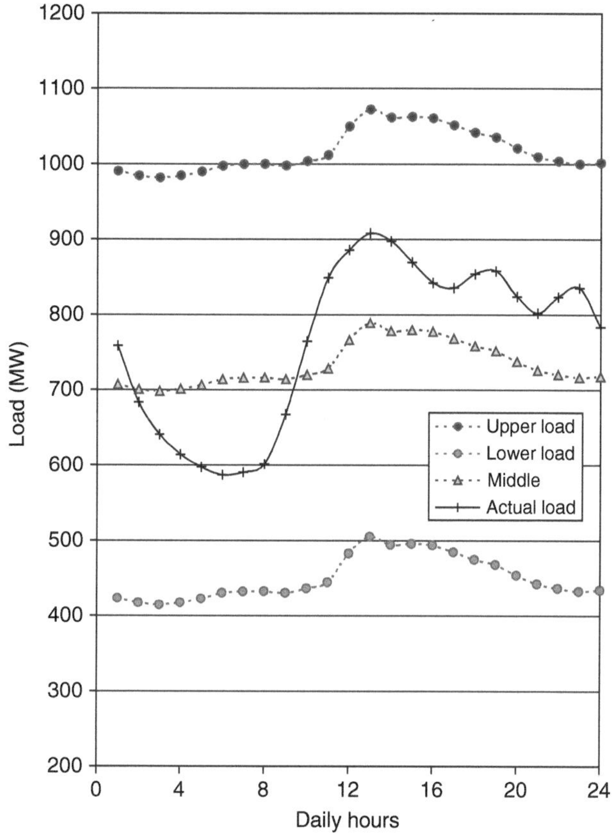

Figure 7.8 Predicted load for a summer weekend day (5% load deviation), model A.

- The estimated load using the fuzzy parameters for all load deviations does not violate either the upper or lower load.
- The actual load violates the upper load in the crisp load case (Figure 7.25), the 5% load deviation case (Figure 7.26), and the 10% load deviation case (Figure 7.27). This violation decreases as the load deviation increases. For example, in Figure 7.28, the actual load does not violate the upper load because the load deviation is increased to 20%, which increases the fuzziness of the load.
- Because the load varies between the upper and lower values, the estimated parameters can sufficiently be used to predict the load for any day in the week in any season. The load parameters must be updated from weekday, weekend day, and from one season to the another.

7.4.3 Load Estimation and Prediction for a Winter Day

The results obtained for the winter weekday are reported in Appendices 7.2 and 7.3. The same conclusions can be made.

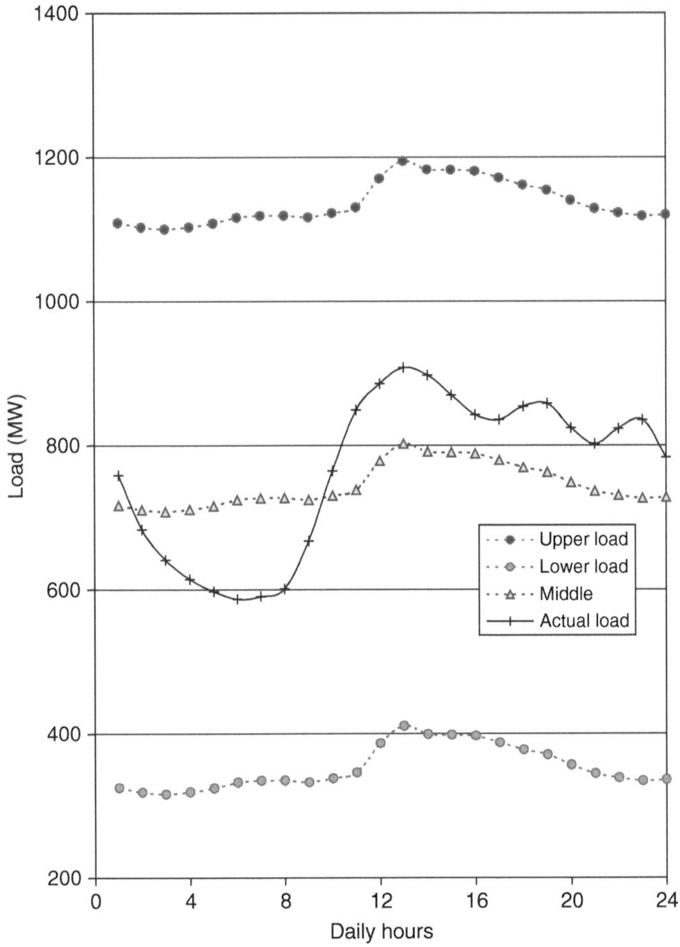

Figure 7.9 Predicted load for a summer weekend day (20% load deviation), model A.

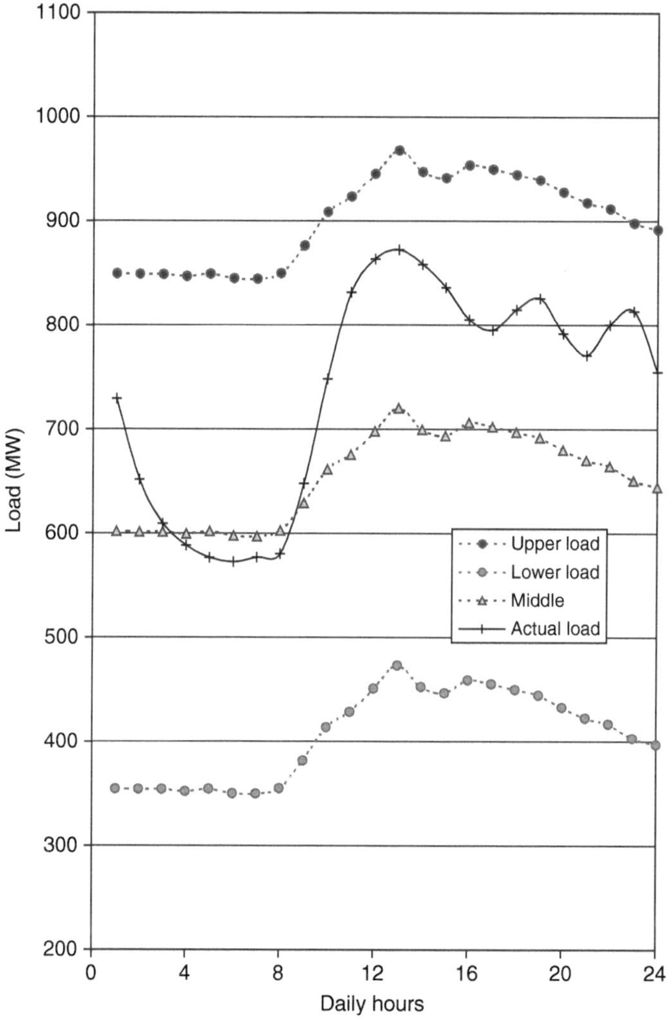

Figure 7.10 Predicted load for a summer weekend day, crisp load, model A.

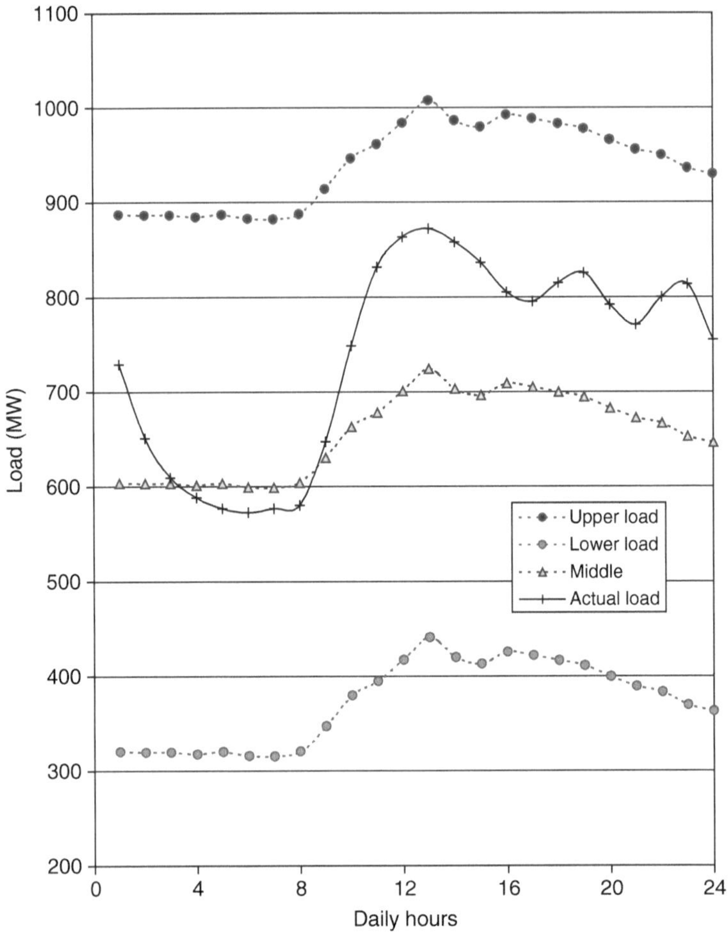

Figure 7.11 Predicted load for a summer weekend day (5% load deviation), model A.

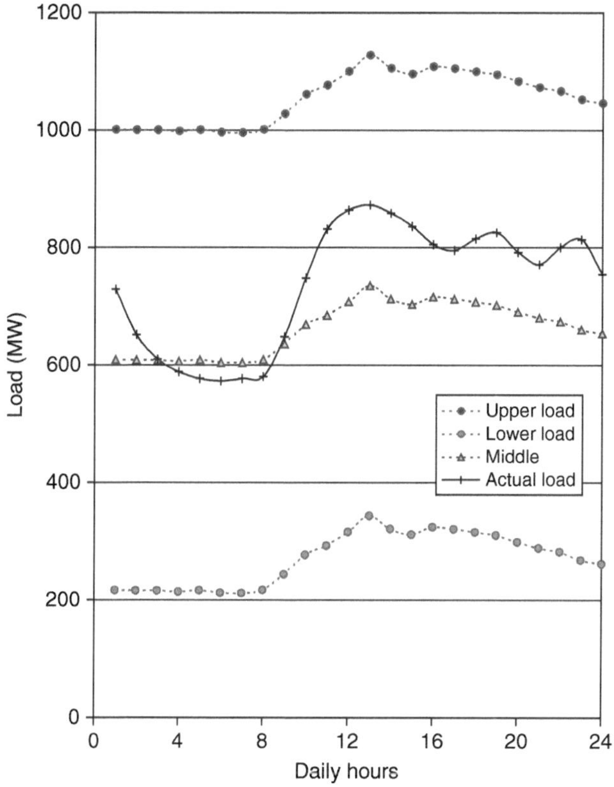

Figure 7.12 Predicted load for a summer weekend day (20% load deviation), model A.

Table 7.3 Fuzzy Parameters for a Summer Day Load, Model B

Parameter	Crisp Load		5% Load Deviation		10% Load Deviation		20% Load Deviation	
	Middle	Spread	Middle	Spread	Middle	Spread	Middle	Spread
A_0	874.32	258.7	875.99	299.306	879.498	340.848	886.038	424.722
A_1	1.594	0.0	1.402	0.0	0.0	0.0	0.0	0.0
A_2	28.95	0.0	28.544	0.0	26.8084	0.0	23.770	0.0
A_3	0.0	0.0	0.355	0.0	1.7189	0.0	3.4214	0.0
A_4	45.81	0.0	45.502	0.0	44.6506	0.0	42.304	0.0
A_5	0.0	14.43	0.0	18.5504	0.00.0	19.411	0.00.0	20.196
A_6	23.40	0.0	23.336	0.0	24.1051	0.0	24.9042	0.0
A_7	13.5	0.0	12.826	0.0	10.7958	0.0	7.3645	0.0
A_8	14.3	0.0	14.083	0.0	12.2520	0.0	11.0160	0.0

Continued

Table 7.3 Fuzzy Parameters for a Summer Day Load, Model B *Continued*

Parameter	Crisp Load		5% Load Deviation		10% Load Deviation		20% Load Deviation	
	Middle	Spread	Middle	Spread	Middle	Spread	Middle	Spread
A_9	108.9	0.0	104.869	0.0	103.782	0.0	100.678	0.0
B_1	66.22	0.0	65.528	0.0	65.1667	0.0	63.921	0.0
B_2	24.80	0.0	23.676	0.0	22.1530	0.0	18.937	0.0
B_3	9.43	0.0	8.582	0.0	9.42273	0.0	9.601	0.0
B_4	29.16	0.0	28.407	0.0	27.5010	0.0	25.045	0.0
B_5	3.93	0.0	2.192	0.0	1.81461	0.0	2.0200	0.0
B_6	1.02	0.0	0.0	0.0	0.0	0.0	0.0	0.0
B_7	54.67	0.0	51.593	0.0	50.220	0.0	47.1926	0.0
B_8	11.62	0.0	11.832	0.0	11.5480	0.0	12.288	0.0
B_9	2.70	0.0	1.7536	0.0	3.2094	0.0	5.8511	0.0

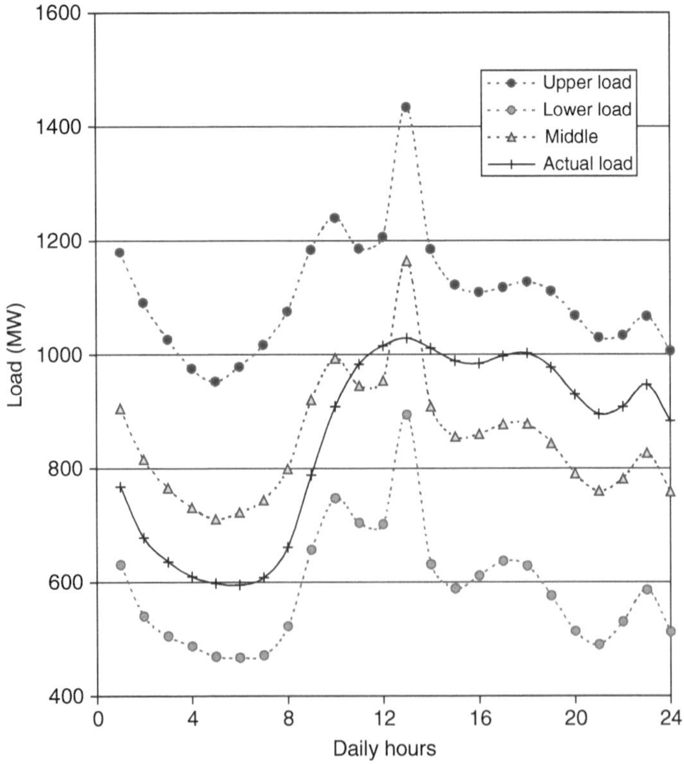

Figure 7.13 Estimated load for a summer day, crisp load, model B.

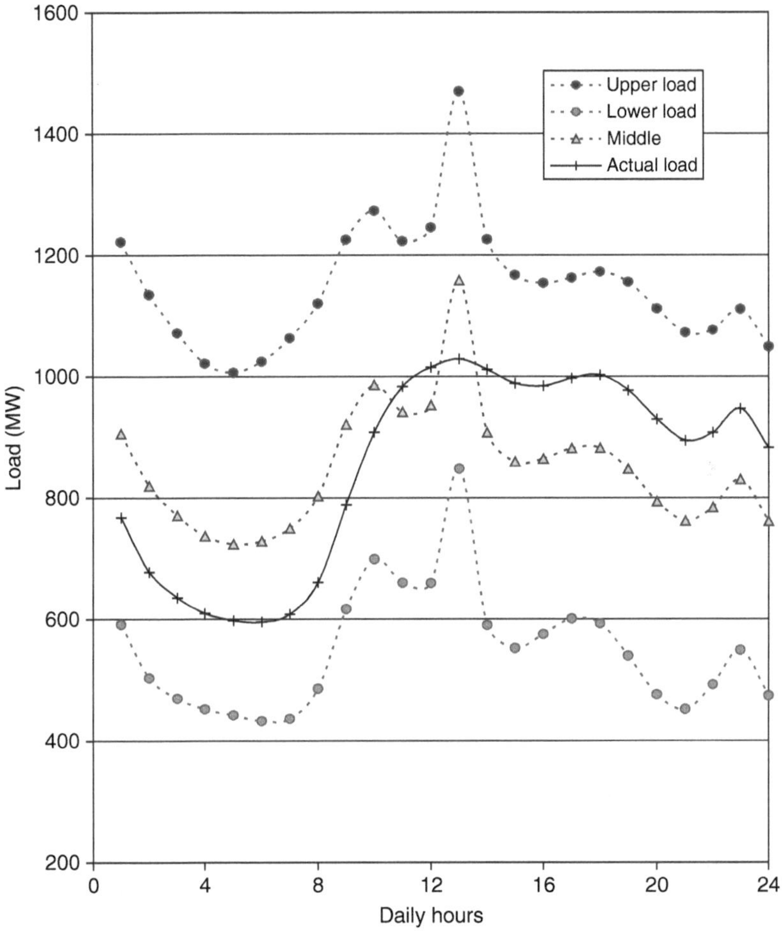

Figure 7.14 Estimated load for a summer day (5% load deviation), model B.

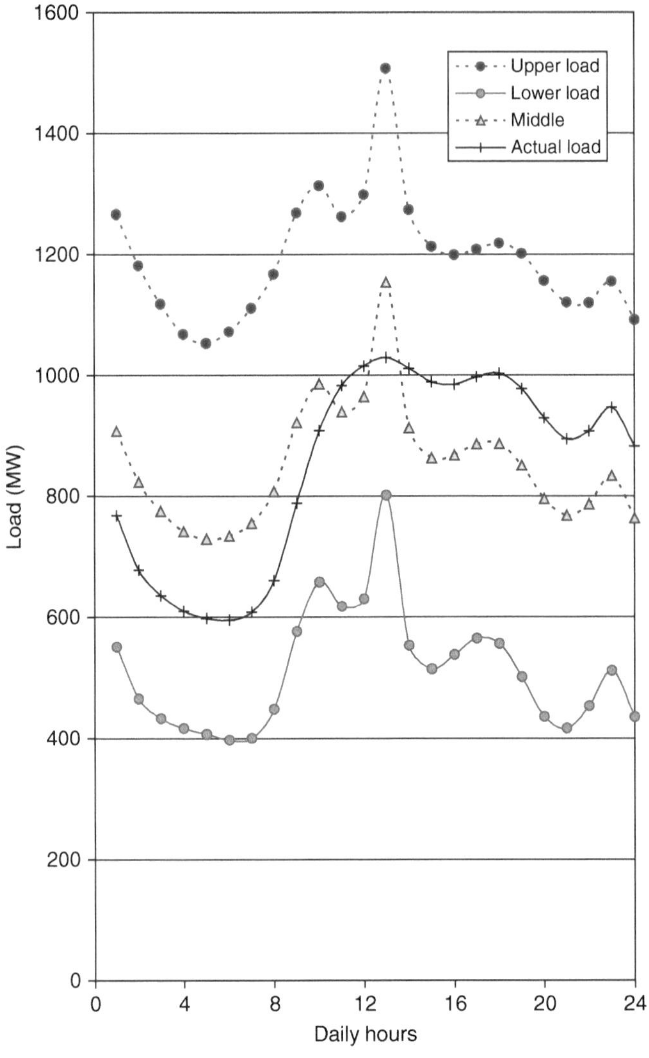

Figure 7.15 Estimated load for a summer day (10% load deviation), model B.

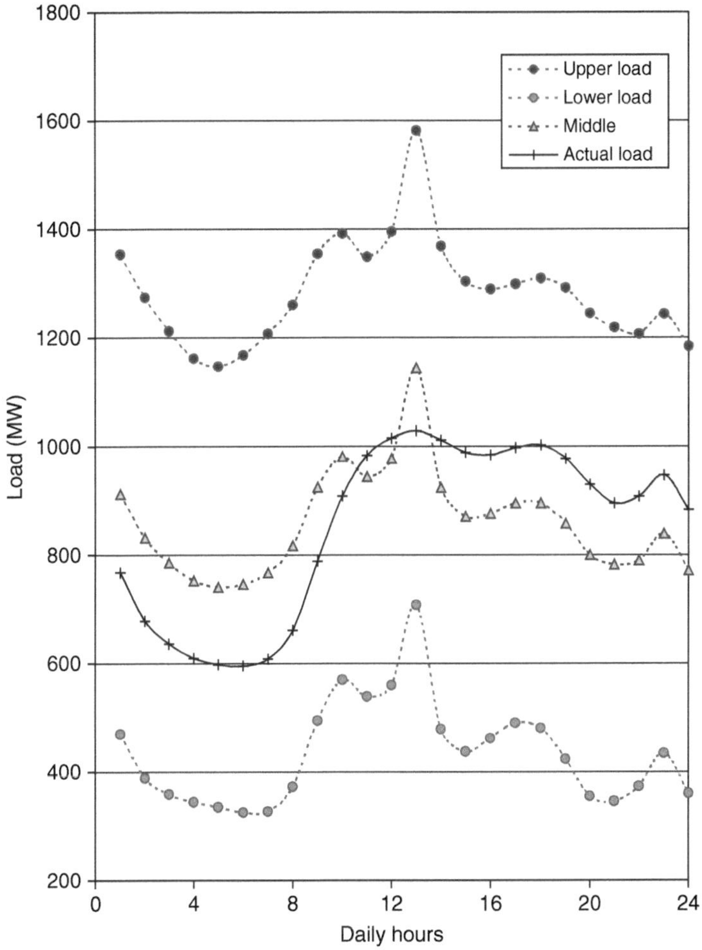

Figure 7.16 Estimated load for a summer day (20% load deviation), model B.

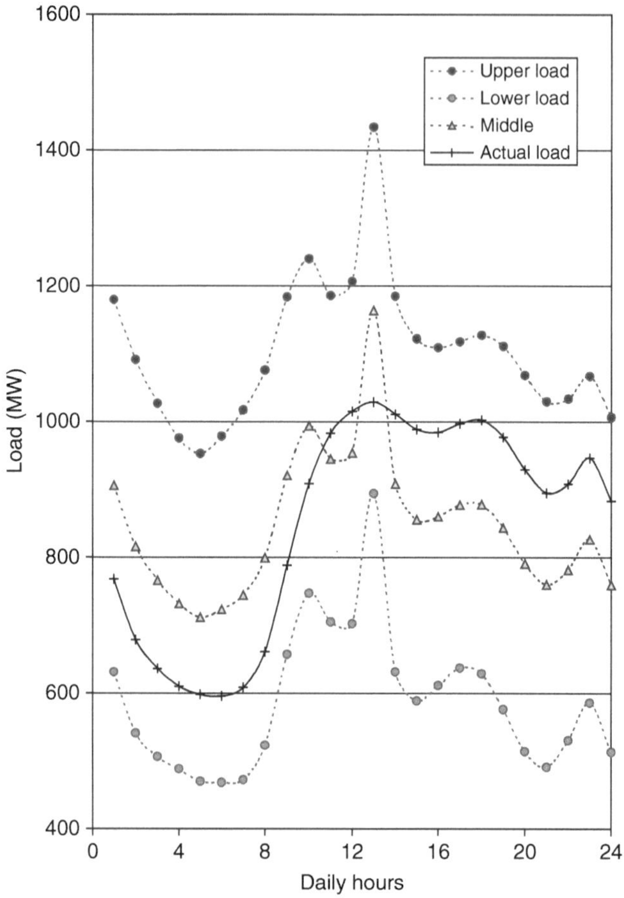

Figure 7.17 Predicted load for a summer day, crisp load, model B.

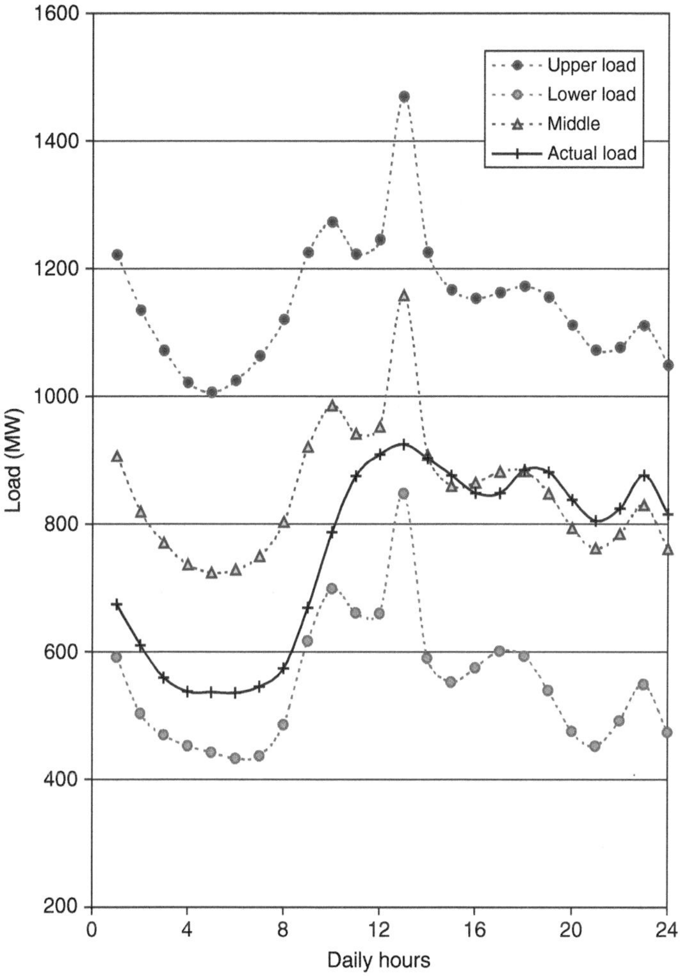

Figure 7.18 Predicted load for a summer day (5% load deviation), model B.

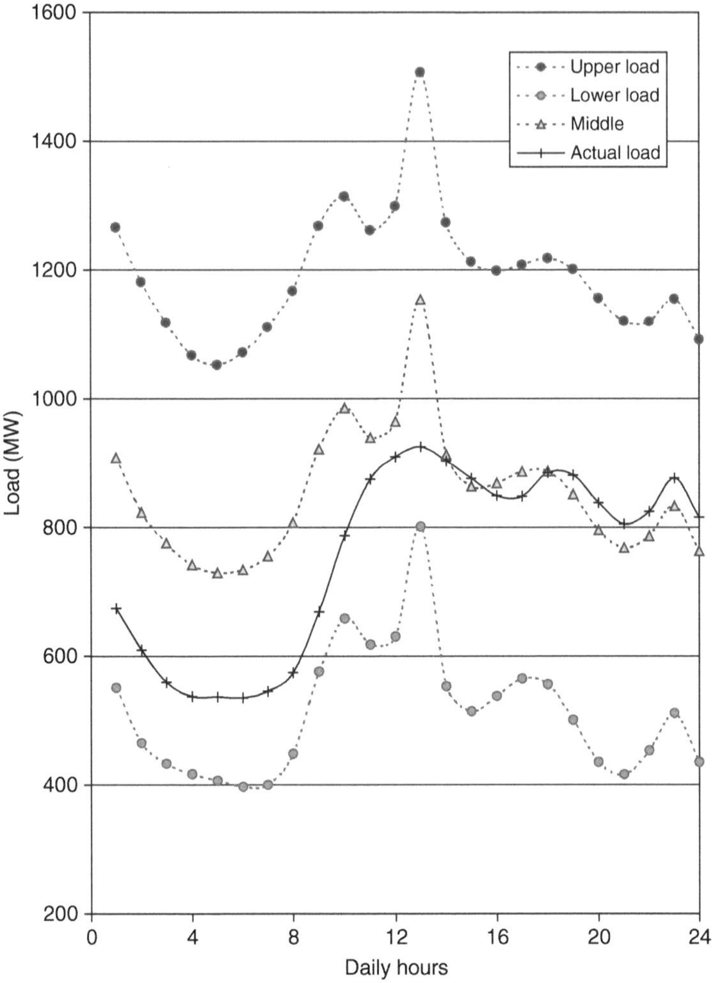

Figure 7.19 Predicted load for a summer day (10% load deviation), model B.

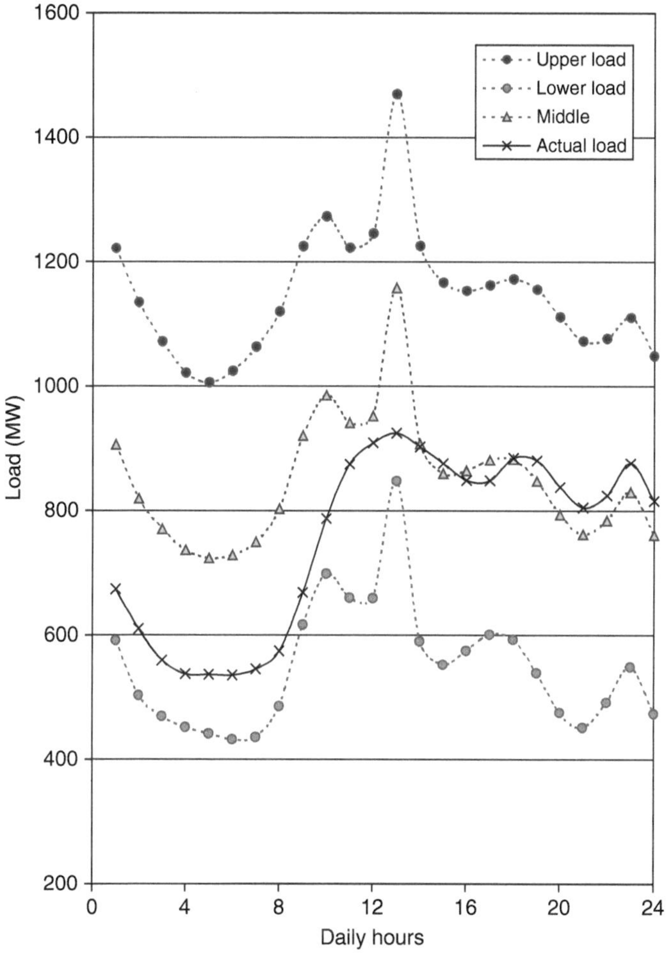

Figure 7.20 Predicted load for a summer day (20% load deviation), model B.

Table 7.4 Fuzzy Parameters for a Summer Day Load, Model C

Parameters	Crisp Load		5% Load Deviation		10% Load Deviation		20% Load Deviation	
	Middle	Spread	Middle	Spread	Middle	Spread	Middle	Spread
A_0	515.54	46.58	520.62	84.5043	520.385	124.141	529.00	205.698
A_1	0.0	0.00	0.00	0.00	0.00	0.00	0.00	0.00
A_2	0.0	0.00	0.408	0.00	2.5126	0.00	3.0646	0.00
A_3	0.0	0.00	0.00	0.00	0.00	0.00	0.00	0.00
A_4	12.14	14.60	17.631	19.00	15.132	12.925	15.346	12.618
A_5	0.0	0.00	0.00	0.00	0.00	0.00	0.00	0.00
A_6	37.0	0.00	36.372	0.00	36.816	0.00	37.101	0.00
A_7	4.59	0.00	8.1702	0.00	8.097	0.00	5.5347	0.00
A_8	0.0	0.00	0.00	0.00	0.0182	0.00	3.8353	0.00
A_9	33.91	0.00	34.568	0.00	33.06	0.00	31.653	0.00
B_1	31.99	0.00	24.101	0.00	24.0741	0.00	19.783	0.00
B_2	0.47	0.00	0.614	0.00	0.447	0.00	0.451	0.00
B_3	0.0	0.00	0.00	0.00	0.00	0.00	0.00	0.00
B_4	13.85	0.00	10.476	0.00	9.667	0.00	7.02	0.00
B_5	0.0	0.00	0.00	0.00	0.00	0.00	0.00	0.00
B_6	31.41	0.00	27.631	0.00	25.8138	0.00	21.295	0.00
B_7	52.04	0.00	51.294	0.00	48.9959	0.00	45.185	0.00
B_8	16.86	12.87	16.531	17.918	17.1895	21.881	18.279	21.5500
B_9	20.42	0.00	16.403	0.00	20.160	0.00	23.135	0.00
C_0	16.12	0.00	14.576	0.00	16.620	0.00	15.600	0.00
C_1	0.0	0.00	0.00	0.00	0.000	0.00	0.00	0.00
C_2	1.13	0.00	1.578	0.00	0.2810	0.00	1.955	0.00
C_3	13.39	0.00	12.710	0.00	13.565	0.00	12.066	0.00

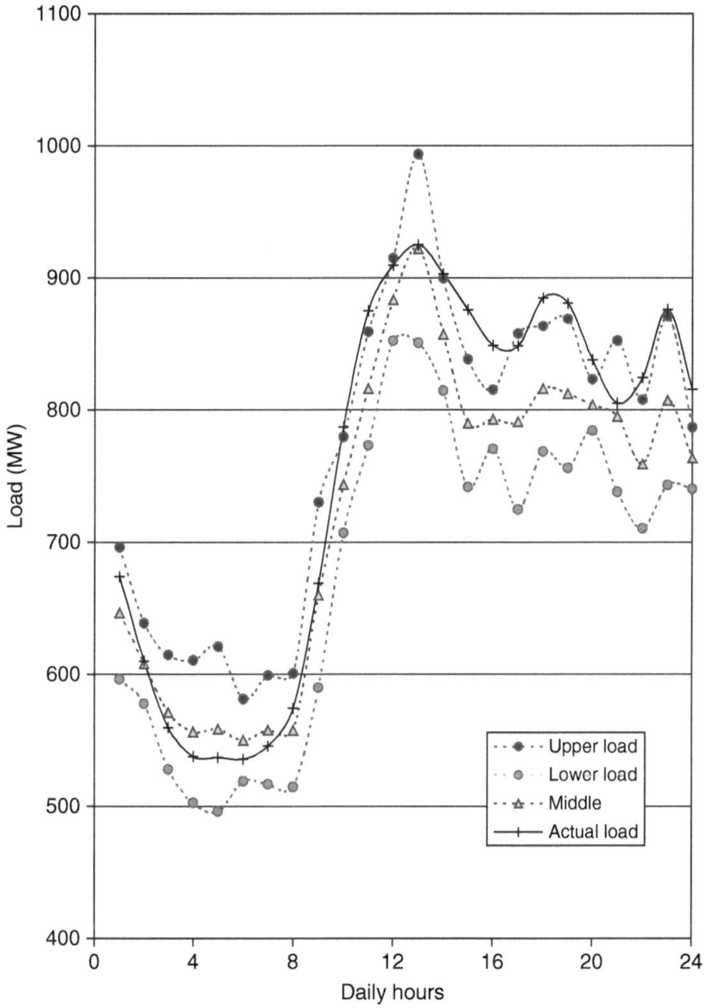

Figure 7.21 Estimated load for a summer day, crisp load, model C.

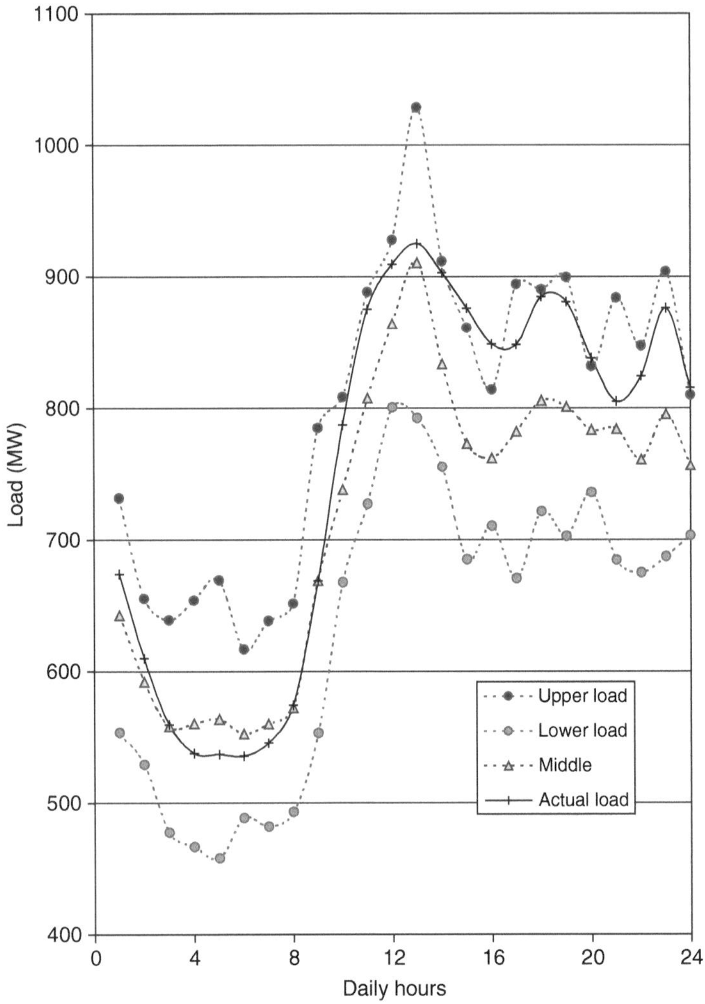

Figure 7.22 Estimated load for a summer day (5% load deviation), model C.

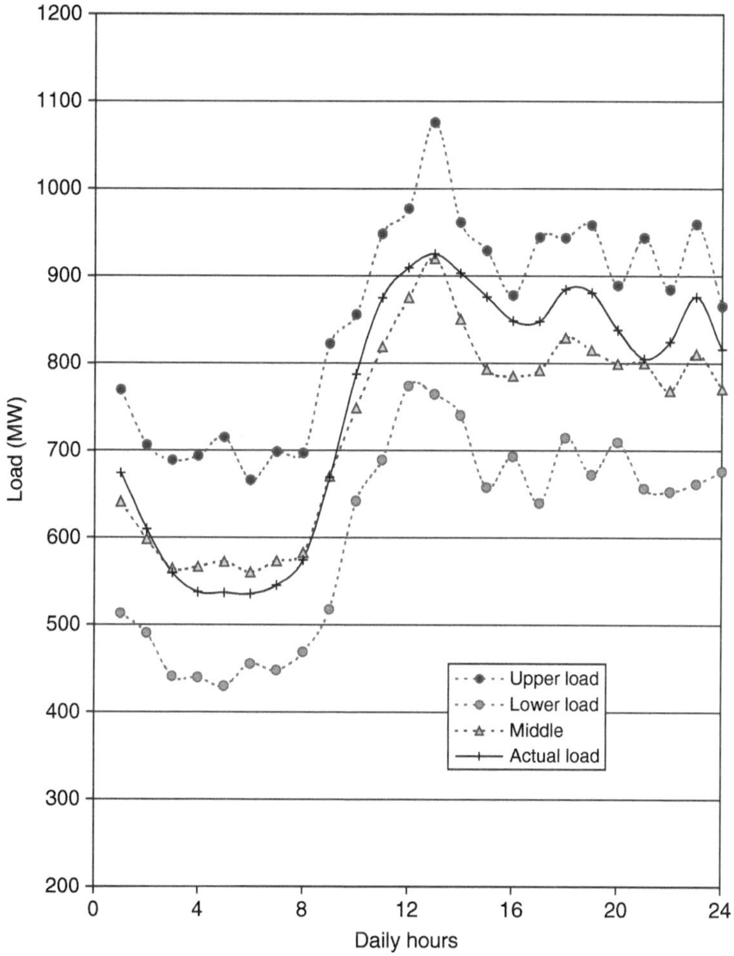

Figure 7.23 Estimated load for a summer day (10% load deviation), model C.

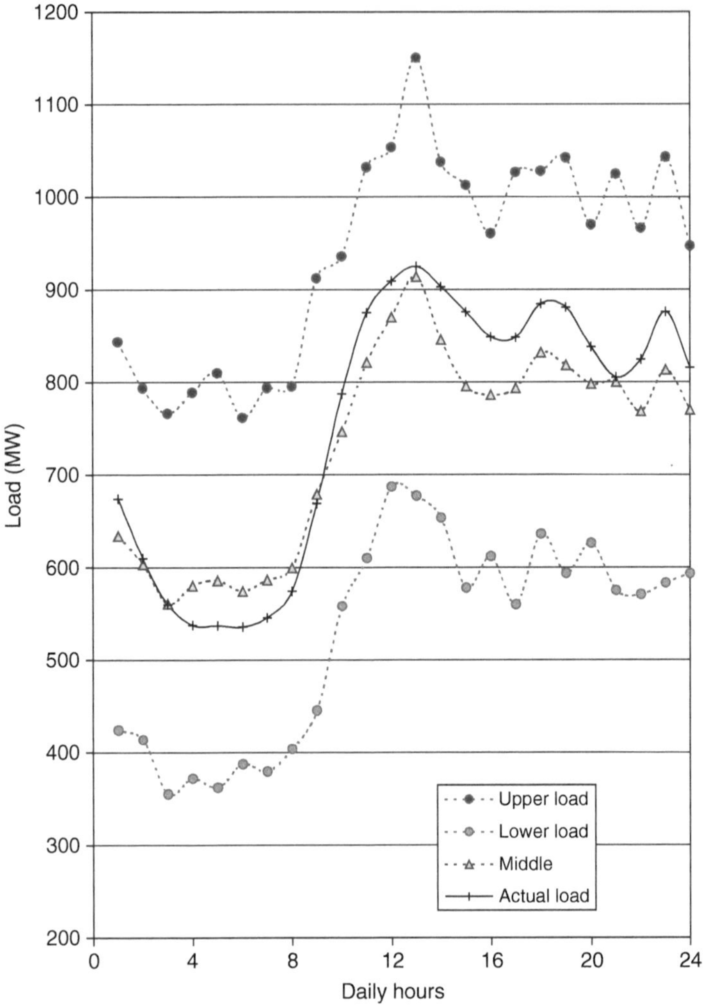

Figure 7.24 Estimated load for a summer day (20% load deviation), model C.

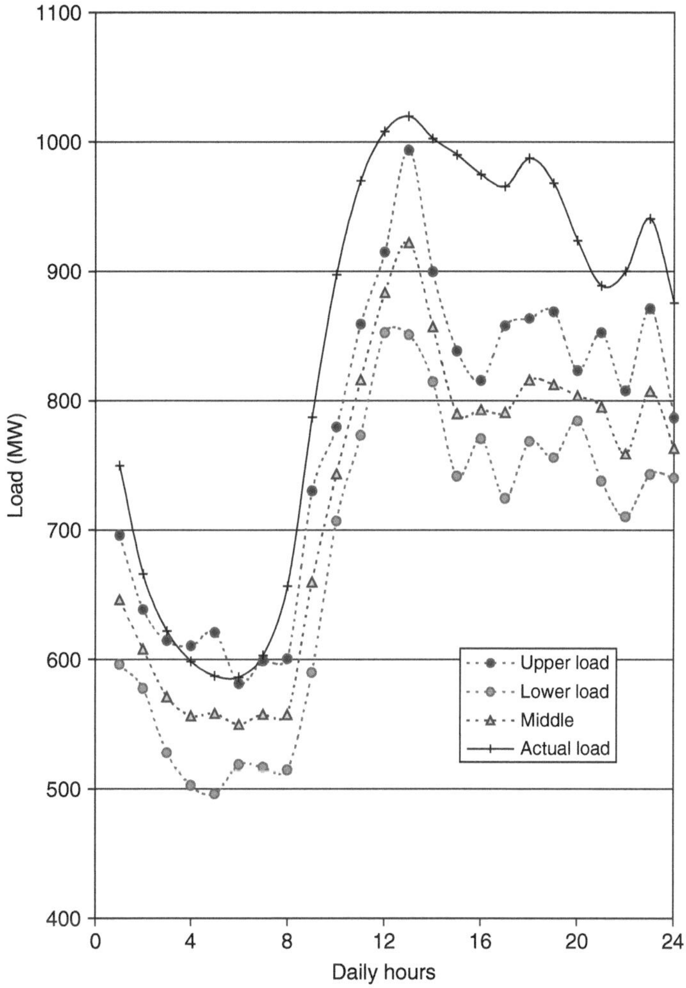

Figure 7.25 Predicted load for a summer day, crisp load, model C.

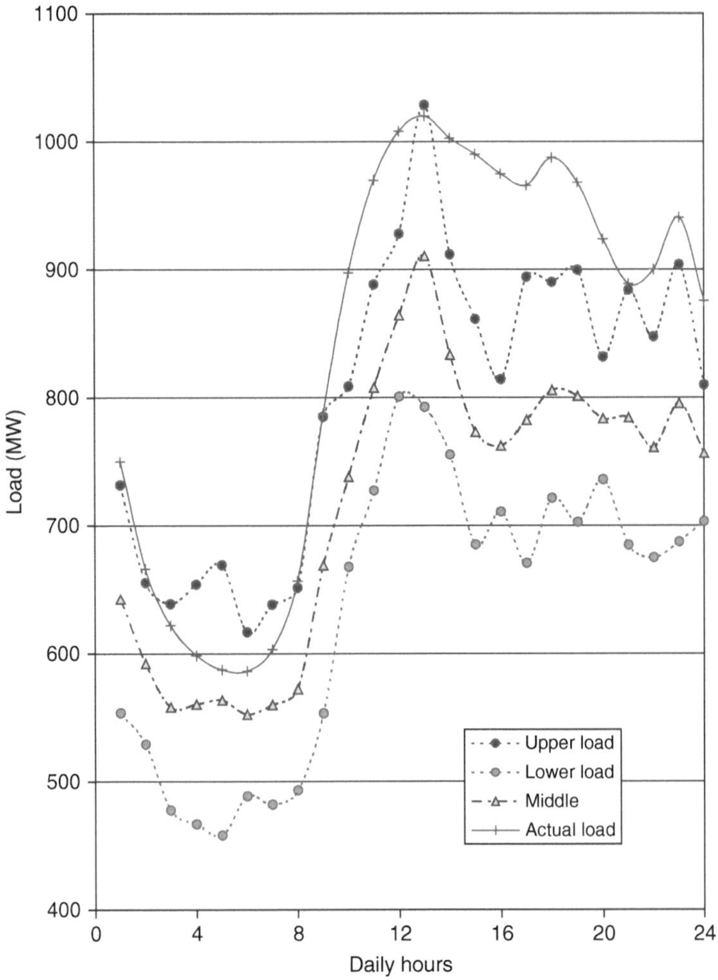

Figure 7.26 Predicted load for a summer day (5% load deviation), model C.

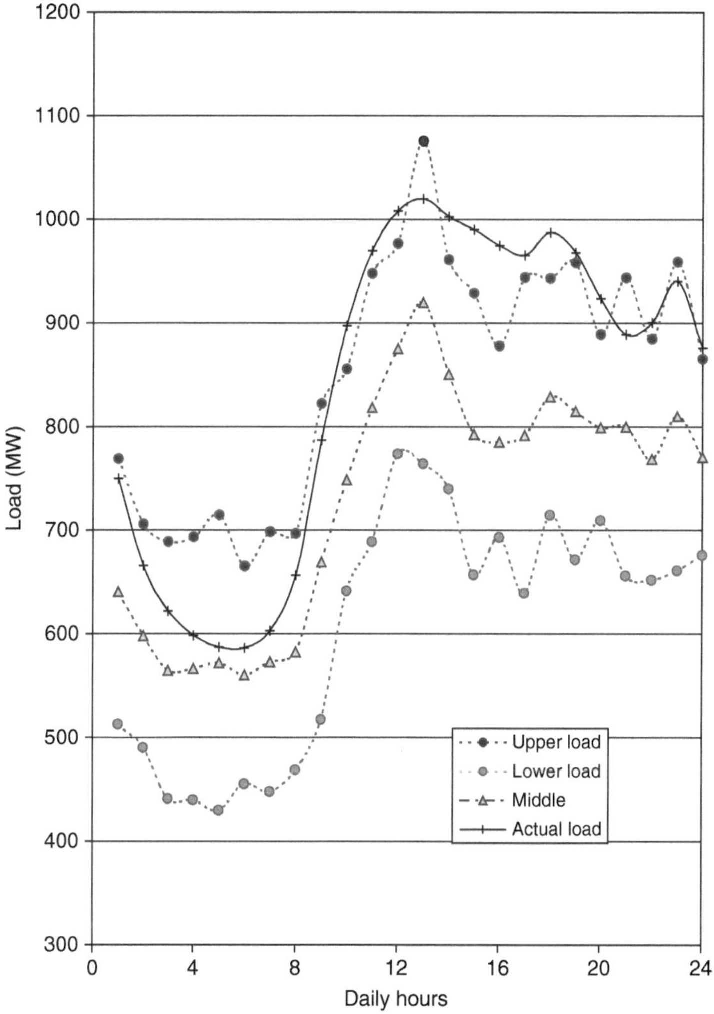

Figure 7.27 Predicted load for a summer day (10% load deviation), model C.

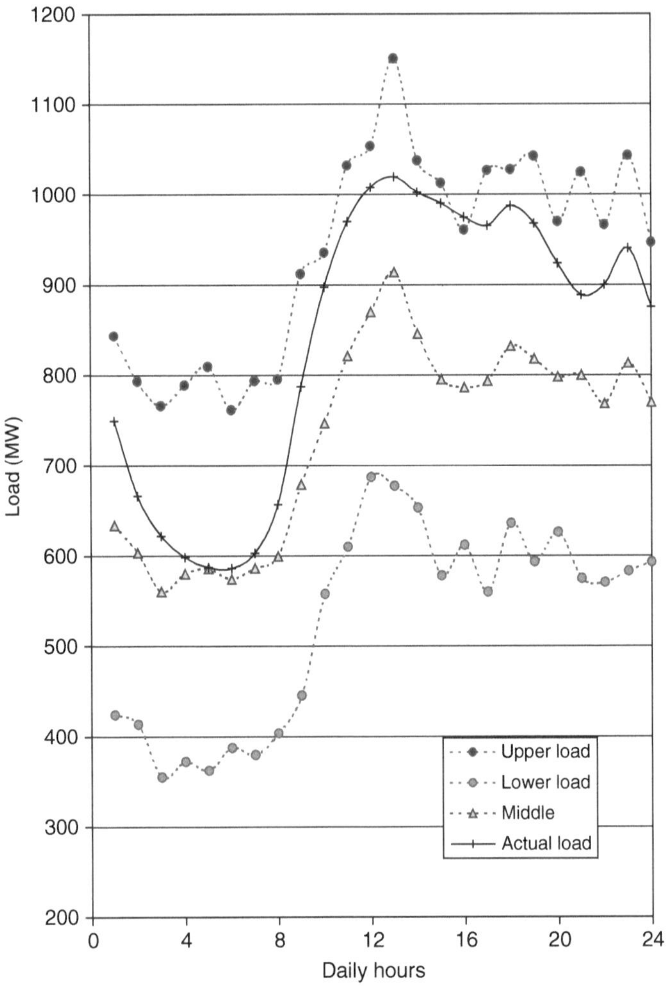

Figure 7.28 Predicted load for a summer day (20% load deviation), model C.

7.5 Conclusion

In this chapter, the fuzzy short-term load-forecasting problem is solved. The three models developed in Chapter 4 are implemented to predict the load. The three models are used to estimate the load power at any day in any season, based on fuzzy optimization rules. The predicted load lies between the upper and lower limits. This chapter showed that the actual load never violates these limits.

Appendix 7.1 Winter Load Forecasting: Fuzzy Case Model A

Table A7.1 Estimated Load for a Winter Weekday (20% Load Deviation), Model A

Daily Hours	Actual Load (MW)	Upper Load (MW)	Middle Load (MW)	Lower Load (MW)
1	735.9	1315.764	786.363	256.962
2	650.6	1336.382	806.9808	277.5798
3	613.1	1328.587	799.1863	269.7853
4	599.6	1333.742	804.3407	274.9397
5	604.8	1338.393	808.9922	279.5912
6	617.1	1349.205	819.804	290.403
7	635.1	1360.268	830.8672	301.4662
8	731.5	1354.234	824.8327	295.4317
9	915.8	1358.382	828.9814	299.5804
10	1001.8	1376.737	847.3363	317.9353
11	1013	1379	849.5992	320.1982
12	1014.6	1380.76	851.3592	321.9582
13	1020.9	1397.229	867.8283	338.4273
14	995.1	1406.91	877.5085	348.1075
15	979.7	1436.202	906.8009	377.3999
16	965.5	1454.934	925.5329	396.1319
17	975.1	1452.545	923.1443	393.7433
18	1029.7	1443.242	913.8412	384.4402
19	1024.8	1453.048	923.6472	394.2462
20	968.3	1438.842	909.441	380.04
21	955.2	1452.671	923.2699	393.8689
22	960	1491.015	961.614	432.213
23	950.7	1511.884	982.4832	453.0822
24	858.3	1566.572	1037.171	507.7695

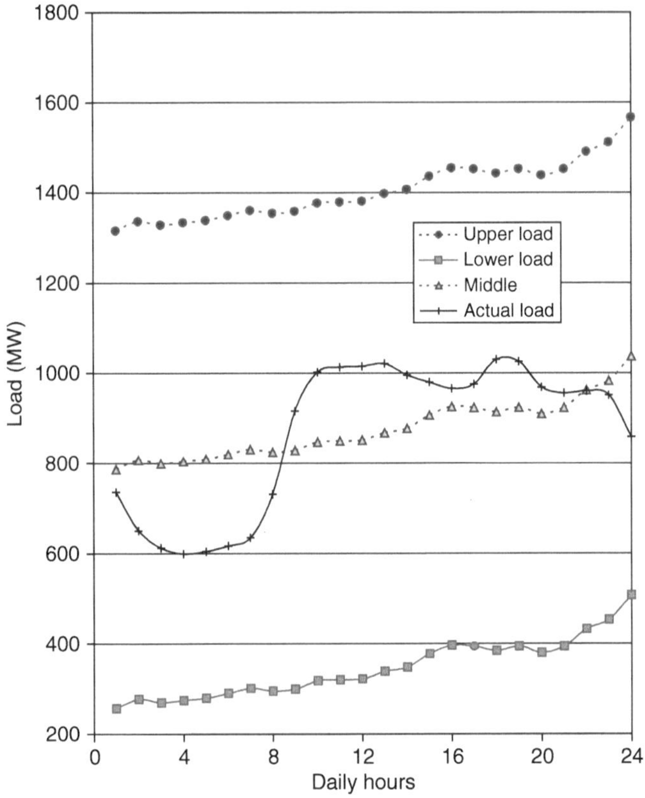

Figure A7.1 Estimated load for a winter weekday (20% load deviation), model A.

Table A7.2 Predicted Load for a Winter Weekday (20% Load Deviation), Model A

Daily Hours	Actual Load (MW)	Upper Load (MW)	Middle Load (MW)	Lower Load (MW)
1	748.8	1433.813	904.4123	375.0113
2	55.9	1434.945	905.5437	376.1427
3	621.5	1438.716	909.3153	379.9143
4	606.2	1434.693	905.2924	375.8914
5	604.1	1433.31	903.9094	374.5084
6	606.6	1430.419	901.0179	371.6169
7	625	1437.208	907.8066	378.4056
8	723.9	1432.305	902.9037	373.5027
9	913.8	1441.105	911.7039	382.3029
10	1004.4	1464.489	935.0875	405.6865
11	1026.9	1460.843	931.4417	402.0406

Continued

Table A7.2 Predicted Load for a Winter Weekday (20% Load Deviation), Model A *Continued*

Daily Hours	Actual Load (MW)	Upper Load (MW)	Middle Load (MW)	Lower Load (MW)
12	1025.6	1476.306	946.905	417.504
13	1021.9	1495.289	965.8884	436.4874
14	992.4	1485.232	955.8309	426.4299
15	972.1	1491.266	961.8654	432.4644
16	946.2	1501.701	972.3	442.899
17	949.4	1508.49	979.0888	449.6878
18	986.3	1510.753	981.3517	451.9507
19	966.8	1536.273	1006.872	477.4715
20	913.5	1557.394	1027.993	498.5922
21	877.1	1554.628	1025.227	495.8264
22	889.4	1547.085	1017.684	488.2833
23	935.5	1543.942	1014.541	485.1404
24	875.6	1523.953	994.5522	465.1512

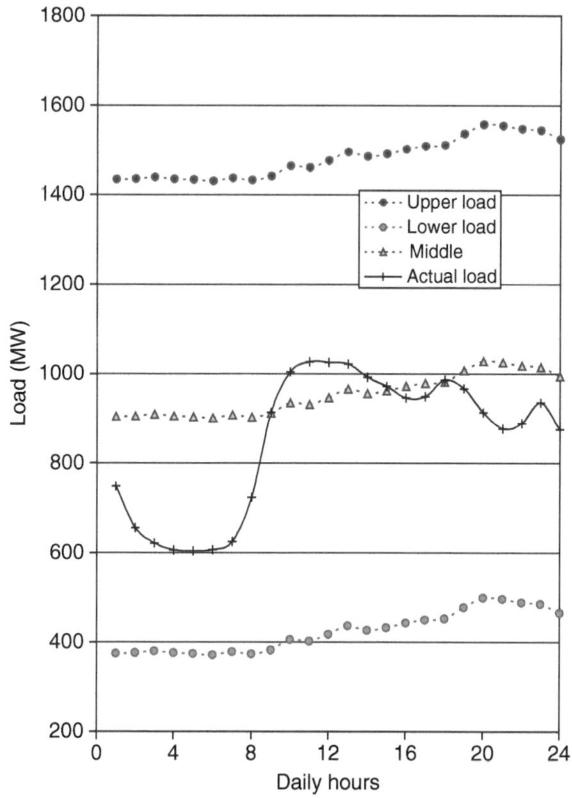

Figure A7.2 Predicted load for a winter weekday (20% load deviation), model A.

Table A7.3 Estimated Load for a Winter Weekday (5% Load Deviation), Model A

Daily Hours	Actual Load (MW)	Upper Load (MW)	Middle Load (MW)	Lower Load (MW)
1	735.9	1171.989	767.3909	362.7927
2	650.6	1194.385	789.7866	385.1884
3	613.1	1185.848	781.2501	376.6518
4	599.6	1191.447	786.849	382.2508
5	604.8	1196.578	791.9793	387.3811
6	617.1	1208.244	803.6458	399.0475
7	635.1	1220.145	815.5464	410.9482
8	731.5	1213.567	808.9683	404.3701
9	915.8	1218.228	813.6301	409.0318
10	1001.8	1238.197	833.5988	429.0006
11	1013	1240.624	836.0258	431.4276
12	1014.6	1242.264	837.6659	433.0677
13	1020.9	1259.998	855.3998	450.8016
14	995.1	1270.451	865.8528	461.2545
15	979.7	1302.363	897.7643	493.166
16	965.5	1322.842	918.2436	513.6454
17	975.1	1320.457	915.8586	511.2603
18	1029.7	1310.515	905.9163	501.3181
19	1024.8	1320.925	916.3272	511.729
20	968.3	1305.3	900.7019	496.1037
21	955.2	1319.778	915.1799	510.5816
22	960	1361.398	956.7994	552.2012
23	950.7	1384.028	979.4294	574.8312
24	858.3	1443.742	1039.143	634.545

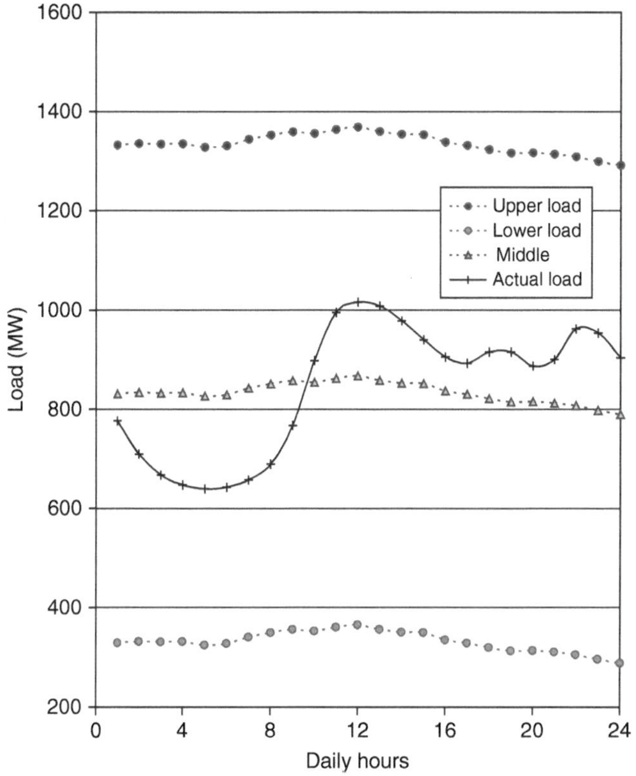

Figure A7.3 Estimated load for a winter weekday (5% load deviation), model A.

Table A7.4 Predicted Load for a Winter Weekday (5% Load Deviation), Model A

Daily Hours	Actual Load (MW)	Upper Load (MW)	Middle Load (MW)	Lower Load (MW)
1	748.8	1299.978	895.3793	490.781
2	655.9	1301.191	896.5928	491.9945
3	621.5	1305.342	900.7439	496.1457
4	606.2	1300.957	896.3585	491.7602
5	604.1	1299.509	894.9107	490.3125
6	606.6	1296.337	891.7388	487.1405
7	625	1303.618	899.0198	494.4215
8	723.9	1298.253	893.6552	489.0569
9	913.8	1307.727	903.1289	498.5307
10	1004.4	1333.018	928.4202	523.822
11	1026.9	1329.144	924.5454	519.9471
12	1025.6	1345.94	941.3422	536.7439

Continued

Table A7.4 Predicted Load for a Winter Weekday (5% Load Deviation), Model A *Continued*

Daily Hours	Actual Load (MW)	Upper Load (MW)	Middle Load (MW)	Lower Load (MW)
13	1021.9	1366.654	962.0558	557.4575
14	992.4	1355.69	951.0922	546.494
15	972.1	1362.269	957.6703	553.0721
16	946.2	1373.424	968.8262	564.228
17	949.4	1380.705	976.1072	571.509
18	986.3	1383.132	978.5342	573.936
19	966.8	1410.893	1006.295	601.6963
20	913.5	1433.757	1029.159	624.5605
21	877.1	1430.862	1026.263	621.665
22	889.4	1422.878	1018.279	613.6812
23	935.5	1419.79	1015.192	610.5933
24	875.6	1398.139	993.5407	588.9425

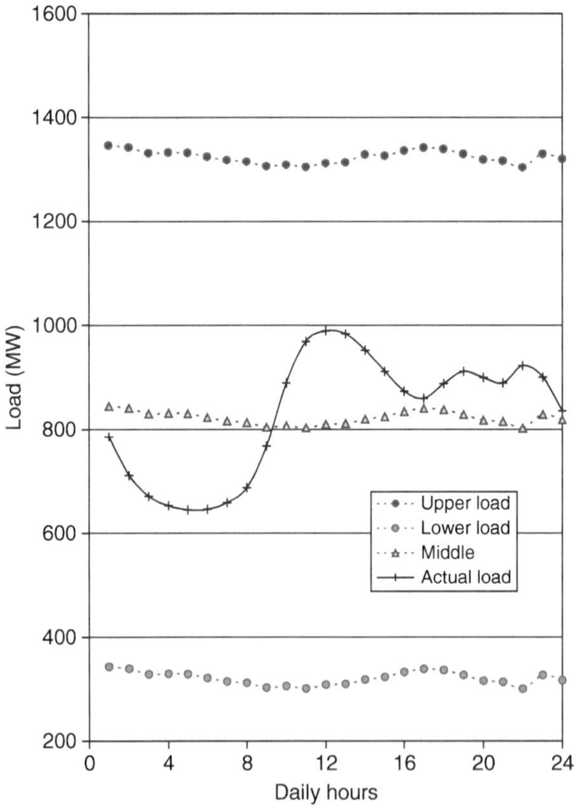

Figure A7.4 Predicted load for a winter weekday (5% load deviation), model A.

Table A7.5 Estimated Load for a Winter Weekend Day (20% Load Deviation), Model A

Daily Hours	Actual Load (MW)	Upper Load (MW)	Middle Load (MW)	Lower Load (MW)
1	776.8	1332.376	830.7912	329.2061
2	710	1335.332	833.7467	332.1616
3	667.1	1334.108	832.5232	330.9381
4	647.2	1334.858	833.2725	331.6873
5	639.3	1327.748	826.1624	324.5773
6	642.8	1330.898	829.3125	327.7274
7	657.2	1343.858	842.2724	340.6873
8	689.3	1352.656	851.0706	349.4855
9	767.5	1359.129	857.5435	355.9583
10	898	1355.821	854.2363	352.6512
11	995.1	1363.775	862.1894	360.6043
12	1016.2	1368.613	867.0279	365.4428
13	1008.1	1359.703	858.1177	356.5326
14	977.9	1354.002	852.4168	350.8317
15	940.1	1353.171	851.5858	350.0007
16	905.1	1338.23	836.6452	335.0601
17	892.8	1331.781	830.1959	328.6108
18	915.4	1323.021	821.4362	319.851
19	915.1	1316.243	814.6578	313.0727
20	887	1316.993	815.408	313.8228
21	900.2	1314.049	812.4639	310.8788
22	961.4	1308.722	807.1371	305.552
23	953.1	1299.203	797.6177	296.0326
24	903.7	1291.464	789.8784	288.2933

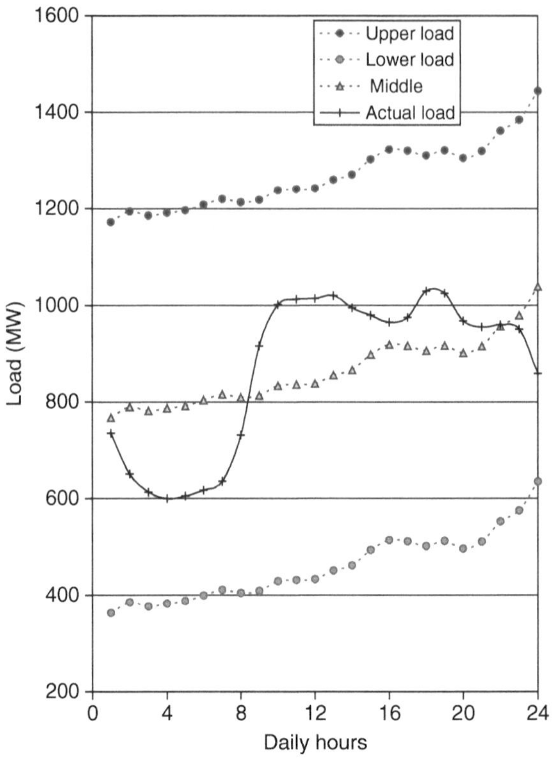

Figure A7.5 Estimated load for a winter weekend day (20% load deviation), model A.

Table A7.6 Predicted Load for a Winter Weekend Day (20% Load Deviation), Model A

Daily Hours	Actual Load (MW)	Upper Load (MW)	Middle Load (MW)	Lower Load (MW)
1	786	1346.21	844.6251	343.0399
2	711.3	1342.504	840.9189	339.3338
3	670.9	1331.717	830.1323	328.5472
4	653	1332.811	831.2263	329.6411
5	645.1	1332.12	830.5352	328.95
6	646	1324.543	822.9579	321.3728
7	659	1317.92	816.3345	314.7494
8	687.6	1315.135	813.5496	311.9645
9	767.7	1306.185	804.6002	303.0151
10	889.4	1308.836	807.251	305.6659
11	968.7	1304.631	803.0461	301.461
12	989.2	1311.826	810.2408	308.6557
13	983.5	1313.216	811.6306	310.0454

Continued

Table A7.6 Predicted Load for a Winter Weekend Day (20% Load Deviation), Model A *Continued*

Daily Hours	Actual Load (MW)	Upper Load (MW)	Middle Load (MW)	Lower Load (MW)
14	952.3	1321.447	819.8619	318.2768
15	911.5	1326.293	824.7076	323.1225
16	873.4	1336.028	834.4432	332.8581
17	859.6	1341.86	840.2751	338.69
18	888	1339.287	837.7015	336.1164
19	911.5	1329.935	828.35	326.7649
20	899.8	1319.202	817.6172	316.0321
21	889.4	1316.834	815.2491	313.664
22	922.6	1303.98	802.3948	300.8097
23	900.4	1329.948	828.3627	326.7776
24	835.8	1320.157	818.5723	316.9872

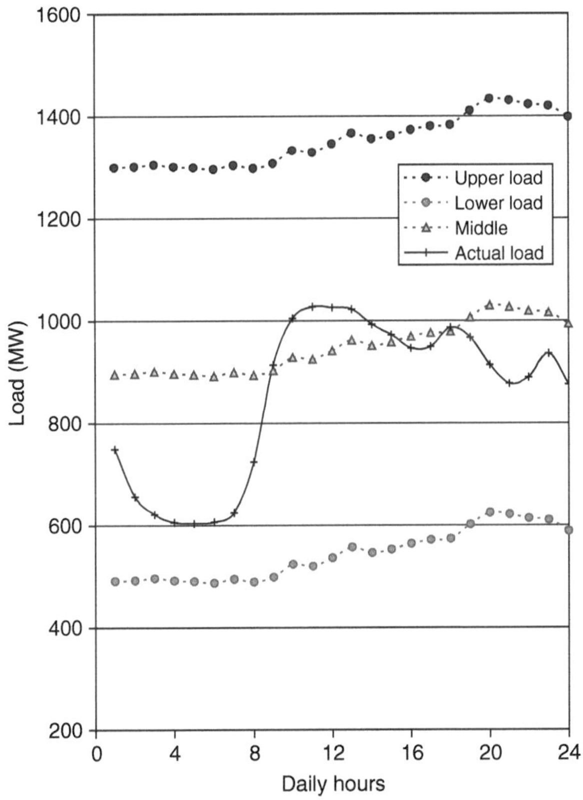

Figure A7.6 Predicted load for a winter weekend day (5% load deviation), model A.

Table A7.7 Estimated Load for a Winter Weekend Day (5% Load Deviation), Model A

Daily Hours	Actual Load (MW)	Upper Load (MW)	Middle Load (MW)	Lower Load (MW)
1	776.8	1197.809	818.4391	439.0692
2	710	1201.219	821.8492	442.4793
3	667.1	1200.542	821.1721	441.8021
4	647.2	1200.075	820.7051	441.3351
5	639.3	1192.373	813.0028	433.6329
6	642.8	1195.785	816.4152	437.0453
7	657.2	1209.825	830.4552	451.0853
8	689.3	1217.553	838.1832	458.8132
9	767.5	1223.011	843.6414	464.2714
10	898	1219.795	840.4249	461.0549
11	995.1	1228.388	849.0179	469.648
12	1016.2	1231.29	851.9202	472.5503
13	1008.1	1222.987	843.6172	464.2473
14	977.9	1219.795	840.4249	461.055
15	940.1	1218.644	839.2739	459.9039
16	905.1	1203.098	823.7279	444.358
17	892.8	1197.971	818.6008	439.2308
18	915.4	1187.932	808.5621	429.1921
19	915.1	1182.77	803.4	424.0301
20	887	1184.363	804.9928	425.6229
21	900.2	1180.994	801.6241	422.2542
22	961.4	1175.262	795.8923	416.5223
23	953.1	1166.613	787.2429	407.8729
24	903.7	1157.237	777.8669	398.4969

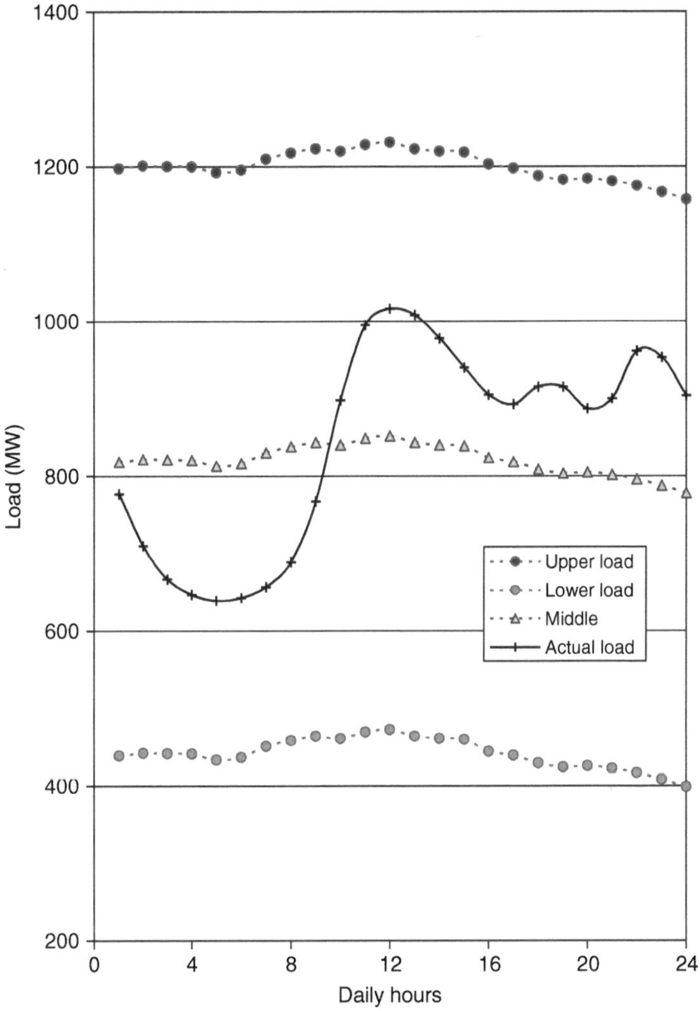

Figure A7.7 Estimated load for a winter weekend day (5% load deviation), model A.

Table A7.8 Predicted Load for a Winter Weekend Day (5% Load Deviation), Model A

Daily Hours	Actual Load (MW)	Upper Load (MW)	Middle Load (MW)	Lower Load (MW)
1	786	1210.677	831.3076	451.9377
2	711.3	1205.9	826.5298	447.1599
3	670.9	1196.521	817.1506	437.7807
4	653	1198.223	818.8534	439.4835
5	645.1	1197.44	818.0698	438.6999
6	646	1189.704	810.3345	430.9646
7	659	1184.206	804.8357	425.4658
8	687.6	1180.378	801.0081	421.6381
9	767.7	1171.985	792.6147	413.2448
10	889.4	1173.958	794.5875	415.2176
11	968.7	1169.659	790.2896	410.9196
12	989.2	1177.446	798.0762	418.7062
13	983.5	1179.021	799.6508	420.2808
14	952.3	1185.808	806.4385	427.0686
15	911.5	1189.627	810.2576	430.8876
16	873.4	1200.271	820.9011	441.5312
17	859.6	1207.217	827.8467	448.4767
18	888	1204.591	825.2209	445.851
19	911.5	1194.832	815.4623	436.0924
20	899.8	1184.677	805.3068	425.9369
21	889.4	1180.792	801.4216	422.0517
22	922.6	1171.357	791.9873	412.6174
23	900.4	1198.294	818.9243	439.5544
24	835.8	1184.288	804.9178	425.5479

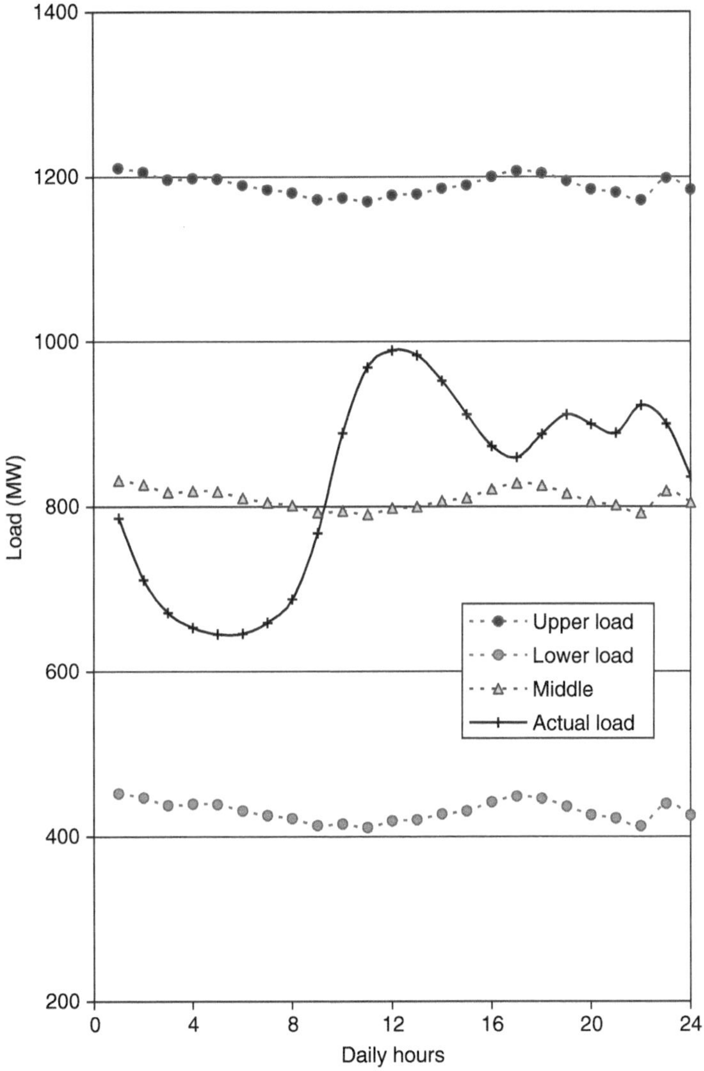

Figure A7.8 Predicted load for a winter weekend day (5% load deviation), model A.

Appendix 7.2 Winter Load Forecasting: Fuzzy Case Model B

Table A7.9 Predicted Load for a Winter Weekday (20% Load Deviation), Model B

Daily Hours	Actual Load (MW)	Upper Load (MW)	Middle Load (MW)	Lower Load (MW)
1	776.8	1354.124	912.2347	470.3455
2	710	1274.516	831.8185	389.1215
3	667.1	1212.3	785.7601	359.2201
4	647.2	1161.758	753.1927	344.6273
5	639.3	1147.11	741.1701	335.2302
6	642.8	1167.671	746.5836	325.4966
7	657.2	1207.364	767.4948	327.6252
8	689.3	1261.066	816.9554	372.8446
9	767.5	1354.7	924.7267	494.7534
10	898	1392.019	981.03	570.041
11	995.1	1348.908	943.9782	539.0481
12	1016.2	1395.226	977.572	559.9182
13	1008.1	1581.876	1144.632	707.3882
14	977.9	1368.704	923.9869	479.2703
15	940.1	1304.163	870.7567	437.3499
16	905.1	1289.748	875.9318	462.1153
17	892.8	1299.183	894.657	490.1308
18	915.4	1309.851	895.4289	481.0066
19	915.1	1292.027	857.8129	423.5983
20	887	1244.799	799.8808	354.9622
21	900.2	1218.812	782.3753	345.9391
22	961.4	1206.893	790.0472	373.2014
23	953.1	1243.734	839.006	434.2779
24	903.7	1183.662	772.0673	360.4725

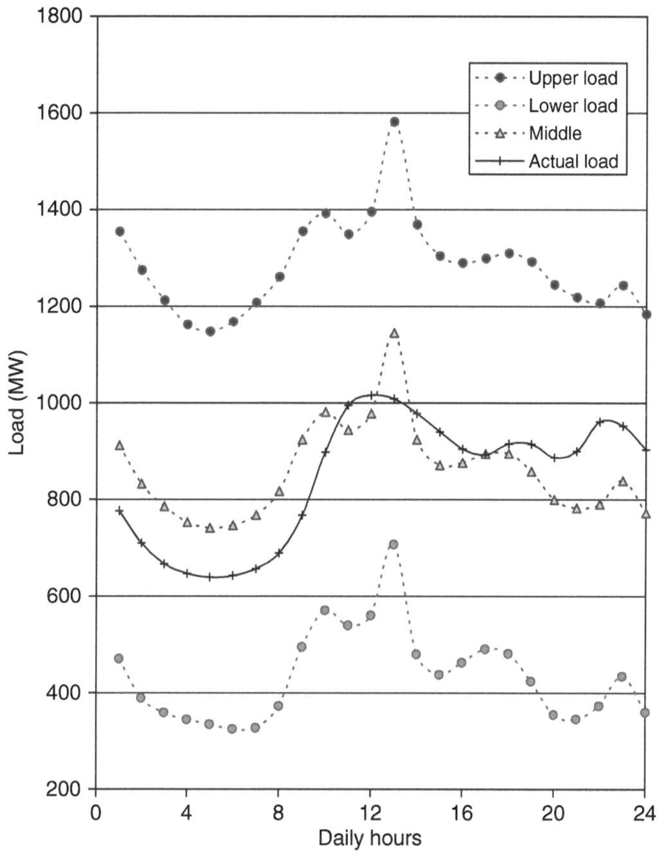

Figure A7.9 Predicted load for a winter weekday (20% load deviation) using the parameters of a summer day, model B.

Appendix 7.3 Winter Load Forecasting: Fuzzy Case Model C

Table A7.10 Estimated Load for a Winter Day (0% Load Deviation), Model C

Daily Hours	Actual Load (MW)	Upper Load (MW)	Middle Load (MW)	Lower Load (MW)
1	1117.6	1117.352	1049.314	981.2758
2	1006.4	1002.796	959.6642	916.5328
3	943.6	947.7949	893.3424	838.8899
4	871.17	930.072	847.0901	764.1083
5	813	886.8589	806.3678	725.8766
6	869.7	935.3588	884.3027	833.2465
7	914.8	944.6049	899.8886	855.1723

Continued

Table A7.10 Estimated Load for a Winter Day (0% Load Deviation), Model C *Continued*

Daily Hours	Actual Load (MW)	Upper Load (MW)	Middle Load (MW)	Lower Load (MW)
8	978.7	1053.5	981.386	909.2726
9	1157.3	1169.818	1082.987	996.1563
10	1223.8	1199.176	1134.987	1070.799
11	1216.8	1190.083	1147.857	1105.631
12	1284.3	1223.936	1165.634	1107.332
13	1258.6	1203.112	1118.093	1033.073
14	1207.8	1053.874	976.5532	899.232
15	1155.4	975.6566	927.5439	879.4313
16	1110.6	913.4152	866.661	819.9069
17	1094.1	954.7243	878.988	803.2518
18	1113.3	1023.045	937.1199	851.1946
19	1186.4	1046.106	985.9926	925.8795
20	1139	1003.872	961.873	919.8738
21	1152.3	1055.361	993.2103	931.0594
22	1226.5	1098.469	1012.091	925.7128
23	1198.4	1031.638	957.7131	883.7883
24	1111.8	1026.037	980.4147	934.7927

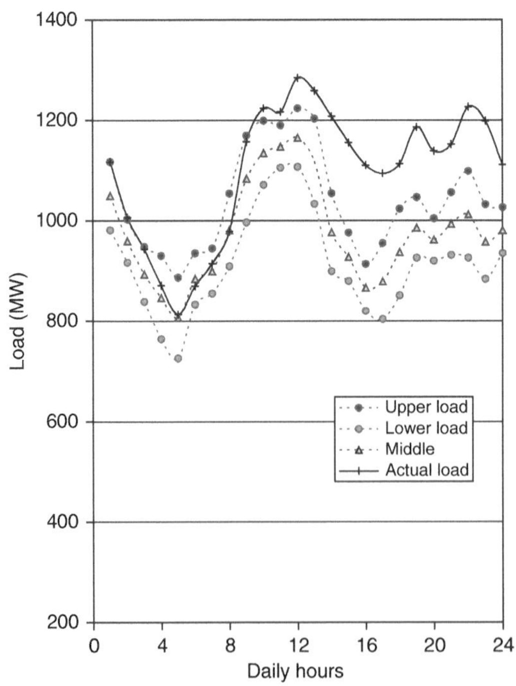

Figure A7.10 Estimated load for a winter day (0% load deviation), model C.

Table A7.11 Predicted Load for a Winter Day (0% Load Deviation), Model C

Daily Hours	Actual Load (MW)	Upper Load (MW)	Middle Load (MW)	Lower Load (MW)
1	883.7	1117.352	1049.314	981.2758
2	806.5	1002.796	959.6642	916.5328
3	779	947.7949	893.3424	838.8899
4	772.4	930.072	847.0901	764.1083
5	780.2	886.8589	806.3678	725.8766
6	795.6	935.3588	884.3027	833.2465
7	843.3	944.6049	899.8886	855.1723
8	966.8	1053.5	981.386	909.2726
9	1145.8	1169.818	1082.987	996.1563
10	1225.8	1199.176	1134.987	1070.799
11	1220.9	1190.083	1147.857	1105.631
12	1188.1	1223.936	1165.634	1107.332
13	1174.1	1203.112	1118.093	1033.073
14	1130.2	1053.874	976.5532	899.232
15	1108.7	975.6566	927.5439	879.4313
16	1082.2	913.4152	866.661	819.9069
17	1105	954.7243	878.988	803.2518
18	1148.7	1023.045	937.1199	851.1946
19	1146.9	1046.106	985.9926	925.8795
20	1120.2	1003.872	961.873	919.8738
21	1128.4	1055.361	993.2103	931.0594
22	1164.8	1098.469	1012.091	925.7128
23	1126.5	1031.638	957.7131	883.7883
24	1026.5	1026.037	980.4147	934.7927

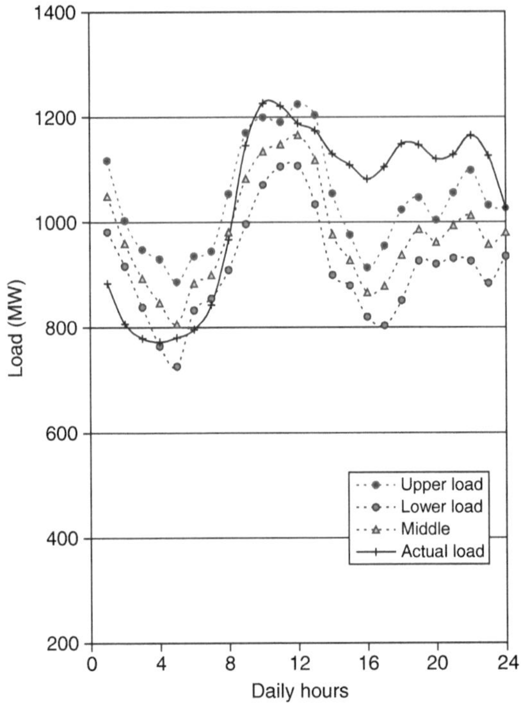

Figure A7.11 Predicted load for a winter day (0% load deviation), model C.

Table A7.12 Estimated Load for a Winter Day (5% Load Deviation), Model C

Daily Hours	Actual Load (MW)	Upper Load (MW)	Middle Load (MW)	Lower Load (MW)
1	1117.6	1173.625	1049.911	926.197
2	1006.4	1055.183	961.504	867.825
3	943.6	1007.615	898.3017	788.9886
4	871.17	998.3513	855.9619	713.5724
5	813	952.1627	815.037	677.9114
6	869.7	993.3805	889.2535	785.1266
7	914.8	1000.306	901.8365	803.3668
8	978.7	1111.747	983.4883	855.2292
9	1157.3	1228.142	1083.423	938.7039
10	1223.8	1254.239	1133.437	1012.635
11	1216.8	1239.701	1144.984	1050.267
12	1284.3	1274.589	1163.387	1052.185
13	1258.6	1262.633	1118.719	974.8044
14	1207.8	1116.402	979.9797	843.5577

Continued

Table A7.12 Estimated Load for a Winter Day (5% Load Deviation), Model C *Continued*

Daily Hours	Actual Load (MW)	Upper Load (MW)	Middle Load (MW)	Lower Load (MW)
15	1155.4	1030.421	930.3224	830.2241
16	1110.6	969.5775	871.4403	773.3031
17	1094.1	1019.414	885.0286	750.6428
18	1113.3	1087.438	941.9798	796.5219
19	1186.4	1101.045	987.7631	874.4816
20	1139	1056.232	962.2513	868.2704
21	1152.3	1114.093	995.4145	876.7358
22	1226.5	1160.676	1016.154	871.6326
23	1198.4	1092.559	962.5981	832.6372
24	1111.8	1082.281	982.8972	883.5136

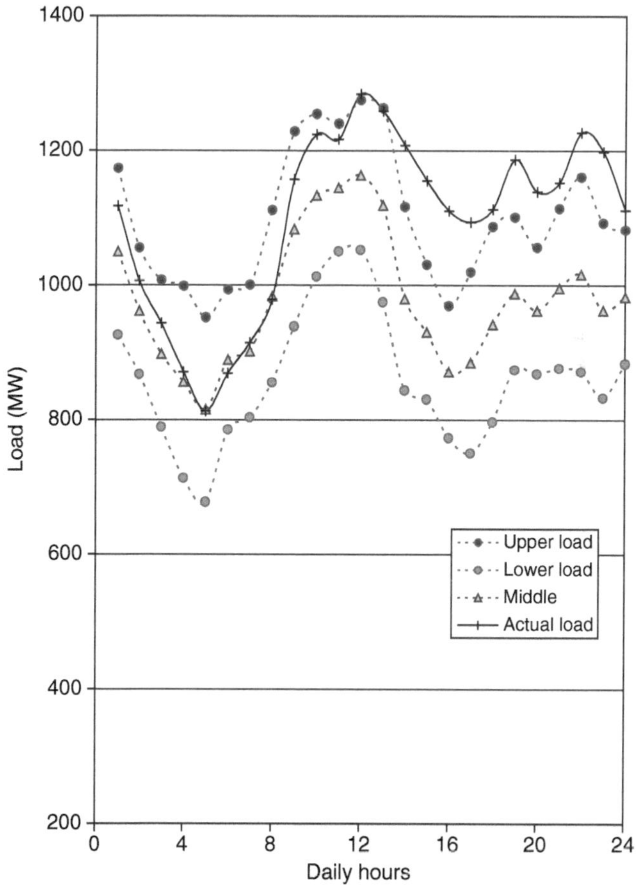

Figure A7.12 Estimated load for a winter day (5% load deviation), model C.

Table A7.13 Predicted Load for a Winter Day (5% Load Deviation), Model C

Daily Hours	Actual Load (MW)	Upper Load (MW)	Middle Load (MW)	Lower Load (MW)
1	883.7	1173.625	1049.911	926.197
2	806.5	1055.183	961.504	867.825
3	779	1007.615	898.3017	788.9886
4	772.4	998.3513	855.9619	713.5724
5	780.2	952.1627	815.037	677.9114
6	795.6	993.3805	889.2535	785.1266
7	843.3	1000.306	901.8365	803.3668
8	966.8	1111.747	983.4883	855.2292
9	1145.8	1228.142	1083.423	938.7039
10	1225.8	1254.239	1133.437	1012.635
11	1220.9	1239.701	1144.984	1050.267
12	1188.1	1274.589	1163.387	1052.185
13	1174.1	1262.633	1118.719	974.8044
14	1130.2	1116.402	979.9797	843.5577
15	1108.7	1030.421	930.3224	830.2241
16	1082.2	969.5775	871.4403	773.3031
17	1105	1019.414	885.0286	750.6428
18	1148.7	1087.438	941.9798	796.5219
19	1146.9	1101.045	987.7631	874.4816
20	1120.2	1056.232	962.2513	868.2704
21	1128.4	1114.093	995.4145	876.7358
22	1164.8	1160.676	1016.154	871.6326
23	1126.5	1092.559	962.5981	832.6372
24	1026.5	1082.281	982.8972	883.5136

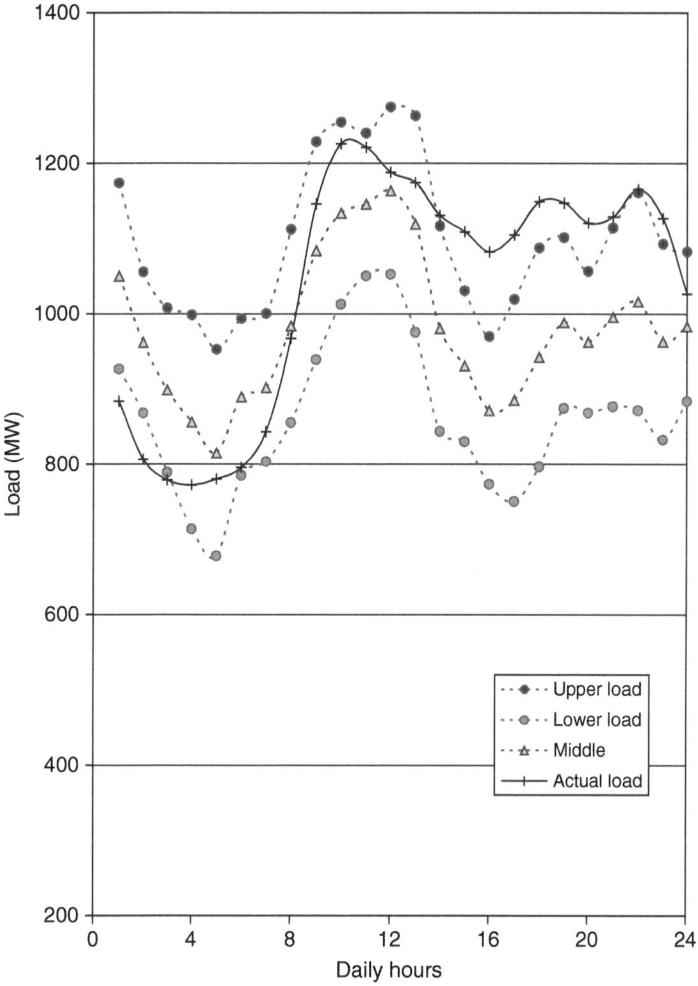

Figure A7.13 Predicted load for a winter day (5% load deviation), model C.

Table A7.14 Estimated Load for a Winter Day (10% Load Deviation), Model C

Daily Hours	Actual Load (MW)	Upper Load (MW)	Middle Load (MW)	Lower Load (MW)
1	1117.6	1229.683	1051.127	872.5706
2	1006.4	1111.278	967.0101	822.7422
3	943.6	1071.517	904.9275	738.3383
4	871.17	1066.356	865.0006	663.6457
5	813	1012.47	822.2262	631.9821
6	869.7	1048.724	890.5163	732.3086
7	914.8	1059.512	903.2762	747.0405
8	978.7	1167.856	984.9442	802.0327
9	1157.3	1284.741	1085.945	887.1494
10	1223.8	1315.252	1136.116	956.9797
11	1216.8	1294.429	1144.169	993.9083
12	1284.3	1326.631	1164.099	1001.566
13	1258.6	1320.511	1119.519	918.528
14	1207.8	1180.547	984.0671	787.587
15	1155.4	1087.571	935.0442	782.517
16	1110.6	1025.568	876.0086	726.4489
17	1094.1	1083.584	889.7103	695.8363
18	1113.3	1148.609	944.8365	741.0643
19	1186.4	1152.285	987.6144	822.9443
20	1139	1113.382	965.0074	816.6323
21	1152.3	1177.339	999.8724	822.4058
22	1226.5	1222.244	1022.885	823.5271
23	1198.4	1153.569	969.8036	786.0383
24	1111.8	1141.033	984.1565	827.2798

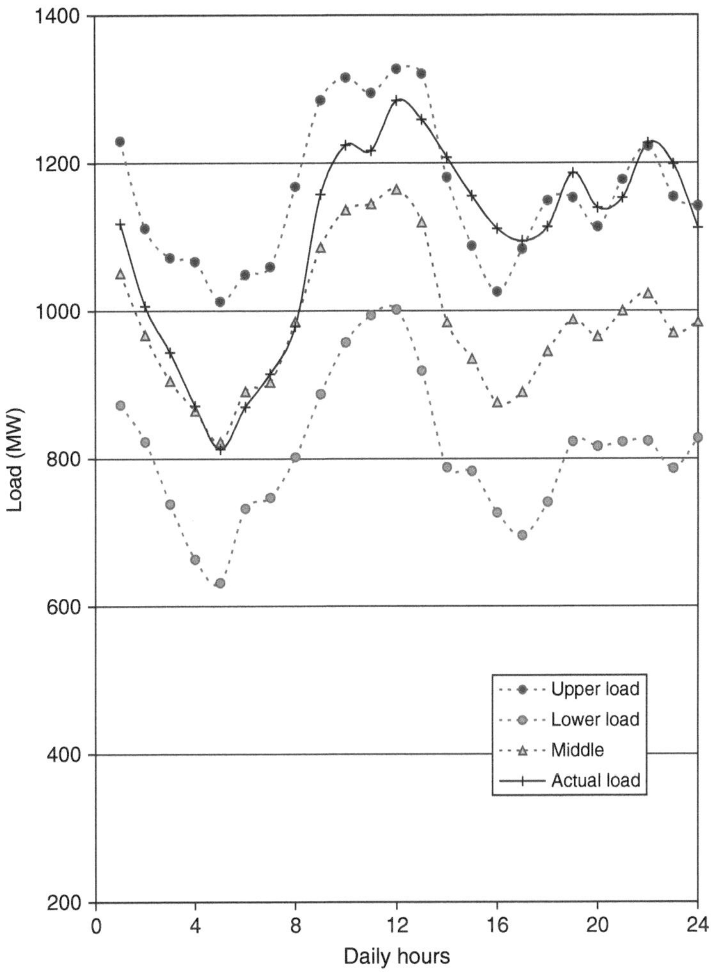

Figure A7.14 Estimated load for a winter day (10% load deviation), model C.

Table A7.15 Predicted Load for a Winter Day (10% Load Deviation), Model C

Daily Hours	Actual Load (MW)	Upper Load (MW)	Middle Load (MW)	Lower Load (MW)
1	883.7	1229.683	1051.127	872.5706
2	806.5	1111.278	967.0101	822.7422
3	779	1071.517	904.9275	738.3383
4	772.4	1066.356	865.0006	663.6457
5	780.2	1012.47	822.2262	631.9821
6	795.6	1048.724	890.5163	732.3086
7	843.3	1059.512	903.2762	747.0405
8	966.8	1167.856	984.9442	802.0327
9	1145.8	1284.741	1085.945	887.1494
10	1225.8	1315.252	1136.116	956.9797
11	1220.9	1294.429	1144.169	993.9083
12	1188.1	1326.631	1164.099	1001.566
13	1174.1	1320.511	1119.519	918.528
14	1130.2	1180.547	984.0671	787.587
15	1108.7	1087.571	935.0442	782.517
16	1082.2	1025.568	876.0086	726.4489
17	1105	1083.584	889.7103	695.8363
18	1148.7	1148.609	944.8365	741.0643
19	1146.9	1152.285	987.6144	822.9443
20	1120.2	1113.382	965.0074	816.6323
21	1128.4	1177.339	999.8724	822.4058
22	1164.8	1222.244	1022.885	823.5271
23	1126.5	1153.569	969.8036	786.0383
24	1026.5	1141.033	984.1565	827.2798

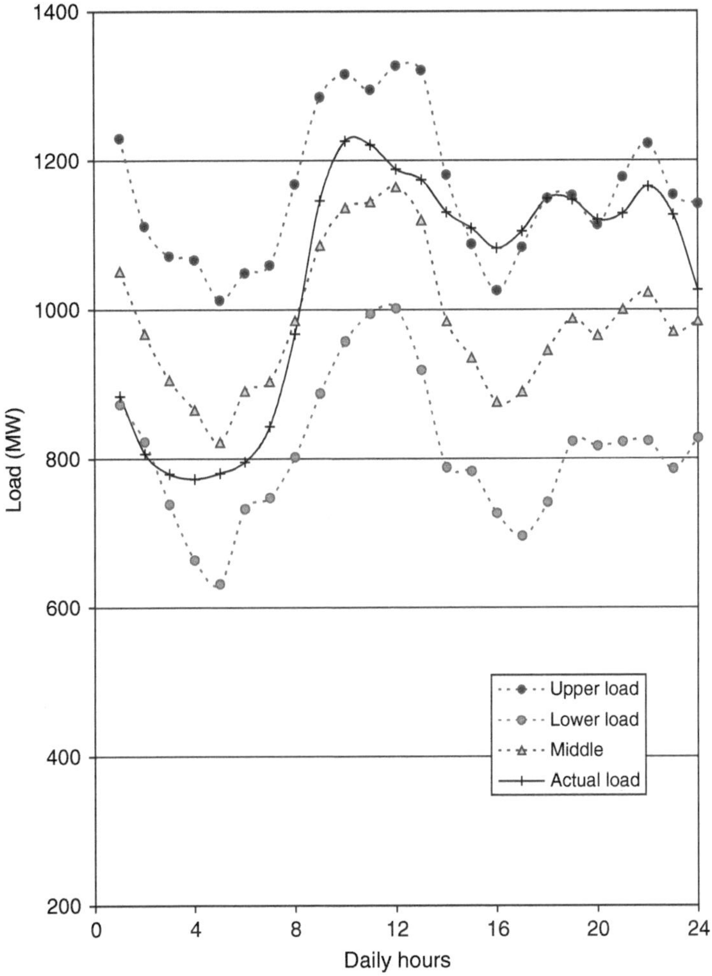

Figure A7.15 Predicted load for a winter day (10% load deviation), model C.

Table A7.16 Estimated Load for a Winter Day (20% Load Deviation), Model C

Daily Hours	Actual Load (MW)	Upper Load (MW)	Middle Load (MW)	Lower Load (MW)
1	1117.6	1341.807	1050.709	759.6112
2	1006.4	1226.967	974.957	722.9469
3	943.6	1196.721	915.1665	633.6118
4	871.17	1198.373	879.9313	561.4897
5	813	1136.893	835.7927	534.6922
6	869.7	1164.667	894.9226	625.1779
7	914.8	1173.425	901.817	630.2086
8	978.7	1278.405	983.1068	687.8086
9	1157.3	1398.555	1087.813	777.0703
10	1223.8	1434.492	1139.156	843.8197
11	1216.8	1402.996	1139.749	876.5016
12	1284.3	1425.355	1154.119	882.8824
13	1258.6	1430.091	1113.992	797.8934
14	1207.8	1302.322	987.6617	673.0016
15	1155.4	1204.581	942.3026	680.0242
16	1110.6	1147.248	889.0203	630.7924
17	1094.1	1214.996	903.5651	592.1339
18	1113.3	1269.902	949.7	629.4979
19	1186.4	1259.544	986.0875	712.6309
20	1139	1223.917	963.7767	703.6367
21	1152.3	1294.554	1000.43	706.3053
22	1226.5	1339.801	1027.696	715.5914
23	1198.4	1269.075	973.8017	678.5282
24	1111.8	1254.081	982.1103	710.1401

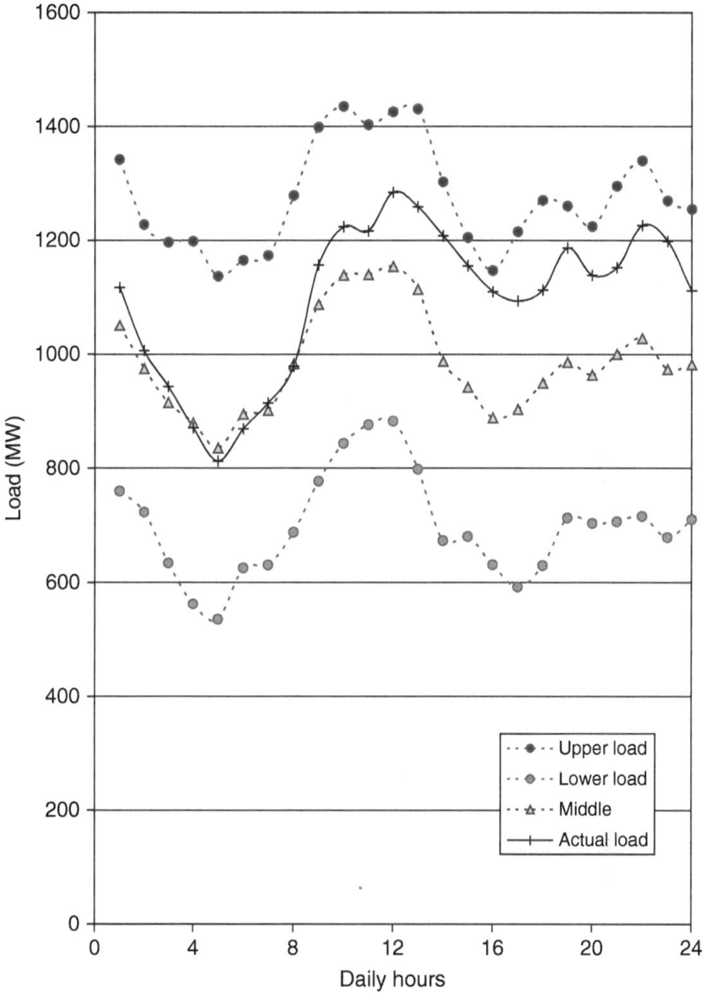

Figure A7.16 Estimated load for a winter day (20% load deviation), model C.

Table A7.17 Predicted Load for a Winter Day (20% Load Deviation), Model C

Daily Hours	Actual Load (MW)	Upper Load (MW)	Middle Load (MW)	Lower Load (MW)
1	883.7	1344.002	1052.904	761.8068
2	806.5	1228.994	976.9838	724.9737
3	779	1204.884	923.3297	641.775
4	772.4	1202.089	883.647	565.2054
5	780.2	1137.512	836.412	535.3115
6	795.6	1164.442	894.6974	624.9526
7	843.3	1176.578	904.9697	633.3613
8	966.8	1276.66	981.3616	686.0634
9	1145.8	1388.196	1077.454	766.7114
10	1225.8	1423.288	1127.952	832.6163
11	1220.9	1400.012	1136.765	873.5177
12	1188.1	1429.521	1158.285	887.0485
13	1174.1	1434.482	1118.383	802.2847
14	1130.2	1308.965	994.305	679.6449
15	1108.7	1209.479	947.2005	684.9222
16	1082.2	1147.586	889.358	631.1302
17	1105	1213.589	902.1577	590.7264
18	1148.7	1271.591	951.389	631.1869
19	1146.9	1257.911	984.4548	710.9983
20	1120.2	1223.917	963.7767	703.6367
21	1128.4	1294.61	1000.486	706.3616
22	1164.8	1336.986	1024.881	712.7765
23	1126.5	1269.019	973.7454	678.4718
24	1026.5	1253.011	981.0406	709.0704

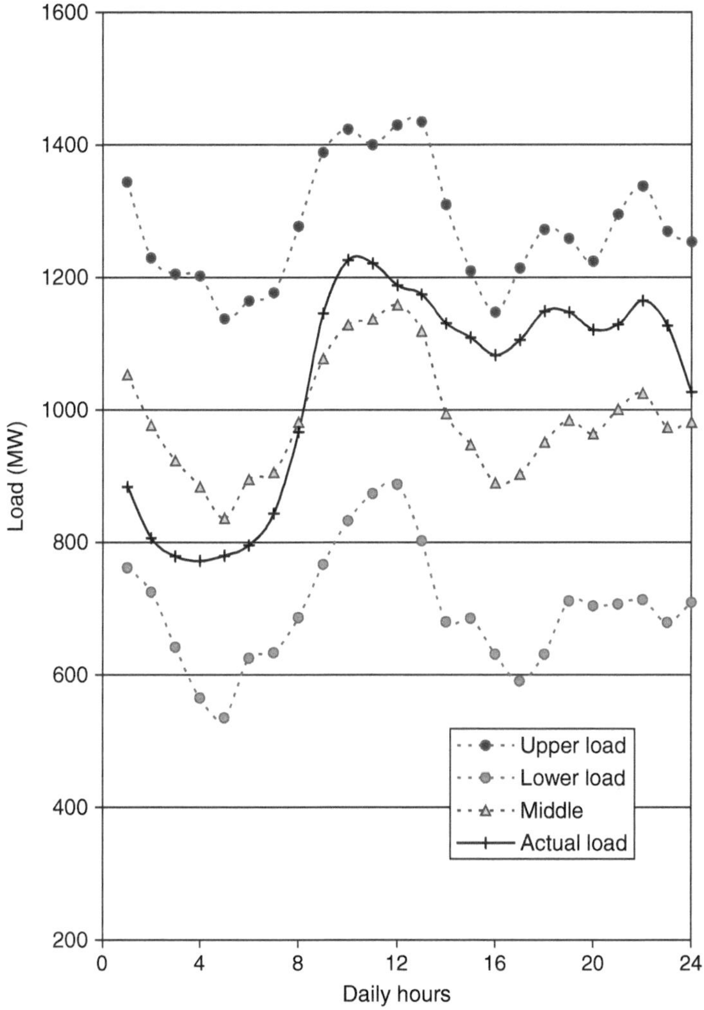

Figure A7.17 Predicted load for a winter day (20% load deviation), model C.

8 Dynamic Electric Load Forecasting

8.1 Objectives

The objectives of this chapter are

- Presenting approaches for long-term and mid term electric power load forecasting.
- Implementing strong short-term correlations of daily (24 hours) and yearly (52 weeks) load behavior to predict future load demand.
- Forecasting weekly average load profiles for 24 hours of a day with a lead time from several weeks to few years.
- Using simple (first-order) linear regression models of previous (one year) data augmented with annual load growth to predict future load demand.
- Introducing the Kalman filtering algorithm with moving window weather, for both crisp and fuzzy loads, to estimate optimal load forecast parameters.
- Using recursive least error squares estimation, as a dynamic estimation, to estimate optimal load-forecast parameters.

8.2 Introduction

Accurate long-term and mid term electric load forecasting plays an essential role for electric power system planning. It corresponds to load forecasting with lead times long enough to plan for long-term and mid term maintenance, construction scheduling for developing new-generation facilities, purchasing of generating units, and developing of transmission and distribution systems. The accuracy of the long-term load forecast has a significant effect on developing future generation and distribution planning. An extensive overestimation of load demand will result in substantial investment for the construction of excess power facilities, whereas underestimation will result in customer discontentment. The time horizon for long-term and mid-term forecasting ranges between a few weeks to several years. Unfortunately, it is difficult to forecast load demand accurately over a planning period of this length. This fact is due to the uncertain nature of the forecasting process. A large number of influential factors characterize and directly or indirectly affect the underlying forecasting process; all of them are uncertain and uncontrollable. Therefore, any long-term forecast, by nature, is *inaccurate!*

Most of the electric load-forecasting methods are dedicated to short-term (a few minutes to 24 hours) forecasting, but not as many are dedicated to long-term (1 to 10 years) or intermediate-term (a few days to several months) load forecasting. Generally, load-forecasting methods can be classified into two broad categories: parametric methods and artificial intelligence–based methods. The artificial intelligence

DOI: 10.1016/B978-0-12-381543-9.00008-7

methods are further classified into neural network–based methods and fuzzy logic–based methods. The parametric methods are based on relating load demand to its affecting factors by a mathematical model. The model parameters are estimated using statistical techniques on historical data of load and its affecting factors. Parametric load-forecasting methods generally can be categorized under three approaches: regression methods, time-series prediction methods, and gray dynamic methods.

The great importance of long-term and mid term load forecasting for electric power utility planning and its economic consequences is encouraging development of forecasting approaches in electric power research to improve its accuracy. Since the 1980s, many techniques have been developed to improve long-term and mid-term forecasting accuracy. Regression models utilize the strong correlation of load with load-affecting factors such as weather. A method of mathematical modeling for global forecasting based on regression analysis was used to forecast load demand up to year 2000. Long-term forecasting based on linear and linear-log regression models of six predetermined sectors has been developed. The time-series models—autoregressive (AR), moving average (MA), and autoregressive moving average (ARMA)—are popular and widely accepted by power utilities at present. They require a massive amount of historical data to produce optimal models. Gray system theory is successfully used to develop dynamic load-forecasting models [1–9].

By nature, long-term electric load forecasting is a complex problem. Among other factors, its accuracy is extremely influenced by the weather as well as social behavior of the community of that load. These factors are difficult to predict for a long-term load-forecasting time horizon. Conversely, short-term forecasting, although affected by weather and daily social habits, is small enough to predict load with high accuracy. Some short-term forecasting algorithms report to have results with a mean absolute error of less than 1%. Consequently, short-term correlation of daily (24 hours) and yearly (52 weeks) load demand of a previous year is utilized to construct a one-year load-demand behavior. The load trends obtained thus far are adjusted with the annual load growth (ALG) to project load demand for the next year. Daily and yearly correlations are modeled as simple linear regressions on weekly average load (WAL) for the 24 hours and 52 weeks, resulting in (24×52) simple linear regression equations. Daily regression is used to depict the relation between the loads at each hour with the hour prior to it, and weekly regression relates the average weekly load with the week prior to it [10–17].

8.3 Load Regression Models

The mid term and long-term electric load demand as a function of time has a complex nonlinear behavior [1]. It depends on a number of complex factors such as daily and seasonal weather, national economic growth, and social habits. All these factors depend on time in a complex way. Therefore, a single mid term and long-term electric load-demand model that accommodates most of these factors will have high nonlinear characteristics and may not give accurate prediction results. The decomposition of the problem into multiple simple (first-order) linear regression models, to capture the global nonlinear behavior of the load, is implemented in this section. Each of the linear regression models extracts the *short-term* correlation of a certain set of data. Then a recursive iterative algorithm is used to tie up the short-term results to capture the global load prediction.

One year's worth of data is arranged into a two-dimensional (2D) layout with 24 columns representing 24 hours of a day and 52 rows representing 52 weeks of the year. Figure 8.1 illustrates the 2D layout of such load data.

Special consideration is taken for the load variation during weekends. Accordingly, weekends are treated separately but in the exact same manner as the working days. The $L(i, k)$ cell in Figure 8.1 is the average load of the working days of the ith week at the kth hour. With this setup of the load data, obvious great intrinsic correlations exist between successive columns as well as between successive rows, as illustrated in Figures 8.2 and 8.3, respectively. These two figures, and all subsequent results, are based on the load demand of one of the largest electric power utilities in Canada for the years 1994 and 1995.

Figure 8.2 shows the load correlation between hour 1 AM and hour 2 AM throughout the whole year 1994. The correlation factor was calculated as 0.997. Similarly, Figure 8.3 shows the load correlation of week 1 and week 2 of year 1994, with a correlation factor of 0.985. The strong correlation is maintained over the entire year for all 24 hours of the day, as illustrated in Figures 8.4 and 8.5.

The persistent correlation of the prevalent load patterns suggests the use of short-term simple linear regression models for successive hours [see equation (8.1a)] and another set for successive weeks [see equation (8.1b)]. This results in 24×52 simple

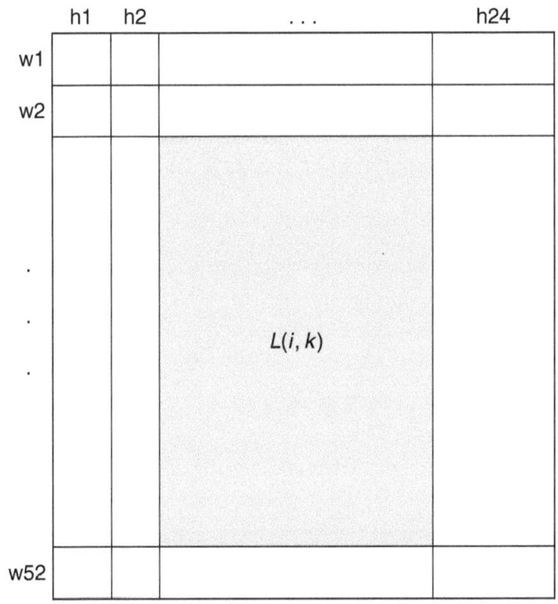

Figure 8.1 Two-dimensional layout of load data.

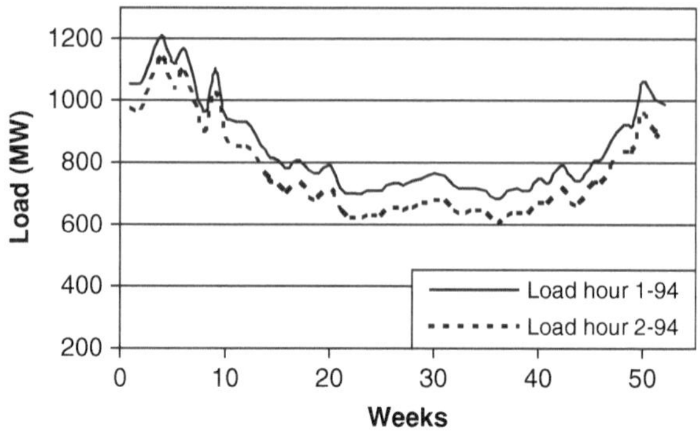

Figure 8.2 Comparing weekly average load of hours 1 and 2, 1994.

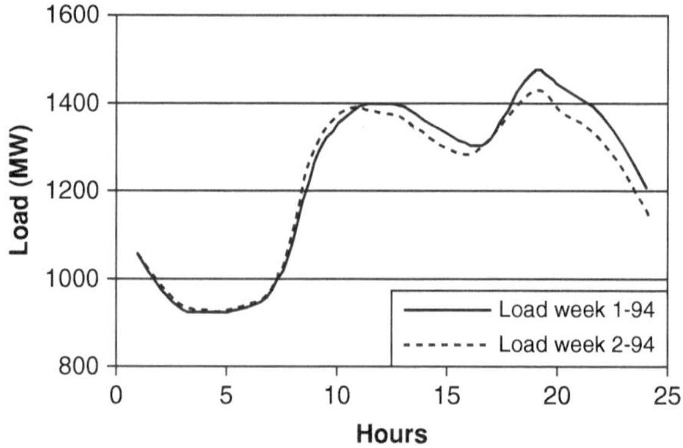

Figure 8.3 Comparing weekly average load of weeks 1 and 2, 1994.

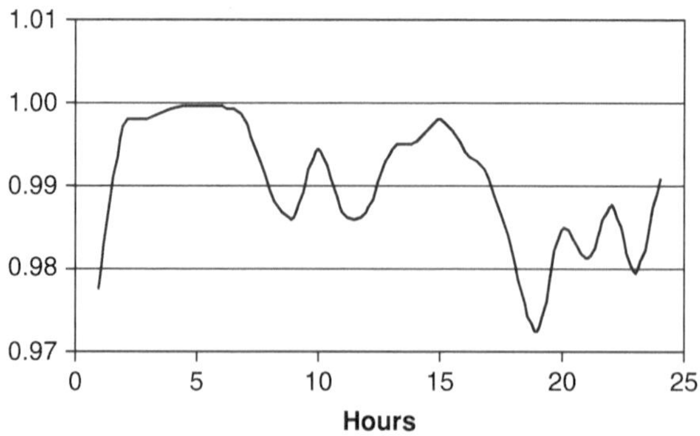

Figure 8.4 Correlation factor for successive hours over 52 weeks of 1994.

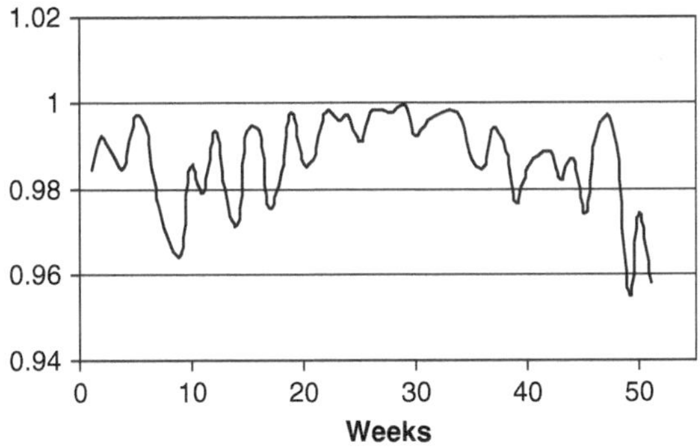

Figure 8.5 Correlation factor for successive weeks over 24 hours of 1994.

linear regression models, which are used to draw the shape of the 2D load behavior contour for one year.

$$L(i,k) = a(k)\ L(i,k-1) + b(k) \quad k = 1,\ldots,24 \tag{8.1a}$$

$$L(i,k) = c(i)\ L(i-1,k) + d(i) \quad i = 1,\ldots,52 \tag{8.1b}$$

where

$a(k)$ and $b(k)$ are regression parameters at the kth hour; $k = 1,2,\ldots,24$, which are estimated using the load pairs $[L(i, k), L(i, k-1)]$ for all $i = 1,2,\ldots,52$, by the least squares method;
$L(i, k)$ and $L(i, k-1)$ are the weekly average load at hours k and $k-1$, respectively, for all weeks $i = 1,\ldots,52$, with the initial condition $L(i, 0) = L(i-1, 24)$;
$c(i)$ and $d(i)$ are regression parameters of the ith week; $i = 1,2,\ldots,52$, which are estimated using the load pairs $[L(i, k), L(i-1, k)]$ for all $k = 1,2,\ldots,24$, by the least squares method;
$L(i, k)$ and $L(i-1, k)$ are the weekly average load in the ith and $(i-1)$th weeks, respectively, for all hours $k = 1,\ldots,24$, with the initial condition $L(0, k) = [L(52, k)$ of the previous year].

8.4 Estimating the Next Year's Load Contour

The first-order regression models developed in the preceding section are used to project the load trends for next year. Figures 8.6 and 8.7 demonstrate the fact that successive years have nearly identical load behavior contours. The load contours of the previous year (1994) coupled with the annual load growth (described in Section 8.5) is utilized to predict next year's load (1995). Each regression model depicts a local relation of the load contours of the two years. The 24 linear regression models of equation (8.1a) relate the load demand of successive hours of a day. They model the daily behavior of the load. The seasonal behavior of the load is modeled by the 52 linear regression models of equation (8.1b).

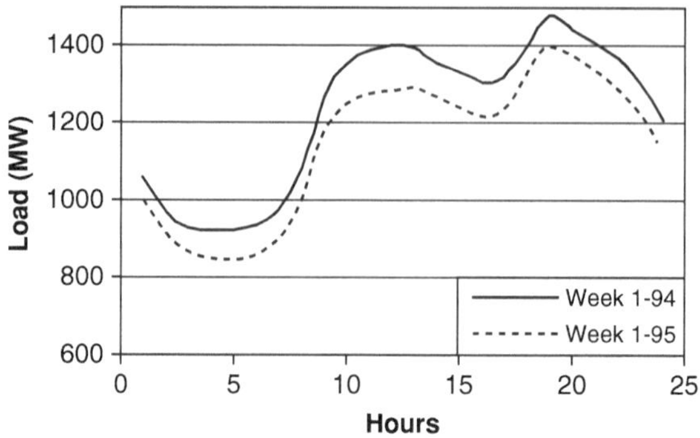

Figure 8.6 Comparing weekly average load of the first weeks of 1994 and 1995.

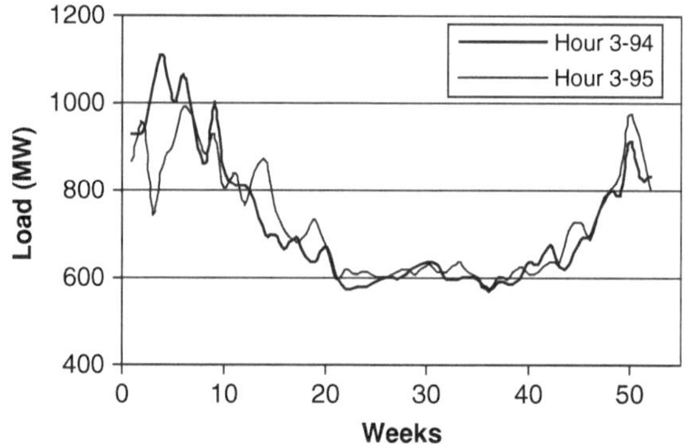

Figure 8.7 Comparing weekly average load of hour 3 of 1994 and 1995.

A recursive procedure used to estimate next year's load contour utilizing regression models of the previous year is as follows:

1. Estimating for the first week the weekly average load: This process corresponds to estimating the first row of the next year's load; refer to Figure 8.8(a). Using equation (8.1b), we calculate $\hat{L}(1,k)$:

$$\hat{L}(1,k) = c(k)\ \hat{L}(0,k) + d(k) \quad k = 1,2,\ldots,24 \tag{8.2}$$

where $\hat{L}(1,k)$ is the estimated weekly average load of the first week at the kth hour; $\hat{L}(0,k)$ is set to $L(52, k)_{\text{last-year}}$, which is the weekly average load of last year's 52nd week; and $[c(k), d(k)]$ is a pair of regression coefficients of the kth hour, obtained from equation (8.1b) using last year's data.

2. Estimating for the first hour the weekly average load: This process corresponds to estimating the first column of the next year's load; refer to Figure 8.8(b). Using equation (8.1a), we calculate $\hat{L}(i,1)$:

$$\hat{L}(i,1) = a(i)\ \hat{L}(i,0) + b(i) \quad i = 2,3,\ldots,52 \tag{8.3}$$

where $\hat{L}(i,1)$ is the estimated weekly average load of the first hour in the ith week; $\hat{L}(i,0)$ is set to $L(i-1, 24)$, which is the weekly average load of the 24th hour of the previous

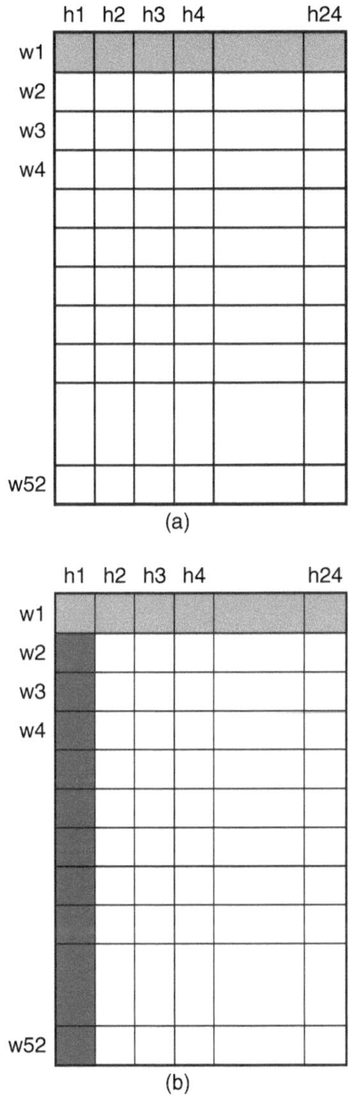

Figure 8.8 (a) First week (row) load estimation resulting from the first iteration. (b) First hour (column) load estimation resulting from the second iteration.

week; and $[a(i), b(i)]$ is a pair of regression coefficients of the ith week, obtained from equation (8.1a) using last year's data.

3. Estimating for the second week the weekly average load: This process corresponds to estimating the second row of the next year's load; refer to Figure 8.8(c). Using equation (8.1b), we calculate $\hat{L}(2,k)$:

$$\hat{L}(2,k) = c(k)\ \hat{L}(1,k) + d(k) \quad k = 2, 3, \ldots, 24 \tag{8.4}$$

where $\hat{L}(2,k)$ is the estimated weekly average load of the second week at the kth hour; and $\hat{L}(1,k)$ is obtained using equation (8.2).

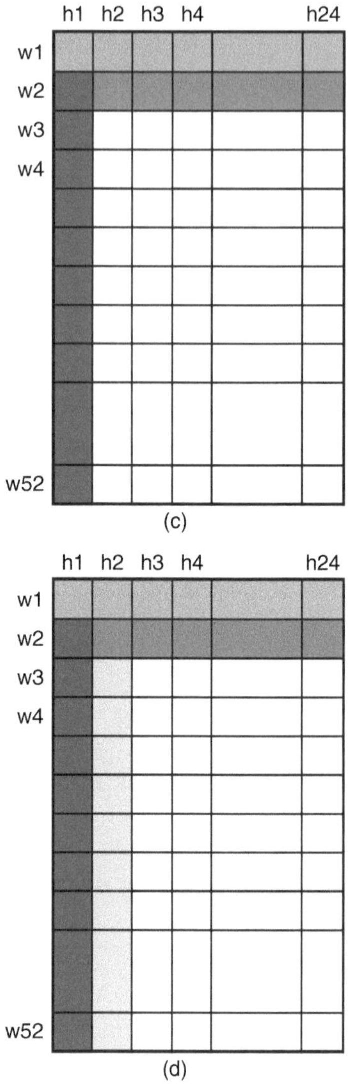

Figure 8.8 (c) Second week (row) load estimation resulting from the third iteration. (d) Second hour (column) load estimation resulting from fourth iteration.

4. Estimating for the second hour the weekly average load: This process corresponds to estimating the second column of the next year's load; refer to Figure 8.8(d). Using equation (8.1a), we calculate $\hat{L}(i,2)$:

$$\hat{L}(i,2) = a(i)\ \hat{L}(i,1) + b(i) \quad i = 3, 4, \ldots, 52 \tag{8.5}$$

where $\hat{L}(i,2)$ is the estimated weekly average load of the second hour in the ith week; and $\hat{L}(i,1)$ is obtained using equation (8.3).

5. The recursive iterations are repeated until $i = k = 24$.

The preceding procedure produces a two-dimensional contour of the load behavior for one year based on regression coefficients of the previous year. The load contour will then be augmented by the annual load growth to account for the load change between successive years.

8.5 Annual Load Growth

To maximize the accuracy of next year's load-demand estimation, we estimate and employ annual load growth as an adjusting factor. It is evident that load demand has a very strong dependence on time. Typical load profiles of successive years reveal very strong correlation at certain periodic time intervals. For example, refer to Figure 8.7; the two load curves at a certain hour over the whole year for two successive years retain the same shape. Moreover, there is, on average, a clear load increase over the previous year. This increase amounts to an annual load growth at that hour as a function of time (weeks) throughout the whole year. The load growth is modeled as the difference between the load curves of two successive years as a function of time.

Practical load profiles show that second-order models will not be sufficient to pick up the annual load growth variations. Third-order models or higher must be used. Models with orders 3, 4, 5, and 6 were tested to best fit load profiles. It was found that models with orders higher than third order were very sensitive to round-off errors and produce "very" incorrect results. A third-order polynomial is utilized to model the load as a function of time at the kth hour as a function of the load of the previous hour. The regression model is as follows:

$$L(i, k) = \beta_0(k) + \beta_1(k)\, L(i, k-1) + \beta_2(k)\, L^2(i, k-1) + \beta_3(k)\, L^3(i, k-1) \tag{8.6}$$

where $\beta_j(k), j = 0, 1, 2, 3$ are regression variables at the kth hour, and $k = 1, 2, \ldots, 24$, which are determined using the load pairs $[L(i, k), L(i, k-1)$, for all $i = 1, 2, \ldots, 52]$ by the least squares method. The initial values $L(i, 0)$ are set to $L(i-1, 24)$. The two curves that approximate the relationship between $L(i, k)$ and $L(i, k-1)$ corresponding to the load behavior of the two years in Figure 8.7 are shown in Figure 8.9. The annual load growth curve is obtained by subtracting the approximate curve for 1995 (estimated data using regression models) from the approximate curve for 1994 (actual data), as shown in Figure 8.10.

Next, the procedure for evaluating the annual load growth is as follows; we assume that the annual load growth is calculated between 1994 and 1995.

1. Using equation (8.6), we determine the regression coefficients (24 sets) for 24 hours for the actual data from year 1994. The coefficients define 24 approximate curves of the weekly average load, one curve per hour.
2. We repeat the calculations of the preceding step for the 1995 estimated data obtained using the regression models from Section 8.3.

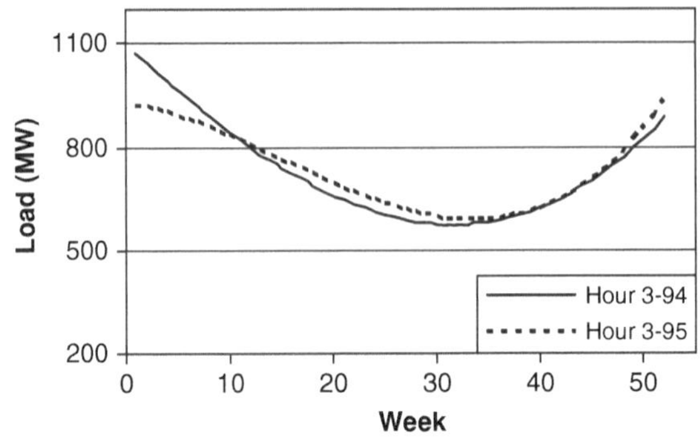

Figure 8.9 Approximate curves of load of hour 3 of 1994 and 1995.

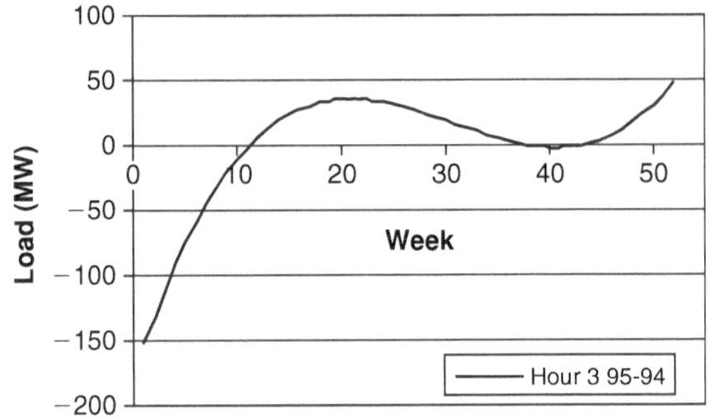

Figure 8.10 Annual load growth variation during 52 weeks of a year.

3. We define the annual load growth as the difference of the approximate load curves of 1995 and 1994 from steps 2 and 1, respectively:

$$\text{Annual Load Growth}\,(i) = L(i,k)_{(95)} - L(i,k)_{(94)} \quad k=1,2,\ldots,24,\, i=1,2,\ldots,52 \tag{8.7}$$

For each hour, the annual load growth is added to the 1995 estimated data obtained using the regression models from Section 8.3 to produce the final prediction results.

8.6 Examples

To verify the effectiveness of the proposed load-demand forecasting technique, we used load data from one of the largest utility companies in Canada for the years 1994 and 1995. Regression models are obtained from 1994 data and used to project load demand for 1995.

8.6.1 Multiple Regression Models Results

Using equation (8.1a), we calculate 24 sets of regression coefficients. Table 8.1 shows the first seven of these sets as a sample. It also lists the correlation factors of successive hours (columns) of the 1994 load data. Similarly, using equation (8.1b), we

Table 8.1 Correlation Factors and Regression Coefficients for Seven Hours of 1994

1994	**Hour 1**	**Hour 2**	**Hour 3**	**Hour 4**	**Hour 5**	**Hour 6**	**Hour 7**
***k* = hour of the day**	**1**	**2**	**3**	**4**	**5**	**6**	**7**
Correlation Factor	0.978	0.997	0.998	0.999	0.999	1.000	0.998
$a(k)$	0.973	0.994	1.014	1.022	1.025	1.024	1.049
$b(k)$	−89.311	−76.835	−49.053	−31.009	−21.580	−11.659	−6.003

Table 8.2 Correlation Factors and Regression Coefficients for Seven Weeks of 1994

1994	***i* = Week Number**	**Correlation Factor**	***c*(*i*)**	***d*(*i*)**
Week 1	1	0.985	0.918	80.911
Week 2	2	0.993	0.964	137.674
Week 3	3	0.987	0.953	123.455
Week 4	4	0.985	0.983	−86.209
Week 5	5	0.997	1.025	43.987
Week 6	6	0.994	0.909	−5.718
Week 7	7	0.976	1.161	−252.143

calculate 52 sets of regression coefficients. Table 8.2 shows the first seven of these sets as a sample, together with the correlation factors of successive weeks (rows) of the 1994 load data.

8.6.2 Estimating 1995 Year Load Contour

The mean absolute percentage error (MAPE) with respect to the actual load is used to measure the effectiveness of the estimated results. For n estimated load values, the MAPE error is given by the equation

$$\mathrm{MAPE} = \frac{100}{n}\sum_{i=1}^{n}\frac{\left|\hat{L}_{est,i} - L_{act,i}\right|}{L_{act,i}} \tag{8.8}$$

where $\hat{L}_{est,i}$ and $L_{act,i}$ are the estimated and actual ith load values, respectively.

The recursive procedure outlined in Section 8.4 is used to project the shape of a 1995 load contour. The regression coefficients determined in Section 8.6.1—namely, [$c(i)$, $d(i)$] and [$a(k)$, $b(k)$]—are alternatively used to estimate a row and a column, respectively, of the 1995 contour described in Figure 8.1. The procedure is carried out for 24 iterations, converging to the actual 1995 load. Figure 8.11(a) shows a sample of the MAPE error convergence for each hour over the 24 iterations. As shown,

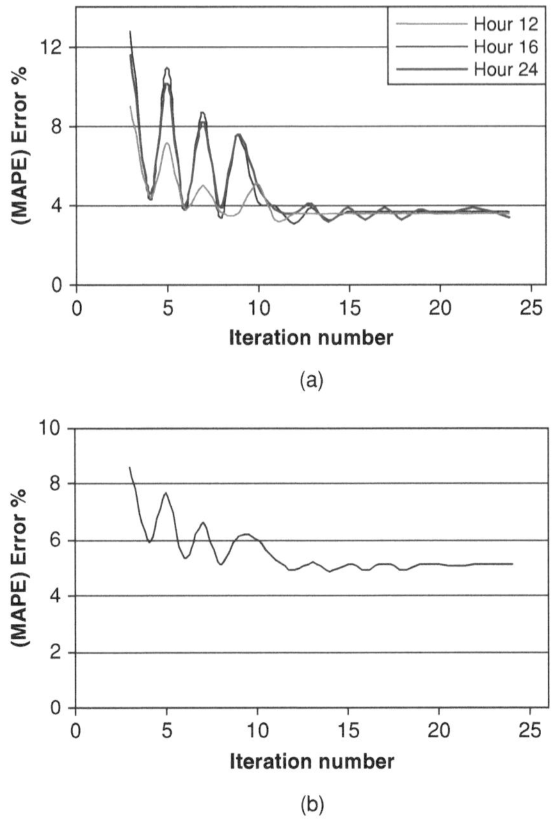

Figure 8.11 (a) Regression estimation (MAPE) error over 52 weeks of 1995. (b) Overall regression estimation (MAPE) error over 52 weeks of 1995.

the error for each hour converges to its minimum. Figure 8.11(b) shows the convergence of the overall MAPE error for the whole year, which was found to be 5.12%.

8.6.3 Annual Load Growth Results

The annual load growth is evaluated and used to augment the estimated load contours. The third-order polynomial load models described in equation (8.6) are used to calculate the annual load growth for each hour of the day. Figure 8.9 shows the approximate fitted curves for hour 3 of 1994 and 1995, and Figure 8.10 shows the annual load growth for that hour. The annual load growth curves for all hours follow almost the same shape with very minimal variations, as illustrated by Figures 8.12 and 8.13. During almost the first 10 weeks, the annual load growth is negative. This trend is accounted for by the unexpectedly low load demand during these weeks in 1995, as shown in Figure 8.7. The low power consumption in these weeks of 1995 was mainly due to above-normal high temperature. The model naturally responds to the given data. It will react differently to different data from different utilities. To reduce the dependency of the annual load growth on uncontrollable

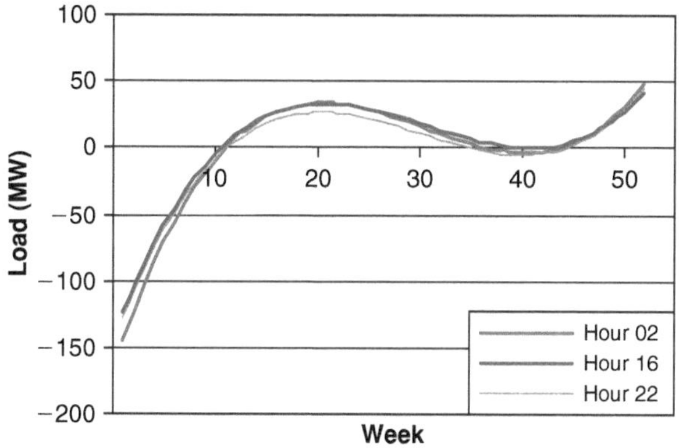

Figure 8.12 Annual load growth throughout 52 weeks of the year.

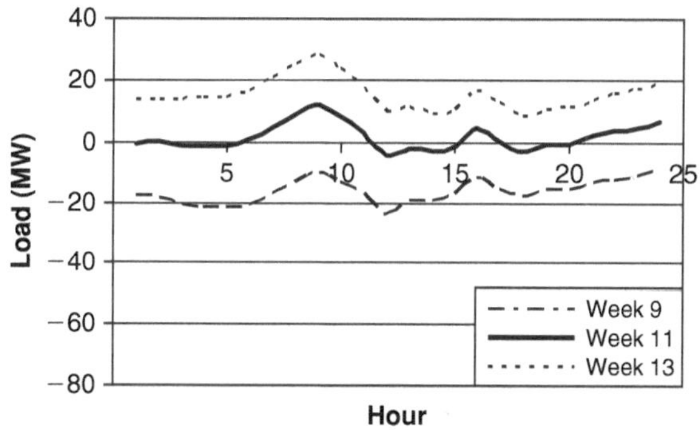

Figure 8.13 Annual growth variations during hours of a day.

short-term weather variations, we can calculate the average of the annual growth over several years.

Figure 8.14 shows a sample of the estimated weekly average load curves for some weeks together with MAPE error over 24 hours of the day. Similarly, Figure 8.15 presents a sample of the weekly average load for some hours varying over 52 weeks of the year. Table 8.2 shows a sample of estimated results. Introducing annual load growth improved the estimation results obtained in Section 8.6.2. The resulting overall MAPE is 3.8 with a standard deviation of 4.14.

8.6.4 Remarks

This part of the chapter demonstrated long-term and mid term electric load-forecasting techniques for forecasting hourly daily load demand for a lead time of several weeks to a few years. This type of forecasting is achieved utilizing short-term correlation of load behavior together with its annual growth. First, using historic data over a specific

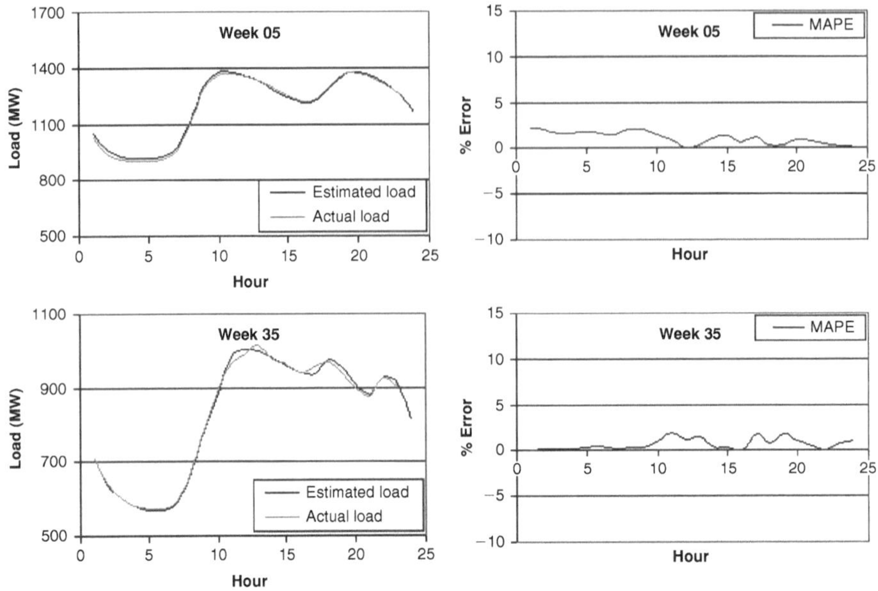

Figure 8.14 Comparison of a sample of estimated and actual load in 1995, during 24 hours.

period of time (one year), we obtained the hourly daily load shape using multiple simple linear regression parametric load models. Second, we employed the parametric models obtained using alternating hourly and weekly load estimations to determine the shape of the load behavior for the next year. Last, we added annual growth load to correct the shape of next year's load. The results indicate that the mean absolute error of the predicted weekly average daily load does not exceed 3.8% of the actual load over a whole year period. With the produced results, the proposed model and forecast technique we used provide a significant advantage compared to those typically seen in the literature for reducing the average absolute error between the forecasted and actual loads over a forecast period of one year ahead.

8.7 Kalman Filtering Algorithm with Moving Window Weather

This section presents a time-varying weather and load model for solving the short-term electric load-forecasting problem. This model utilizes a moving window of current values of weather data as well as recent past history of load and weather data. The load forecasting is based on state space and the Kalman filter approach. A time-varying state space model is used to model the load demand on an hourly basis. The Kalman filter is used recursively to estimate the optimal load-forecast parameters for each hour of the day. The results indicate that the forecasting model produces a robust and accurate load forecast. (See Figure 8.16.)

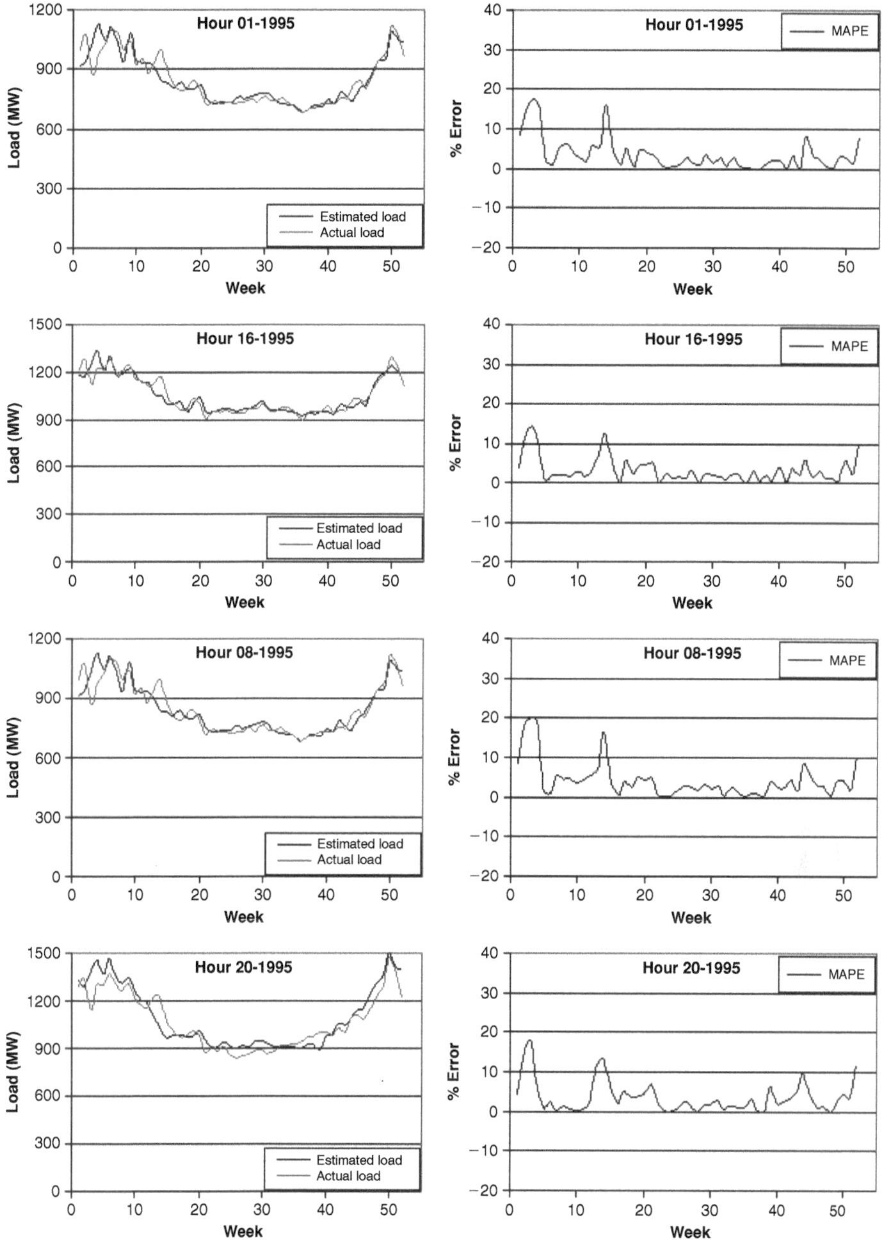

Figure 8.15 Comparison of a sample of estimated and actual load in 1995, throughout 52 weeks.

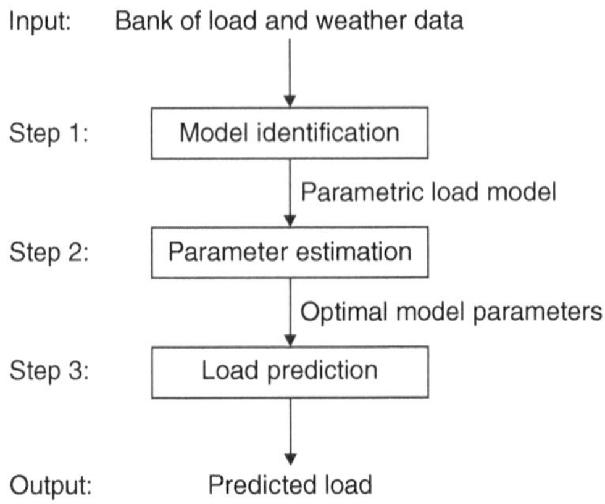

Figure 8.16 Schematic diagram for the load prediction steps.

Hybrid models are used to express the load as a combination of past loads and dominant weather variables to predict future load. Moreover, the Kalman filter is used to estimate the load model parameters.

8.7.1 Load-Forecasting Model

One of the most difficult tasks in load forecasting is the load model identification step [18–32]. Setting up an appropriate load model has a great impact on prediction performance. Model formation entails deciding on the model order and postulating the variables that have an effect on the load. The model order controls the number of variables forming the model. We will show, in the results, the effect of the model order on the prediction performance. Next, we will answer this question: What variables should be used to form the model?

Prevalent weather patterns have a significant impact on the nature of the load profile. The profile shows a very strong dependence of the load on time. At certain periodic time intervals, typical load profiles overcome the existence of the very strong correlation between these intervals. For example, the daily load profile for a specific day retains the same shape more or less as the next day. Moreover, there is a strong correlation in load of successive hours, which cannot be overlooked in designing a load model. Figures 8.17 through 8.19 show these effects. These smooth fluctuations in load are mainly due to the smooth weather changes. The effect of these changes on the same hour on successive days and the changes on the successive hours of the same day have to be imported into the model to maximize prediction accuracy.

Figure 8.17 displays the load fluctuations over 24 hours of seven days of the first week of January 1995 for the Canadian weather data. Each curve represents the load for one day. The curves have almost the same shape, with peaks of loads around 10 AM and 7 PM. For load modeling, this fact strongly suggests expressing the load as a function of the load value of the same hour and/or adjacent hours of the previous day. Figure 8.18 displays the load for one hour over the whole month of January 1995. Five hours, from 1 AM to 5 AM, are plotted one curve per hour.

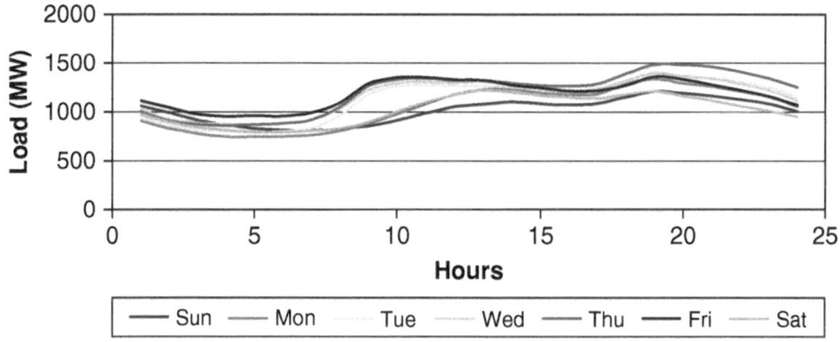

Figure 8.17 Correlation of the load for the first week of days in January 1995.

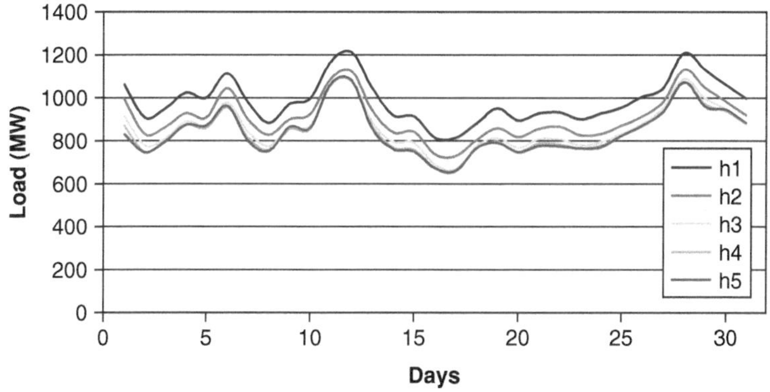

Figure 8.18 The correlation between load curves of the same hour of previous days (hours 1–7), January 1995.

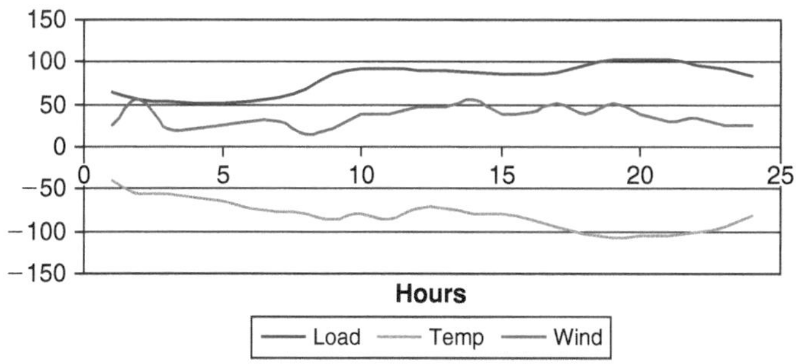

Figure 8.19 The correlation between load demand and weather variables over one day, Thursday, January 5, 1995.

The chart shows the strong correlation of these curves and the strong dependence of the load of a certain hour on the previous hour(s) of the same day. Figure 8.19 shows three curves: load, temperature, and wind-chill factor over 24 hours for Thursday, January 5, 1995. (The load and wind curves have been scaled to display the three curves on one chart.) The impact of weather, particularly temperature, on load variation is obvious. As temperature peaks low, the load peaks high and vice versa. Similarly, the wind-chill factor has an effect on the load-demand variation but with a lesser amount than the temperature. As the wind-chill factor increases at about 7 AM, the load starts rising at about 9 AM.

Based on the preceding facts regarding strong dependence of load on weather and time, the load model is expressed as a linear function of past values of load and dominant weather variables. The winter model is expressed next.

8.7.2 Winter Model

The dynamic variation of load with respect to weather factors is expressed as a time-varying linear model with variables belonging to moving windows of recent past load and weather variables The load at any discrete time instant k, $k = 1, 2, \ldots, 24$, corresponding to 24 hours of one day, can be expressed as

$$\begin{aligned} y(k) = {} & a_0(k) + a_1(k)\ y_1(k) + \cdots + a_{n1}(k)\ y_{n1}(k) \\ & + b_0(k)\ t_0(k) + b_1(k)\ t_1(k) + \cdots + b_{n2}(k)\ t_{n2}(k) \\ & + c_0(k)\ w_0(k) + c_1(k)\ w_1(k) + \cdots + c_{n3}(k)\ w_{n3}(k) \end{aligned} \quad (8.9)$$

where

$y(k) =$ load at time instant k;
$y_i(k) = n_1$ previous load values related to time instant k, defined below;
$t(k) =$ temperature at time instant k;
$t_i(k) = n_2$ previous temperature values related to time instant k, defined below;
$w(k) =$ wind at time instant k;
$w_i(k) = n_3$ previous wind values related to time instant k, defined below;
$a_0(k) =$ base load at time instant k;
$a_i(k) = n_1$ load coefficients at time instant k;
$b_i(k) = n_2$ temperature coefficients at time instant k;
$c_i(k) = n_3$ wind coefficients at time instant k.

The model assumes that the load, temperature, wind, and their coefficients are constant over each discrete time instant k, $k = 1, \ldots, 24$, of the 24 hours of the day. Next, we define the moving window of the load. Temperature and wind windows are treated in exactly the same manner. The load data are arranged in a matrix of 24 columns with rows holding 24 values for one day. The load window associated with the kth time instant is defined in Figure 8.20. This figure illustrates positions of the load values relative to the kth time instant in the data matrix.

The indices are labeled such that

0 represents the current value, $y(k)$;
1 represents the previous hour's value: $y_1(k) = y(k-1)$;
2 represents the hour of the previous day's value: $y_2(k) = y(k-24)$;
3 represents one hour before the hour of the previous day's value: $y_3(k) = y(k-25)$;
4 represents one hour after the hour of the previous day's value: $y_4(k) = y(k-23)$;

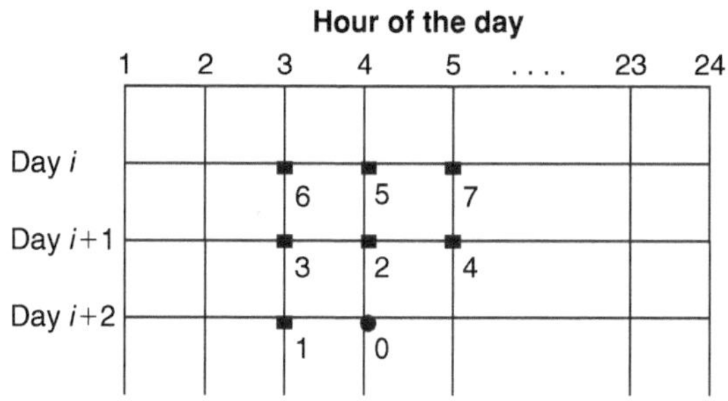

Figure 8.20 Illustration on the data matrix.

5 represents the hour of the two days' back value: $y_5(k) = y(k-48)$;
6 represents one hour before the hour of the two days' back value: $y_6(k) = y(k-49)$;
7 represents one hour after the hour of the two days' back value: $y_7(k) = y(k-47)$.

As shown, although the model sets a window of values for only two days back and only one hour before and one hour after the current hour, it can be easily generalized by stretching the window to include more values. The results, as will be illustrated Section 8.8, show that the size of window considered in Figure 8.20 is sufficient for parameter estimation using the Kalman filter. In fact, a subset of the values of Figure 8.20 produces a sufficiently accurate prediction of the load.

Because coefficients in the model are assumed constant over one hour time instance, parameter estimation is carried out for each of the 24 discrete instances in a day. Accordingly, 24 sets of coefficients are required to be estimated for one day. The estimated coefficients can be plugged into the model to predict hourly loads for the next day.

8.8 Kalman Filter Parameter Estimation

8.8.1 Basic Kalman Filter

In this section, we will address only the necessary equations for the development of the basic recursive discrete Kalman filter. Given the discrete state equations as

$$\begin{aligned} x(k+1) &= A(k)\, x(k) + w(k) \\ z(k) &= C(k)\, x(k) + v(k) \end{aligned} \tag{8.10}$$

where

$x(k)$ is $n \times 1$ system states;
$A(k)$ is an $n \times n$ time-varying state transition matrix;
$z(k)$ is an $m \times 1$ measurement vector;
$C(k)$ is an $m \times n$ time-varying output matrix;
$w(k)$ is an $n \times 1$ system error;
$v(k)$ is an $m \times 1$ measurement error.

The noise vectors $w(k)$ and $v(k)$ are uncorrected white noises that have zero means:

$$E[w(k)] = E[v(k)] = 0 \tag{8.11}$$

no time correlation:

$$E\left[w(i)\ w^T(j)\right] = E\left[v(i)\ v^T(j)\right] = 0,\ \text{for } i = j \tag{8.12}$$

and known covariance matrices (noise levels):

$$\begin{aligned} E\left[w(k)\, w^T(k)\right] &= Q_1 \\ E\left[v(k)\, v^T(k)\right] &= Q_2 \end{aligned} \tag{8.13}$$

where Q_1 and Q_2 are positive semi-definite and positive definite matrices, respectively. The basic discrete-time Kalman filter algorithm is given by the following set of recursive equations. Given a priori estimates of the state vector $\hat{x}(0) = \hat{x}_0$ and its error covariance matrix, $P(0) = P_0$, set $k = 0$, and then recursively compute the following:
Kalman gain:

$$K(k) = \left[A(k)\ P(k)\ C^T(k)\right]\left[C(k)\ P(k)\ C^T(k) + Q_2\right]^{-1} \tag{8.14}$$

New state estimate:

$$\hat{x}(k+1) = A(k)\ \hat{x}(k) + K(k)[z(k) - C(k)\ \hat{x}(k)] \tag{8.15}$$

Error covariance updates:

$$P(k+1) = [A(k) - K(k)\ C(k)]\ p(k)\ [A(k) - K(k)\ C(k)]^T + K(k)\ Q_2\ K^T(k) \tag{8.16}$$

An intelligent choice of the a priori estimate of the state $\hat{x}_0$ and its covariance error P_0 enhances the convergence characteristics of the Kalman filter. A few samples of the output waveform $z(k)$ can be used to get a weighted least squares as an initial value for $\hat{x}_0$ and P_0,

$$\begin{aligned} \hat{x}_0 &= \left[H^T Q_2^{-1} H\right]^{-1} H^T Q_2^{-1} z_0 \\ P_0 &= \left[H^T Q_2^{-1} H\right]^{-1} \end{aligned} \tag{8.17}$$

where

z_0 is an $(mm_1) \times 1$ vector of m_1 measured samples;
H is an $(mm_1) \times n$ matrix.

$$z_0 = \begin{bmatrix} z(1) \\ z(2) \\ \vdots \\ z(m_1) \end{bmatrix} \quad \text{and} \quad H = \begin{bmatrix} C(1) \\ C(2) \\ \vdots \\ C(m_1) \end{bmatrix} \tag{8.18}$$

8.8.2 Prediction of the Kalman Filter Model

In this section we use the weather-load model to form a time-varying discrete dynamic system suited for the Kalman filter. The dynamic system of equation (8.10) is used with the following definitions:

1. State transition matrix, $A(k)$, is a constant identity matrix.
2. The error covariance matrices, Q_1 and Q_2, are constant matrices. Q_1 and Q_2 values are based on some knowledge of the actual characteristics of the process and measurement noises, respectively. Q_1 and Q_2 are chosen to be identity matrices for this simulation; Q_1 would be assigned better value if more knowledge were obtained on the sensor accuracy.
3. The state vector, $x(k)$, consists of N parameters. From equation (8.9), $N = n_1 + n_2 + n_3 + 1$.
4. $C(k)$ is the N elements time-varying row vector, which relates the measured load and weather data to the state vector. Refer to equation (8.19).
5. The observation vector, $z(k)$, for this application is a scalar representing the load at time instant k. Refer to equation (9.19).

The observation equation $z(k) = C(k)\, x(k)$ has the form

$$z(k) = \begin{bmatrix} 1 & y_1 & \cdots & y_{n1} & t_0 & \cdots & t_{n2} & w_0 & \cdots & w_{n3} \end{bmatrix} \begin{bmatrix} a_0 \\ a_1 \\ \vdots \\ a_{n1} \\ a_{n1+1} \\ \vdots \\ a_{n1+n2+1} \\ a_{n1+n2+2} \\ \vdots \\ a_{n1+n2+n3+2} \end{bmatrix} \tag{8.19}$$

where the parameters and load weather values are defined in equation (8.9), and k represents the time instant of the 24 discrete hours of the day, $k = 1, \ldots, 24$.

8.8.3 Examples and Results

To verify the effectiveness of the load model parameters estimation and forecasting demand load, we used a load for one of the largest utility companies in Canada as well as the weather data for the years 1994 and 1995 for the same utility company.

The steps taken to predict the next day's load are as follows:

Step 1 Select the model and model order. For all examples, the moving window of the load and weather data is selected. The model order for all examples is taken to be $N = 10$. Although the next section shows that lower-order models can give the same results, $N = 10$ is taken to inform the results for comparisons and to take almost all effects of weather and load in the model.

Step 2 Select the initial condition for the parameters for the Kalman filter estimation. The initial condition of the parameter vector is fixed arbitrarily to ones in all our results.

Step 3 Set system noise and measurement error covariance matrices Q_1 and Q_2 of equation (8.13). In all examples, they are set to unity matrices with the appropriate dimension, depending on the order taken in step 1.

Step 4 Run the Kalman filter for the first hour of the day and save the resulting coefficient to be used later for predicting the first hour of any day that needs to be predicted.

Step 5 Use the estimated parameters of the previous hour as the initial condition for estimating the next hour's coefficient using the Kalman filter. Run the Kalman filter and save the estimated parameters.

Step 6 Repeat step 5 for all 24 hours of the day.

Step 7 Using the moving windows as described in step 1, use the 24 sets of estimated parameters, one set per hour, to predict the load 24 hours ahead.

8.8.4 Order of the Load Model

Through the use of the load model of equation (8.9) and state model of equation (8.10), several program runs are made with different model order values: $N = 2$ to 7. Figure 8.21 shows the effect of the model order and the moving window size on load prediction. In the figure, a measure M and a percentage P are used to measure the effectiveness of load prediction. The measure M is taken as the Euclidean norm error between the actual and predicted loads over the 24 hours of a day. P is the mean absolute percentage error with respect to the actual load. For 24 hours, M and P are given by the equations

$$M = \sqrt{d_1^2 + \cdots + d_{24}^2} \tag{8.20}$$

$$P = \left(\frac{|d_1|}{y_{a1}} + \cdots + \frac{|d_{24}|}{y_{a24}} \right) \frac{100}{24} \tag{8.21}$$

where $d_k = y_{pk} - y_{ak}$, $k = 1, \ldots, 24$; and y_{pk} and y_{ak} are the predicted and actual loads at instant k, respectively.

The results depicted in Figure 8.21 show that the second row of each set of runs has the minimum M and P. This observation suggests the importance of including the load of the same hour of the previous day, y_2, as defined in Figure 8.20. The observation also suggests that increasing the model order has little effect on improving the prediction error. But for prediction purposes, low-order models can be very sensitive to weather changes. Therefore, it is advisable to use a sufficiently high-order model to take into account severe changes of weather on the load. The outcome also shows that using only weather effects in the model results in high prediction error. All the preceding observations suggest using a hybrid model with sufficiently high order to take care of the unexpected and uncontrolled weather and load fluctuations. The model order used in all the program runs in the examples is $N = 10$.

8.8.5 One-Hour Prediction

In this subsection, the results of predicting one hour (1 AM, February 28, 1995) are discussed. The load and weather data of the first hour of the day (1 AM) of the

	1	y_1	y_2	y_3	y_4	y_5	y_6	y_7	t_0	t_1	t_2	t_3	t_4	t_5	t_6	w_0	w_1	w_2	w_3	w_4	w_5	w_6	M	P
$N=2$	1	1																					370.21	4.84
	1		1																				48.23	0.63
	1				1																		337.85	4.52
	1								1														917.79	11.72
	1											1											764.53	9.80
	1																	1					1294.54	17.47
	1																				1		1268.98	16.94
$N=3$	1	y_1	y_2	y_3	y_4	y_5	y_6	y_7	t_0	t_1	t_2	t_3	t_4	t_5	t_6	w_0	w_1	w_2	w_3	w_4	w_5	w_6	M	P
	1	1	1																				77.74	1.11
	1		1		1																		47.53	0.57
	1						1					1											349.20	4.07
	1					1							1										233.28	2.63
	1									1				1									781.28	9.84
	1														1		1						858.19	10.90
	1																		1		1		1230.23	16.43
$N=4$	1	y_1	y_2	y_3	y_4	y_5	y_6	y_7	t_0	t_1	t_2	t_3	t_4	t_5	t_6	w_0	w_1	w_2	w_3	w_4	w_5	w_6	M	P
	1	1		1	1																		367.86	4.74
	1		1		1			1															47.53	0.57
	1		1				1				1												53.72	0.66
	1				1								1		1								316.51	4.28
	1									1				1					1				802.98	10.38
	1			1						1								1					320.31	3.88
	1											1					1				1		820.26	10.60
$N=5$	1	y_1	y_2	y_3	y_4	y_5	y_6	y_7	t_0	t_1	t_2	t_3	t_4	t_5	t_6	w_0	w_1	w_2	w_3	w_4	w_5	w_6	M	P
	1	1		1			1			1													367.86	4.74
	1		1				1							1			1						53.72	0.66
	1				1					1			1					1					315.13	4.36
	1			1			1						1		1								366.15	4.60
	1							1			1	1				1							378.56	4.87
	1											1						1	1			1	814.56	10.38
	1									1				1						1	1		864.59	11.39
$N=6$	1	y_1	y_2	y_3	y_4	y_5	y_6	y_7	t_0	t_1	t_2	t_3	t_4	t_5	t_6	w_0	w_1	w_2	w_3	w_4	w_5	w_6	M	P
	1	1			1			1			1								1				92.17	1.11
	1		1				1							1				1				1	53.72	0.66
	1			1						1			1				1				1		320.31	3.88
	1						1				1				1		1			1			355.53	4.31
	1					1						1				1			1			1	233.28	2.63
	1									1				1			1				1	1	802.98	10.38
	1	1	1				1		1			1											77.74	1.11
$N=7$	1	y_1	y_2	y_3	y_4	y_5	y_6	y_7	t_0	t_1	t_2	t_3	t_4	t_5	t_6	w_0	w_1	w_2	w_3	w_4	w_5	w_6	M	P
	1	1		1						1		1								1		1	367.86	4.74
	1		1		1		1		1		1									1			47.53	0.57
	1		1			1					1								1		1	1	109.64	1.40
	1				1			1	1			1		1	1								318.7	4.65
	1		1				1	1								1		1			1		53.718	0.663
	1								1		1			1	1					1		1	843.684	11.049
	1	1	1	1	1	1	1																77.742	1.107

Figure 8.21 Effects of load model order and moving window on load prediction.

previous two months (January and February 1995) are used to estimate the load model parameters. Moreover, two extra points between every two actual values of load, as well as weather, are generated using cubic interpolation to boost up Kalman filter iterations. The model of equation (8.19) is used with load, temperature, and wind data windows of sizes $n_1 = 4$, $n_2 = 2$, and $n_3 = 1$, respectively. This entails the use of a window with four load values (y_1, y_2, y_3, and y_4), a window with three temperature values (t_0, t_1, and t_2), and a window with two wind values (w_0 and w_1), as defined in Figure 8.21. The combined load and weather model is then written as

$$y(k) = \begin{bmatrix} 1 & y_1 & y_2 & y_3 & y_4 & t_0 & t_1 & t_2 & w_0 & w_1 \end{bmatrix} \begin{bmatrix} a_0 \\ a_1 \\ a_2 \\ a_3 \\ a_4 \\ a_5 \\ a_6 \\ a_7 \\ a_8 \\ a_9 \end{bmatrix} \tag{8.22}$$

Figure 8.22 presents a sample of the estimated model coefficients of the Kalman filter iterations. As illustrated, all estimated parameters converge to their steady-state value after some transient fluctuations. The first parameter, a_0, requires more iteration to converge; in fact, it continues its fluctuation in the second hour and then converges in the rest of the hours. Figure 8.23 shows the convergence of the 10 parameters for the last hour of the same day. A comparison between the load resulting from the estimated parameters and the actual load is shown in Figure 8.24. The results show how closely the estimated model matches the actual load. Figure 8.25 displays the difference between the estimated and actual loads.

8.8.6 Twenty-Four-Hour Prediction

The one-hour prediction is repeated for 24 hours of one day (February 28, 1995). Table 8.3 presents 24 sets; each set consists of 10 values of estimated parameters for one day. The estimated parameters are used to predict demand load for one day. The results are shown in Table 8.4 and Figures 8.26 to 8.29. The results show that the load percent error, P, as given by equation (8.21), is less than 1% for all 24 predicted loads except for the second hour, which is due to the transient behavior of the a_0 parameter. The variation of each coefficient over 24 hours is illustrated in Figure 8.30.

The dependency of the load model on the load and weather history is evident in the results depicted in Table 8.4. For example, for the first hour, h00, the parameters depend mainly on y_1, y_2 loads; the three temperature values; and the two wind factors. The effects of the load values y_1, y_2, and y_3, and the three temperature values dominate the predicted load in h12. The wind effect has less effect on the predicted load at that hour. The results necessitate the use of a combined load and weather model.

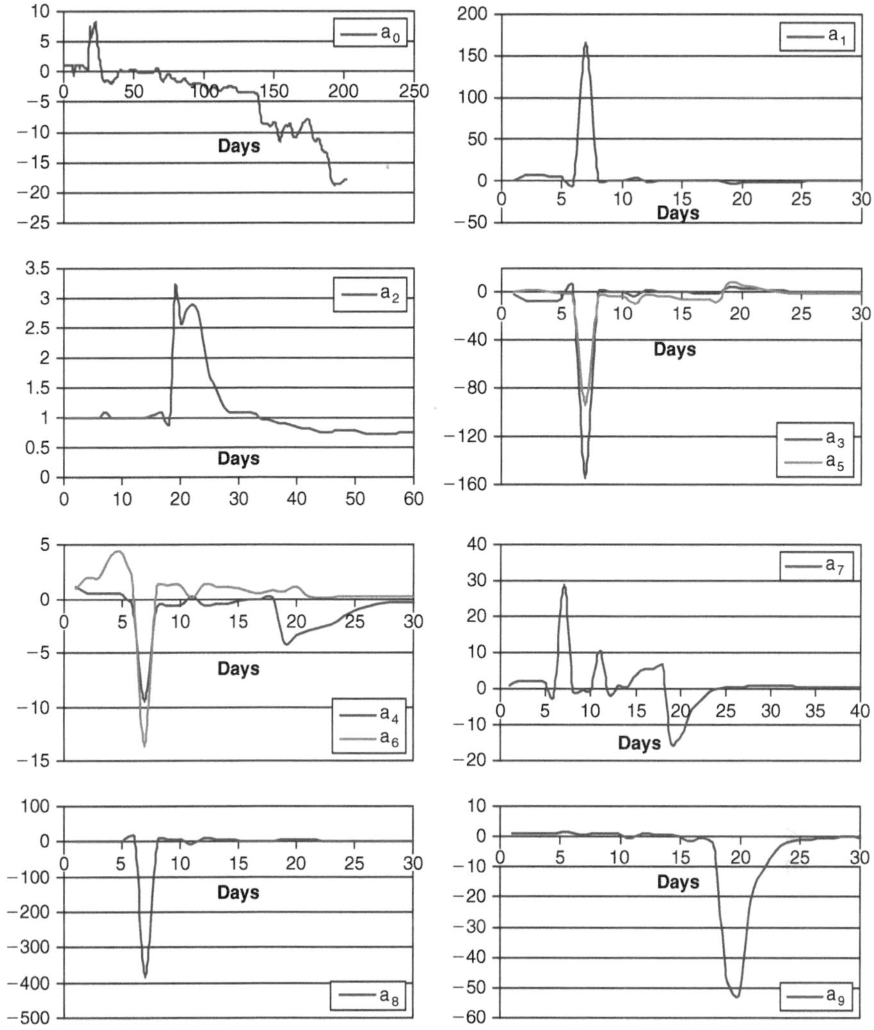

Figure 8.22 Convergence of the Kalman filter for some of the estimated parameters.

8.8.7 Weekdays and Weekends Profiles

Load-demand profiles for weekdays and weekends possess different patterns. Accordingly, treating them separately is expected to produce better results than combining all days as if they have the same profile patterns. The actual and predicted forecasted loads for two months, four months, and weekdays and weekends of four months are provided in Table 8.5. We can note from the results in Table 8.7 that the Euclidean norm error, *M*, and the mean absolute error, *P*, are less for the separated weekdays than for the combined days. The *M* and *P* for the weekends are slightly more than that for the all days' predictions because the number of data used for the weekend case is less than the number

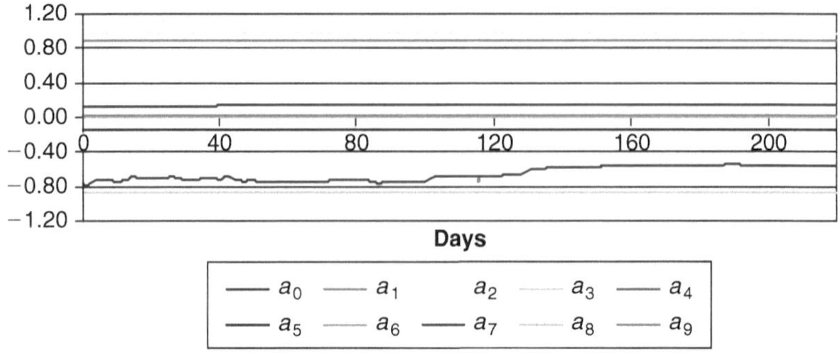

Figure 8.23 Kalman filter parameter estimation of the last hour of the day.

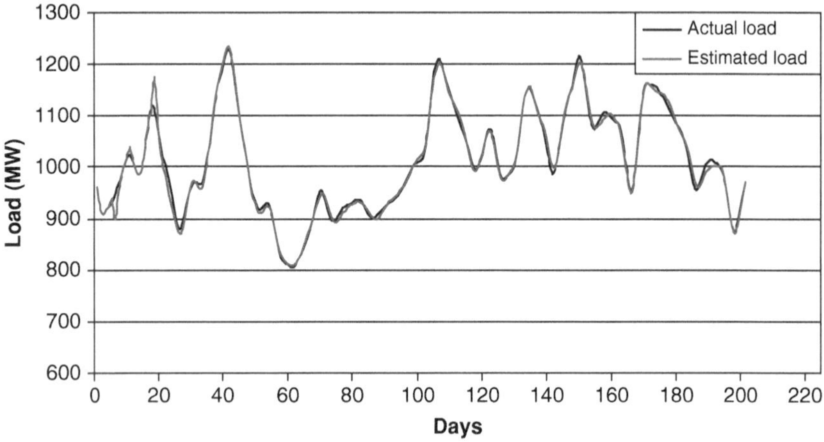

Figure 8.24 Comparison of actual and estimated loads.

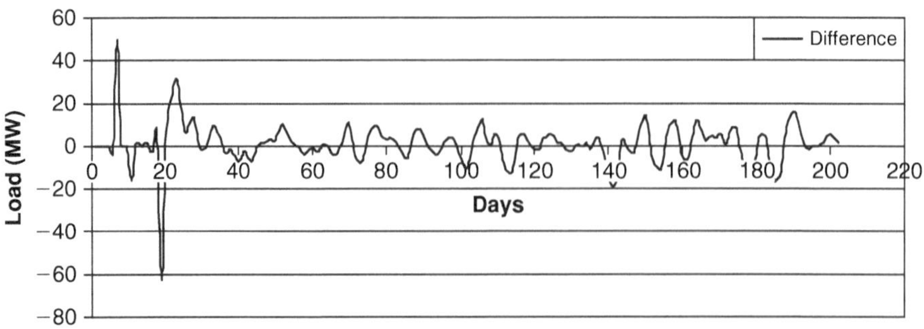

Figure 8.25 Difference between actual and estimated loads.

Table 8.3 Set of 10 Estimated Parameters for Each Hour and Actual and Estimated Loads

Hour	$\underline{a}_0$	a_1	a_2	a_3	a_4	a_5	a_6	a_7	a_8	a_9	Actual	Estimated	Difference
h00	−17.76	0.16	0.86	0.02	−0.05	−1.09	0.23	0.82	0.17	0.18	970.94	969.48	1.47
h01	427.63	−2.70	0.45	3.20	−0.53	−1.14	−0.61	0.77	0.01	0.74	897.94	837.07	60.87
h02	121.93	−0.24	0.60	0.51	−0.04	−1.21	0.14	0.72	0.16	0.15	863.07	848.66	14.41
h03	79.47	0.14	0.58	0.10	0.07	−1.08	0.23	0.62	0.13	0.09	853.19	838.77	14.42
h04	49.89	0.36	0.57	−0.15	0.14	−0.97	0.28	0.54	0.11	0.05	855.92	849.03	6.89
h05	33.24	0.49	0.56	−0.28	0.17	−0.89	0.31	0.47	0.12	0.00	871.83	864.45	7.39
h06	47.12	0.50	0.59	−0.29	0.14	−0.86	0.30	0.42	0.12	0.02	911.33	903.35	7.98
h07	21.14	0.64	0.68	−0.48	0.13	−0.78	0.31	0.40	0.19	−0.08	1027.68	1025.86	1.81
h08	−9.52	0.89	0.92	−0.84	0.04	−0.60	0.29	0.32	0.11	−0.03	1188.25	1194.32	−6.07
h09	−12.76	0.90	0.92	−0.85	0.04	−0.55	0.27	0.29	0.09	−0.05	1283.96	1279.16	4.80
h10	−12.44	0.87	0.92	−0.81	0.04	−0.55	0.26	0.31	0.09	−0.04	1340.34	1327.92	12.42
h11	−10.50	0.85	0.92	−0.80	0.04	−0.54	0.25	0.30	0.07	−0.03	1349.98	1346.29	3.69
h12	−10.54	0.84	0.92	−0.79	0.04	−0.53	0.25	0.29	0.06	−0.02	1372.39	1360.78	11.60
h13	−9.08	0.84	0.93	−0.79	0.04	−0.52	0.24	0.29	0.07	−0.03	1329.03	1326.98	2.05
h14	−8.28	0.84	0.93	−0.79	0.04	−0.51	0.23	0.28	0.07	−0.03	1300.70	1296.51	4.20
h15	−7.00	0.84	0.94	−0.80	0.03	−0.49	0.22	0.26	0.06	−0.02	1265.96	1264.92	1.04
h16	−5.47	0.84	0.95	−0.80	0.02	−0.47	0.22	0.26	0.06	−0.02	1268.11	1264.78	3.33
h17	−4.82	0.84	0.95	−0.81	0.02	−0.46	0.21	0.25	0.05	−0.02	1330.20	1326.26	3.94
h18	−4.54	0.84	0.95	−0.81	0.02	−0.46	0.20	0.26	0.06	−0.03	1346.51	1355.39	−8.89
h19	−4.31	0.84	0.95	−0.81	0.02	−0.46	0.20	0.26	0.06	−0.03	1354.54	1365.99	−11.45
h20	−3.94	0.84	0.95	−0.81	0.02	−0.46	0.20	0.26	0.06	−0.03	1336.76	1341.37	−4.61
h21	−3.75	0.84	0.95	−0.81	0.02	−0.45	0.19	0.26	0.06	−0.03	1293.78	1295.13	−1.35
h22	−1.54	0.87	0.96	−0.85	0.02	−0.20	0.02	0.17	0.00	0.04	1241.41	1243.63	−2.22
h23	−0.92	0.87	0.97	−0.85	0.01	−0.19	0.01	0.18	−0.01	0.04	1146.63	1148.44	−1.81
	1	y_1	y_2	y_3	y_4	t_0	t_1	t_2	w_0	w_1			

Table 8.4 Difference and Error Percentage of the Predicted Load to the Actual Load

Hour	Actual	Predicted	Difference	Percent
0	990.4	990.9	–0.48	–0.048
1	915.4	882.2	33.17	3.623
2	883.2	875.5	7.73	0.876
3	873.8	866.1	7.70	0.881
4	875.9	873.2	2.65	0.303
5	890.9	888.2	2.72	0.305
6	930.4	926.0	4.36	0.469
7	1046.4	1041.1	5.34	0.511
8	1205.8	1202.9	2.86	0.237
9	1301.1	1300.0	1.10	0.084
10	1356.1	1357.9	–1.76	–0.130
11	1364.8	1370.8	–6.01	–0.440
12	1384.6	1388.5	–3.87	–0.280
13	1338.4	1345.6	–7.18	–0.536
14	1312.1	1312.8	–0.69	–0.053
15	1277.4	1280.4	–2.99	–0.234
16	1279.8	1281.6	–1.80	–0.141
17	1342.0	1342.6	–0.56	–0.042
18	1356.8	1362.9	–6.06	–0.447
19	1365.7	1368.2	–2.46	–0.180
20	1347.9	1350.2	–2.31	–0.172
21	1306.5	1308.0	–1.52	–0.116
22	1252.1	1254.3	–2.15	–0.172
23	1157.4	1158.0	–0.65	–0.056

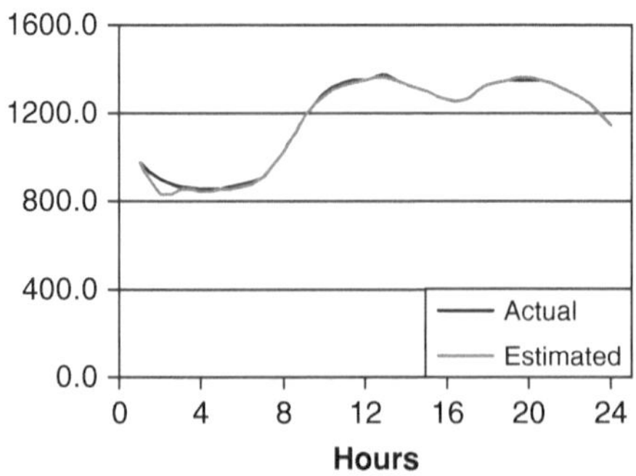

Figure 8.26 Actual and estimated loads over 24 hours.

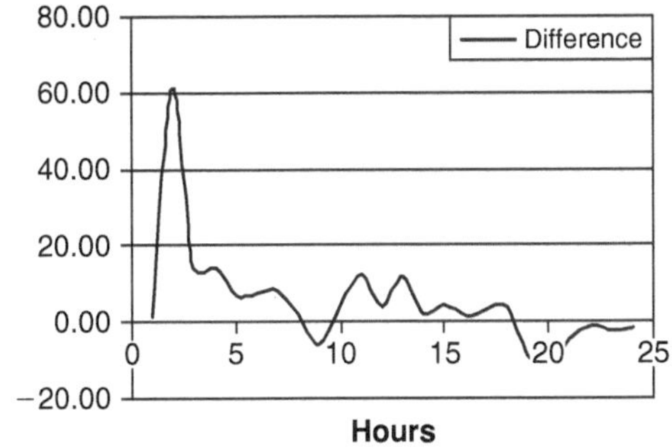

Figure 8.27 Difference between actual and estimated loads over 24 hours.

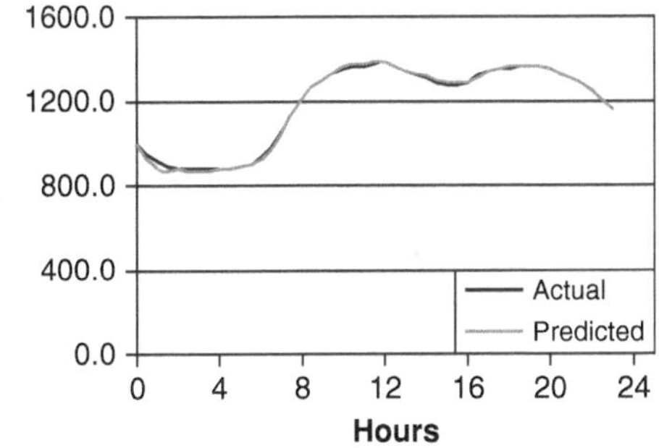

Figure 8.28 Actual and predicted loads over 24 hours.

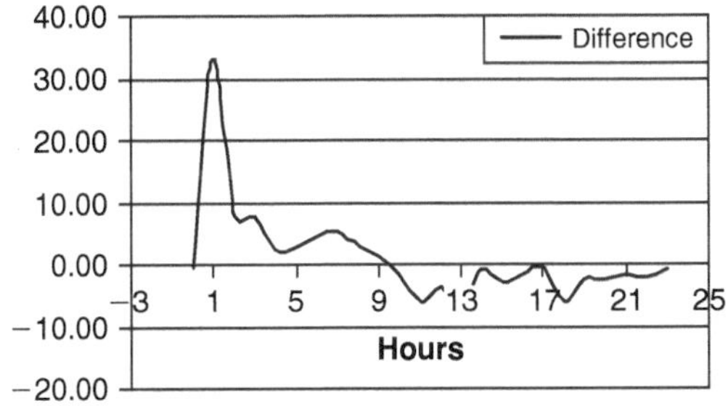

Figure 8.29 Difference between actual and predicted loads over 24 hours.

of data used for the former three cases. To increase the accuracy, we must increase the number of data. This task can be achieved in two possible ways. First, we can increase the actual past data provided for the Kalman filter estimation. However, this option would take the weather and load profiles to another season, which would be summer in this case. Most weather and load profiles experience major changes with the season's change, which has a great effect on the prevailing predicting model. The second way is to generate extra points between actual weather and load data using the interpolation technique. Tables 8.6 and 8.7 and Figure 8.31 depict the effect of introducing points using the

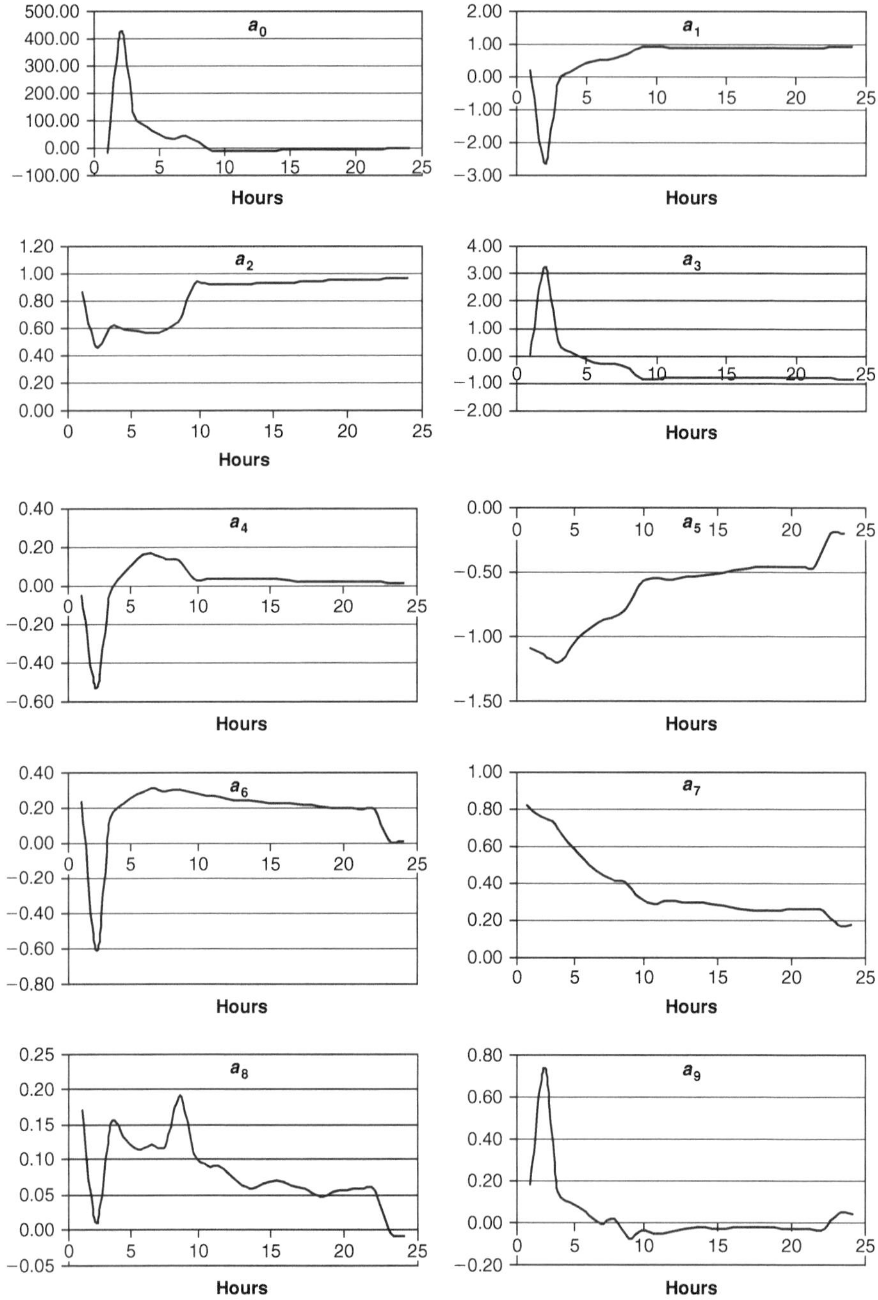

Figure 8.30 Variation of each of the 10 estimated parameters over 24 hours.

Table 8.5 Actual and Predicted Forecasted Loads for Two Months, Four Months, Weekdays, and Weekends of Four Months

Types of Days	All Days				All Days			Weekdays			Weekend			
No. of Months	**2 (Jan-95 to Feb-95)**				**4 (Nov-94 to Feb-95)**			**4 (Nov-94 to Feb-95)**			**4 (Nov-94 to Feb-95)**			
Interpolation Points	**2**				**2**			**2**			**2**			
Total Points	**169**				**352**			**250**			**94**			
Hour	**Actual**	**Pred**	**Diff**	**%**	**Pred**	**Diff**	**%**	**Pred**	**Diff**	**%**	**Actual**	**Pred**	**Diff**	**%**
0	1104.3	1112.4	−8.1	−0.73	1103.9	0.4	0.04	1065.7	38.6	3.50	912.7	897.8	14.9	1.63
1	1043.0	1063.6	−20.6	−1.97	1035.1	7.9	0.76	1037.6	5.4	0.52	838.3	828.6	9.7	1.16
2	1019.9	1013.1	6.8	0.67	1010.7	9.2	0.91	1012.3	7.6	0.75	799.0	785.1	13.9	1.75
3	1018.9	1012.9	6.0	0.59	1009.1	9.8	0.96	1010.4	8.5	0.83	780.5	772.6	7.9	1.01
4	1036.3	1023.3	13.0	1.25	1025.2	11.1	1.07	1027.4	8.9	0.86	781.3	771.3	10.0	1.28
5	1048.7	1046.6	2.1	0.20	1043.3	5.4	0.52	1047.0	1.7	0.16	783.0	779.6	3.4	0.43
6	1098.9	1090.2	8.7	0.80	1090.1	8.8	0.80	1095.6	3.3	0.30	808.6	801.2	7.4	0.92
7	1242.0	1210.5	31.5	2.54	1214.8	27.2	2.19	1233.1	8.9	0.72	853.1	848.1	5.0	0.59
8	1426.0	1397.1	28.9	2.03	1394.8	31.2	2.19	1424.1	1.9	0.13	937.4	929.8	7.6	0.81
9	1472.6	1469.1	3.5	0.24	1469.3	3.3	0.23	1473.7	−1.1	−0.08	1068.2	1048.7	19.5	1.83
10	1461.3	1465.2	−3.9	−0.27	1466.1	−4.8	−0.33	1463.6	−2.3	−0.16	1158.3	1142.7	15.6	1.35
11	1431.6	1433.2	−1.6	−0.11	1433.8	−2.2	−0.15	1435.6	−4.0	−0.28	1190.5	1180.8	9.7	0.82
12	1398.7	1370.7	28.0	2.00	1370.5	28.2	2.02	1397.4	1.3	0.09	1194.9	1216.1	−21.2	−1.78
13	1331.9	1333.2	−1.3	−0.10	1334.2	−2.3	−0.17	1334.4	−2.5	−0.19	1166.7	1137.5	29.2	2.51
14	1278.6	1275.4	3.2	0.25	1275.0	3.6	0.28	1279.9	−1.3	−0.10	1122.0	1122.1	−0.1	−0.01

Continued

Table 8.5 Actual and Predicted Forecasted Loads for Two Months, Four Months, Weekdays, and Weekends of Four Months *Continued*

Types of Days		**All Days**			**All Days**			**Weekdays**			**Weekend**			
No. of Months		**2 (Jan-95 to Feb-95)**			**4 (Nov-94 to Feb-95)**			**4 (Nov-94 to Feb-95)**			**4 (Nov-94 to Feb-95)**			
Interpolation Points			**2**			**2**			**2**				**2**	
Total Points			**169**			**352**			**250**				**94**	
Hour	**Actual**	**Pred**	**Diff**	**%**	**Pred**	**Diff**	**%**	**Pred**	**Diff**	**%**	**Actual**	**Pred**	**Diff**	**%**
15	1251.5	1245.2	6.3	0.50	1246.6	4.9	0.39	1255.5	−4.0	−0.32	1075.9	1079.0	−3.1	−0.29
16	1259.7	1252.1	7.6	0.61	1252.3	7.4	0.58	1258.7	1.0	0.08	1065.4	1070.7	−5.3	−0.50
17	1321.5	1315.4	6.1	0.47	1314.9	6.6	0.50	1317.0	4.5	0.34	1108.5	1109.2	−0.7	−0.07
18	1358.1	1360.4	−2.3	− 0.17	1361.0	−2.9	−0.22	1357.8	0.3	0.03	1163.8	1161.5	2.3	0.20
19	1388.6	1398.2	−9.6	− 0.69	1398.4	−9.8	−0.71	1388.9	−0.3	−0.02	1239.8	1223.0	16.8	1.36
20	1382.8	1386.4	−3.6	− 0.26	1386.5	−3.7	−0.27	1379.7	3.1	0.23	1234.5	1226.6	7.9	0.64
21	1348.1	1350.6	−2.5	− 0.19	1350.7	−2.6	−0.19	1345.2	2.9	0.22	1201.5	1196.6	4.9	0.41
22	1286.5	1288.9	−2.4	− 0.19	1289.3	−2.8	−0.22	1284.3	2.2	0.17	1158.1	1152.3	5.8	0.50
23	1182.9	1186.5	−3.6	− 0.31	1186.3	−3.4	−0.29	1186.9	−4.0	−0.34	1087.6	1082.4	5.2	0.48
M and *P* (Equations 8.20, 8.21):			61.2	0.71		57.5	0.67		44.0	0.43			57.6	0.93

Table 8.6 Effect of Increasing Number of Iterations on the Kalman Filter Parameter Estimation and Load Prediction for Sundays

Interpolation Points			**1**			**2**			**3**			**4**	
Total Points			**32**			**48**			**64**			**80**	
Hour	**Actual**	**Pred**	**Diff**	**%**	**Pred**	**Diff**	**%**	**Pred**	**Diff**	**%**	**Pred**	**Diff**	**%**
0	978.1	1116.1	−138.0	−14.11	1002.7	−24.6	−2.52	974.6	3.5	0.35	968.8	9.3	0.95
1	921.8	938.1	−16.3	−1.77	946.3	−24.5	−2.65	942.1	−20.3	−2.20	938.0	−16.2	−1.76
2	886.6	898.4	−11.8	−1.33	903.5	−16.9	−1.90	900.0	−13.4	−1.51	897.0	−10.4	−1.18
3	872.5	876.5	−4.0	−0.45	883.2	−10.7	−1.23	881.7	−9.2	−1.06	880.0	−7.5	−0.86
4	867.2	872.8	−5.6	−0.65	879.3	−12.1	−1.40	878.3	−11.1	−1.28	876.6	−9.4	−1.09
5	875.3	870.9	4.4	0.50	880.6	−5.3	−0.61	881.3	−6.0	−0.68	880.6	−5.3	−0.61
6	893.6	894.3	−0.7	−0.08	898.4	−4.8	−0.54	898.4	−4.8	−0.54	897.9	−4.3	−0.48
7	924.4	917.3	7.1	0.77	923.1	1.3	0.14	923.8	0.6	0.06	923.9	0.5	0.05
8	968.7	981.9	−13.2	−1.36	973.8	−5.1	−0.52	970.5	−1.8	−0.19	969.3	−0.6	−0.06
9	1054.2	1037.8	16.4	1.55	1045.9	8.3	0.79	1049.6	4.6	0.43	1051.3	2.9	0.27
10	1115.9	1104.3	11.6	1.04	1107.0	8.9	0.79	1108.8	7.1	0.64	1109.9	6.0	0.54
11	1130.7	1132.3	−1.6	−0.15	1130.7	0.0	0.00	1129.3	1.4	0.12	1128.7	2.0	0.18
12	1135.1	1140.9	−5.8	−0.51	1135.7	−0.6	−0.05	1133.9	1.2	0.11	1133.3	1.8	0.16
13	1114.2	1114.9	−0.7	−0.06	1114.9	−0.7	−0.07	1113.9	0.3	0.03	1113.3	0.9	0.09
14	1068.0	1081.8	−13.8	−1.29	1076.1	−8.1	−0.76	1072.7	−4.7	−0.44	1070.8	−2.8	−0.26
15	1034.3	1057.4	−23.1	−2.24	1048.4	−14.1	−1.36	1043.5	−9.2	−0.89	1040.7	−6.4	−0.62
16	1048.0	1071.2	−23.2	−2.21	1060.1	−12.1	−1.15	1055.3	−7.3	−0.70	1053.1	−5.1	−0.48
17	1108.9	1121.5	−12.6	−1.13	1114.8	−5.9	−0.53	1112.6	−3.7	−0.34	1111.7	−2.8	−0.25

Continued

Table 8.6 Effect of Increasing Number of Iterations on the Kalman Filter Parameter Estimation and Load Prediction for Sundays *Continued*

Interpolation Points			**1**			**2**			**3**			**4**	
Total Points			**32**			**48**			**64**			**80**	
Hour	**Actual**	**Pred**	**Diff**	**%**	**Pred**	**Diff**	**%**	**Pred**	**Diff**	**%**	**Pred**	**Diff**	**%**
18	1160.8	1169.0	−8.2	−0.71	1163.4	−2.6	−0.22	1161.6	−0.8	−0.07	1161.0	−0.2	−0.02
19	1188.2	1196.7	−8.5	−0.71	1186.1	2.1	0.17	1183.1	5.1	0.43	1182.2	6.0	0.51
20	1173.2	1185.2	−12.0	−1.03	1176.1	−2.9	−0.25	1172.9	0.3	0.03	1171.6	1.6	0.14
21	1130.7	1125.4	5.3	0.47	1125.6	5.1	0.45	1126.1	4.6	0.41	1126.6	4.1	0.37
22	1070.1	1084.4	−14.3	−1.34	1076.0	−5.9	−0.55	1072.5	−2.4	−0.23	1070.8	−0.7	−0.07
23	982.2	988.4	−6.2	−0.64	985.2	−3.0	−0.31	982.9	−0.7	−0.07	981.7	0.5	0.05
M and *P* (Equations 8.20, 8.21):			148.9	1.50		50.1	0.79		34.2	0.53		29.0	0.46

Table 8.7 Effects of Interpolation Points
Weekend Days (Nov-94 to Feb-95)

N	*N* **Pts**	*M*	*P*
1	62	89.3	1.31
2	94	57.6	0.93
3	126	48.6	0.74
4	158	41.6	0.60
5	190	36.0	0.49
6	222	31.6	0.42
7	254	28.6	0.37
8	286	29.6	0.39
9	318	27.4	0.34
10	350	24.7	0.33
11	382	29.0	0.38
12	414	20.7	0.27
13	446	23.6	0.29
14	478	15.2	0.17
15	510	19.7	0.26
16	542	48.1	0.32
17	574	33.2	0.46
18	606	34.7	0.35
19	638	78.4	0.45
20	670	26.9	0.40
21	702	28.8	0.25
22	734	63.2	0.37
23	766	12.4	0.15
24	798	16.9	0.23
25	830	12.8	0.17

cubic interpolation technique. As clearly seen, the prediction error is decreased by increasing the total number of points up to about 10 points. After 10 points, the results are indeterminately fluctuating, which indicates that increasing the number of interpolation points has little or no significance in improving the prediction accuracy.

To illustrate the effect of load-demand pattern on the parameter estimation, only Sunday's data are taken for winter months (November 1994 to February 1995). Because there are only 17 Sundays in the four months, there are not sufficient data for the Kalman filter to produce adequate convergence of the parameters. Extra points have been generated using the cubic interpolation technique. The parameter convergence and the 24-hour load prediction have been improved, as shown in Table 8.5. With one interpolation point, the total number of points used was 32. The prediction error is 14.1% because the Kalman filter did not get enough iteration for vigorous parameter convergence. This error decreases as the number of interpolation points is increased.

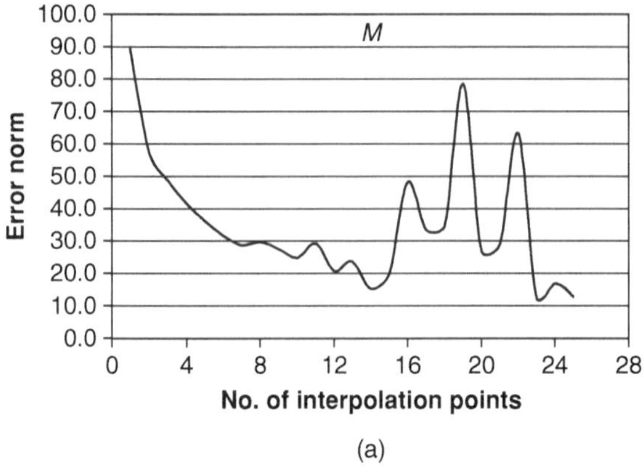

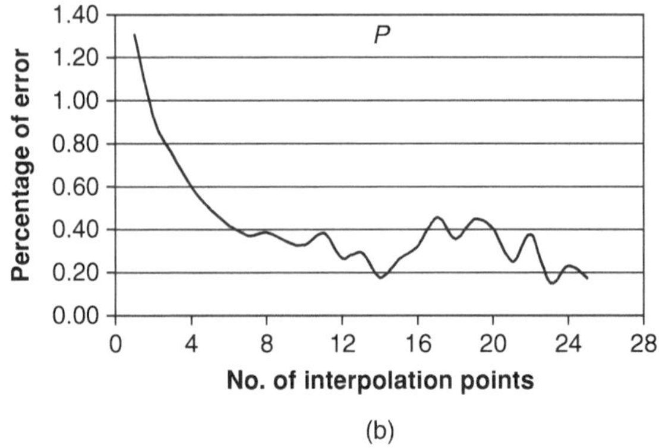

Figure 8.31 (a) *M* and (b) *P* as functions of interpolation points.

8.8.8 Conclusions

This section presented a novel algorithm-forecasting model that depends on a moving window of previous load and weather influential factors. The load at a certain instant of a time is expressed as a linear combination of previous load and weather data with time-varying coefficients. Kalman filter–based estimation was used to estimate the model parameters using previous history of load and weather data. This prediction technique has shown indisputably superior results over other techniques. The results were presented and compared with results based on the ordinary least squares–based techniques. The results showed a less than 1% prediction error for 24 hours, demonstrating superior performance of the presented model and technique. It has been shown, from the predicted load, that the proposed model with the Kalman filtering algorithm produces better and more accurate results compared to those obtained using different algorithms published earlier in the literature.

8.9 Fuzzy Load Forecasting Using the Kalman Filter

Electric load is sensitive to several weather factors, such as temperature, wind speed, humidity, and chill factor [8,9,10,11,17]. The main difficulty of modeling load demand is to come up with a model that comprises only dominant factors and minimizes the prediction error. Due to the increasing pressure on the need for accurate load forecasts, numerous models have been proposed for the short-term load-forecasting problem. Time-series prediction models—autoregressive (AR), moving average (MA), autoregressive moving average (ARMA), and autoregressive integrated moving average (ARIMA)—are the most popular. However, they are all non-weather-sensitive. Regression models that incorporate weather and previous load values are also widely used. Other weather-sensitive and non-weather-sensitive models are used in conjunction with the Kalman filtering technique. Fuzzy sets and fuzzy rule–based inference combined with neural networks and with regression models have been employed for solving the load-forecasting problem. The main reason behind their use is to improve prediction accuracy by importing expert knowledge along with the model. Expert knowledge, in the form of fuzzy rule–based logic, increases model reliability in cases of unusual severe weather conditions and varied holiday activities.

A combined fuzzy Kalman filter approach has been developed. A weather-sensitive fuzzy linear load model is employed that expresses load as a linear combination of weather factors and previous load values. The model utilizes fuzzy coefficients, each of which belongs to a fuzzy set of values represented by the triangular membership function. The coefficients are estimated using the Kalman filter technique in conjunction with expert knowledge fuzzy rule–based logic. Figure 8.32 illustrates the prediction process in a schematic block diagram.

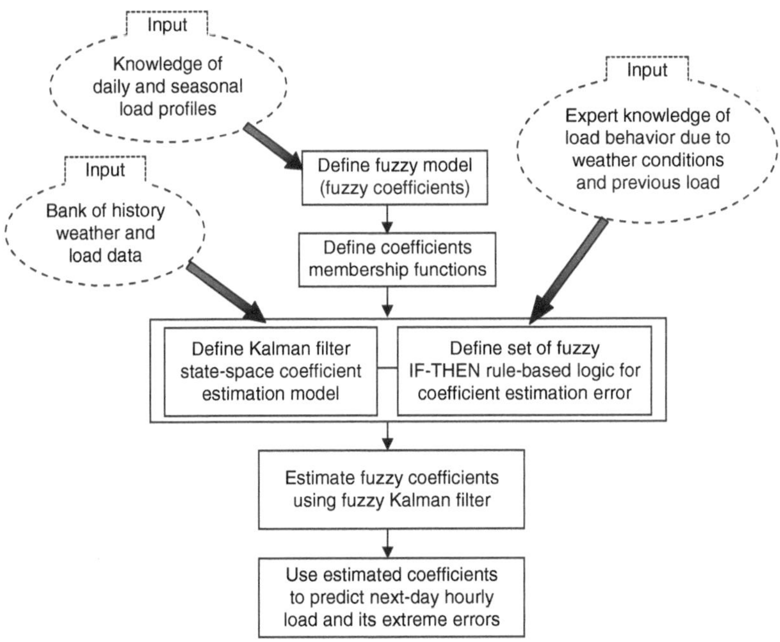

Figure 8.32 Schematic block diagram for the prediction process.

8.9.1 Fuzzy Linear Model

Electric load demand depends on a collection of seasonal, weekly, and daily factors of weather, as well as human habits and social behavior. Any parametric load model that strives to depict the dependence of the load on these factors is subject to inaccurate and imprecise representation, partly due to measuring instrument and human collection errors in the data involved. Therefore, the load model used in this section is a fuzzy parametric one in which the load is defined as a linear combination of weather and load values of the previous hours and days. The load model formulation is discussed in the following section. Moreover, fuzzy rule–based logic, together with Kalman filtering, is used to estimate the model fuzzy parameters.

In this subsection we present an overview of fuzzy linear models. A fuzzy linear model is given by $\underline{Y}$,

$$\underline{Y} = f(x) = \underline{A}_0 + \underline{A}_1 x_1 + \underline{A}_2 x_2 + \cdots + \underline{A}_n x_n \tag{8.23}$$

where $\underline{Y}$ is the dependent fuzzy variable (output), $\underline{x} = \{x_1, x_2, \ldots, x_n\}$ is a set of crisp (not fuzzy) independent variables, and $\{\underline{A}_0, \underline{A}_1, \ldots, \underline{A}_n\}$ is a set of symmetric fuzzy numbers. The membership function of $\underline{A}_i$ is a symmetrical triangle defined by center and spread values, p_i and c_i, respectively (see Figure 8.33). It is expressed as

$$\mu_{A_i}(a_i) = \begin{cases} 1 - \dfrac{|p_i - a_i|}{c_i} & p_i - c_i \le a_i \le p_i + c_i \\ 0 & \text{otherwise} \end{cases} \tag{8.24}$$

Therefore, with $\underline{A}_i = (p_i, c_i)$, the fuzzy function $\underline{Y}$ can be expressed as

$$\underline{Y} = f(x) = (p_0, c_0) + (p_1, c_1)x_1 + \cdots + (p_n, c_n)x_n \tag{8.25}$$

The membership function for the output $\underline{Y}$ is given by

$$\mu_{\underline{Y}}(y) = \begin{cases} \max\left(\min\left[\mu_{\underline{A}_i}(a_i)\right]\right) & \{a|y = f(x)\} \ne \text{nullset} \\ 0 & \text{otherwise} \end{cases} \tag{8.26}$$

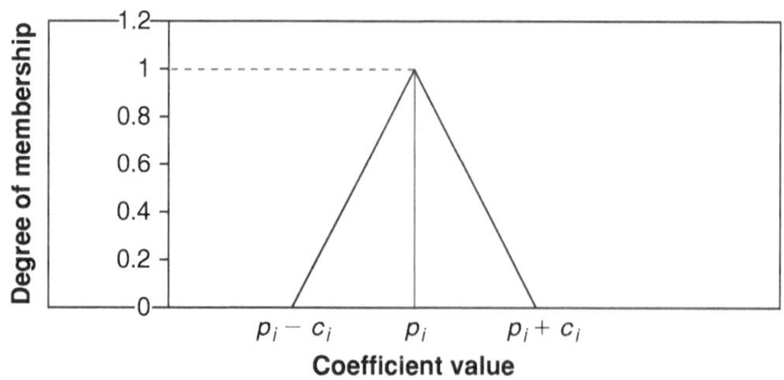

Figure 8.33 Membership function for fuzzy load model coefficients $\underline{A}_i = (p_i, c_i)$.

Substituting equation (8.25) in equation (8.26), we get

$$\mu_{\underline{Y}}(y) = \begin{cases} 1 - \dfrac{\left| y - \sum_{i=1}^{n} p_i x_i \right|}{\sum_{i=1}^{n} c_i |x_i|} & x_i \neq 0 \\ 1 & x_i = 0, \quad y = 0 \\ 0 & x_i = 0, \quad y \neq 0 \end{cases} \tag{8.27}$$

In the next section, the load-forecasting problem will be formulated as a fuzzy linear model.

A crisp load model of the load demand during a certain hour of the day is formulated as a linear function of previous values of load and weather factors. The load (weather) values belong to a moving window defined on a matrix of load (weather) data. The load (weather) matrix is an arrangement of one year's data in 24-hour columns and 365-day rows. In this subsection, a fuzzy load model of the form defined in equation (8.28) is employed. The load at any discrete time hour j of a day, $j = 1, \ldots, 24$, is expressed as

$$\underline{Y}_j = \underline{A}_{j,0} + \underline{A}_{j,1}\, y_{j-1} + \underline{A}_{j,2}\, y_{j-24} + \underline{A}_{j,3}\, y_{j-25} + \underline{A}_{j,4}\, t_j + \underline{A}_{j,5}\, w_j \tag{8.28}$$

where $\underline{A}_{j,i} = \left(p_{j,i}, c_{j,i}\right)$, $i = 0, \ldots, 5$, is the ith fuzzy coefficient belonging to a symmetric triangle defined by equation (8.24), and the load and weather values are defined as follows:

$\underline{Y}_j$ represents the current value of load at hour j of a day;
t_j represents the current value of temperature at hour j of a day;
w_j represents the current value of wind speed at hour j of a day;
y_{j-1} represents the load value of the previous hour;
y_{j-24} represents the load value of the previous day's hour;
y_{j-25} represents the load value of one hour before the previous day's hour.

Thus, 24 models in the form of equation (8.28) are defined for the 24 hours of the day. The load-forecasting problem is to find 24 sets of coefficients of the 24 models, one set per hour. Although the model is defined with only three values for previous load and only two values for weather, experimental results show that increasing the model order will not improve the prediction results. Moreover, the lower the order of the model, the faster the Kalman filter parameter estimation will be. The coefficients in the model are assumed constant over a one-hour time period; parameter estimation is carried out for each of the 24 discrete instances in a day. Accordingly, 24 sets of coefficients are required to be estimated for one day. The estimated coefficients can be plugged into the model to predict hourly loads for the next day. Parameter estimation using the Kalman filter is described in the following section.

8.9.2 Fuzzy Parameter Estimation Using Kalman Filtering

As we said earlier, the Kalman filter algorithm is a well-established technique for estimating the parameters of random processes. It is suitable for estimating stochastic processes, producing optimal estimates in the presence of noise. It formulates the parameter estimation problem as state space dynamic equations that include system as well as measurement-uncorrelated Gaussian white noises. The estimates produced are optimized in the least squares sense by minimizing error equations containing covariance matrices of both noises. Only necessary derivations of the filter and its recursive equations are presented.

8.9.3 Kalman Filter Prediction Model

In this section we use the fuzzy linear load model to form a discrete time-varying dynamic system suited for the Kalman filter. The dynamic system of equation (8.10) is used with the following definitions:

1. State transition matrix, $A(k)$, is a constant identity matrix.
2. The error covariance matrices, Q_1 and Q_2, are constant matrices. Q_1 and Q_2 values are based on some knowledge of the actual characteristics of the process and measurement noises, respectively. Q_1 and Q_2 are chosen to be identity matrices for this simulation.
3. The state vector, $x(k)$, consists of N variables. From equation (8.25), there are 6 fuzzy coefficients to estimate; therefore, $N = 12$—6 central and 6 spread values.
4. The output matrix $C(k)$ is a $2 \times N$ time-varying matrix. The first row relates the measured load and weather data to the state vector, and the second row defines the spread part of the fuzzy coefficients as a function of load and weather values. (Refer to equation (8.28).)
5. The observation vector, $z(k)$, for this application, is a 2×1 vector representing the measured load and defuzzified spread at time instant k. (Refer to equation (8.28).) The defuzzified spread is the output of an inferenced fuzzy rule–based machine.

The observation equation $z(k) = C(k)\, x(k)$ has the form

$$\begin{bmatrix} y_{k,j} \\ e_{k,j} \end{bmatrix} = \begin{bmatrix} 1 & y_{k,j-1} & y_{k,j-24} & y_{k,j-25} & t_{k,j} & w_{k,j} & 0 & 0 & 0 & 0 & 0 & 0 \\ 0 & 0 & 0 & 0 & 0 & 0 & 1 & y_{k,j-1} & y_{k,j-24} & y_{k,j-25} & t_{k,j} & w_{k,j} \end{bmatrix} \begin{bmatrix} p_{j,0} \\ p_{j,1} \\ \vdots \\ p_{j,5} \\ c_{j,0} \\ c_{j,1} \\ \vdots \\ c_{j,5} \end{bmatrix} \tag{8.29}$$

where the parameters and load weather values are defined in equation (8.28), with j representing the time instant of the 24 discrete hours of the day, $j = 1, \ldots, 24$, and k is the Kalman filter iteration index representing consecutive days. The value $e_{k,j}$ is an inferred value of the spread of $\underline{Y}_{k,j}$. It is obtained from a fuzzy rule–based machine that employs a set of propositions on the error of the previously estimated values of the load $y_{k,j}$.

8.9.4 Fuzzy Rule-Based Inference

The estimation of the model fuzzy parameters using Kalman filtering requires the left side of the observation equation, equation (8.28). Specifically, a measured load center (crisp) value of $\underline{Y}_{k,j}$ as well as its spread must be fed, as an observed input, to the recursive Kalman filter on every iteration. The observed value of the center is obviously the measured load value $y_{k,j}$. However, the observed value of the spread is unknown, and some expert knowledge must be used to deduce its value. An inference of the spread value is formed using a rule-based fuzzy logic inference machine described next.

Define the load estimation error as the difference between actual and estimated load values at any hour j of day k:

$$e_{k,j} = y_{k,j} - \hat{y}_{k,j} \tag{8.30}$$

The load spread value is determined using a two-input, one-output fuzzy logic machine (see Figure 8.34).

The fuzzy logic machine accepts fuzzified linguistic fuzzy variables for the inputs. Each linguistic variable belongs to a set of values and is represented by a triangular membership function in the range [0, 1], which corresponds to the degree to which the input belongs to the linguistic class. Five linguistic variables are defined (see also Figure 8.35):

LN = Large Negative
SN = Small Negative
ZE = Zero
SP = Small Positive
LP = Large Positive

The linguistic variables considered here belong to a simple fuzzy set consisting of five variables. Extra variables can be added by including more fuzzy classifiers such as VLN (very large negative), VSN (very small negative), VLP (very large positive), and VSP (very small positive). However, experimental results revealed that increasing the number of classifiers (partitions) did not improve the forecast results. Therefore, five linguistic variables were found to be sufficient for this application. These linguistic variables represent the fuzzy input and desired fuzzy output of the fuzzy logic machine. The sets are used to describe the load error for forming the rule base. We apply disjunctive canonical IF-THEN rules of inference based on our knowledge of

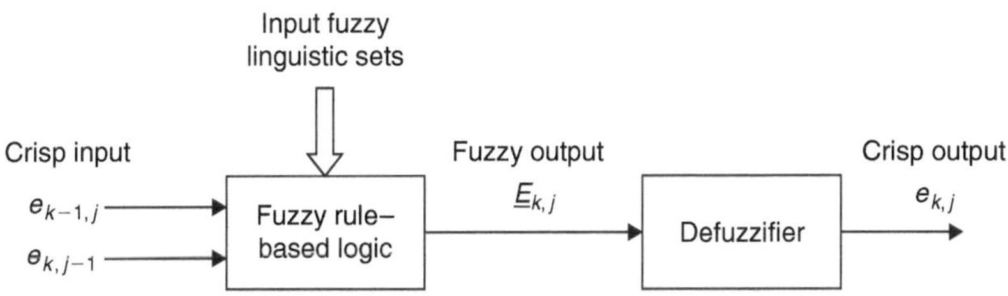

Figure 8.34 Fuzzy logic machine.

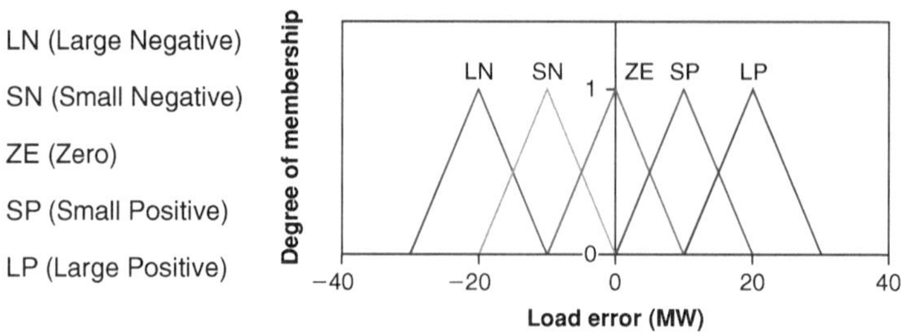

Figure 8.35 Membership functions for fuzzy load error variables.

Table 8.8 Load Error Fuzzy Logic Rules

		$e_{k-1,j}$				
		LN	**SN**	**ZE**	**SP**	**LP**
$e_{k,j-1}$	**LN**	LN	SN	SN	ZE	ZE
	SN	SN	SN	ZE	ZE	SP
	ZE	SN	ZE	ZE	SP	SP
	SP	ZE	ZE	SP	SP	LP
	LP	ZE	SP	SP	LP	LP

the daily weekly load profiles. The IF-THEN rules are employed to make a more accurate inference for the variations of forecast error from the linear load model. Referring to Figure 8.35, for different inputs $e_{k-1,j}$ and $e_{k,j-1}$, we can obtain a set of fuzzy logic rules according to Table 8.8.

Table 8.8 lists the 25 possible canonical fuzzy logic IF-THEN rules used to determine the degree of membership of $e_{k,j}$ depending on the fuzzy inputs $e_{k-1,j}$ and $e_{k,j-1}$. Following are two examples of such rules:

$$\begin{aligned}&\textbf{Rule 1:}\ \text{IF } e_{k-1,j} \text{ is SP AND } e_{k,j-1} \text{ is ZE THEN } e_{k,j} \text{ is SP}\\&\textbf{Rule 2:}\ \text{IF } e_{k-1,j} \text{ is LP AND } e_{k,j-1} \text{ is SP THEN } e_{k,j} \text{ is LP}\end{aligned} \quad (8.31)$$

The rules in Table 8.8 reflect the dominance of $e_{k,j-1}$ over $e_{k-1,j}$ formed from our long experience with the daily and weekly load profiles. Typical load profiles such as the ones in Figures 8.36 and 8.37 show that there is a greater load fluctuation of the same hour over several days (Figure 8.37) than that of the hours of the same day (Figure 8.36). Accordingly, the fuzzy rules of Table 8.8 are established, emphasizing $e_{k,j-1}$ over $e_{k-1,j}$.

The fuzzy inference method used to implement the fuzzy rules was the max-min composition based on the Mamdani implication method of inference. For a set of r disjunctive rules, the aggregated output membership function is given by

$$\mu_{\underline{B}}(e_{k,j}) = \max_i \left[\min\left\{\mu_{\underline{A_1}}(e_{k-1,j}), \mu_{\underline{A_2}}(e_{k,j-1})\right\}\right] \qquad i = 1, 2, \ldots, r \quad (8.32)$$

where $\mu_{\underline{B}}$, $\mu_{\underline{A_1}}$, and $\mu_{\underline{A_2}}$, are output and input membership functions, respectively.

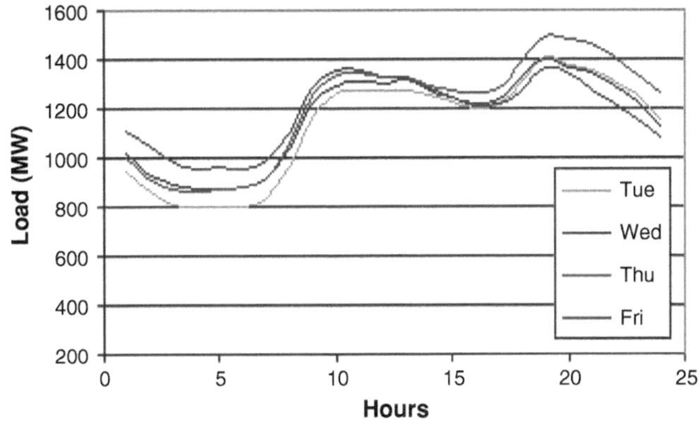

Figure 8.36 Load of first weekdays of January 1995.

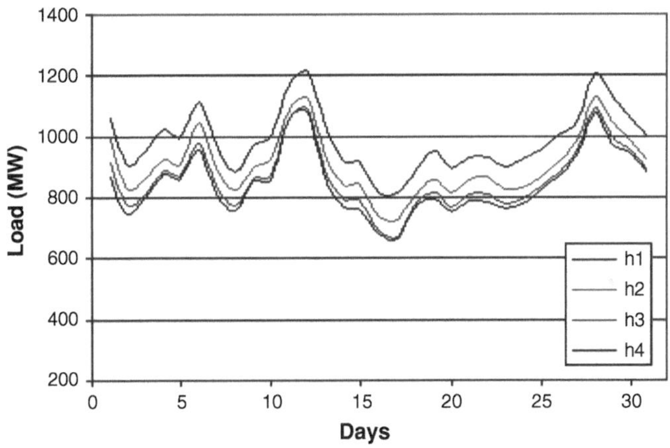

Figure 8.37 Load curves of the same hour of previous days, hours 1–7, January 1995.

A simple numeric and graphical illustration of applying equation (8.32) to only two rules of equation (8.31) is given in Figure 8.38, with $e_{k-1,j}=12$ and $e_{k,j-1}=5$. The first and second rows of Figure 8.38 refer to the two inputs and their minimum (min) fuzzy output of Rule 1 and Rule 2, respectively. The last row shows the maximum (max) of the two (min) fuzzy outputs. The extension for more than two rules is straightforward, as given by equation (8.32). We should notice that the aggregate output membership function in Figure 8.39 defines a set of values [0, 30] for the output, but it does not define the output's crisp (center) value.

We next demonstrate the computation of the crisp output. After obtaining the output membership function using the max-min method, we evaluate the output crisp

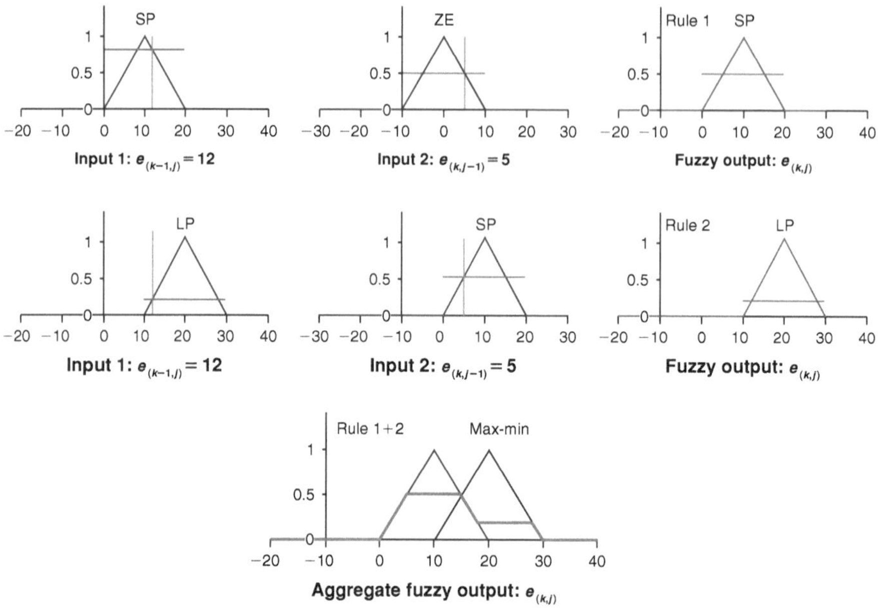

Figure 8.38 Graphical (max-min) inference method.

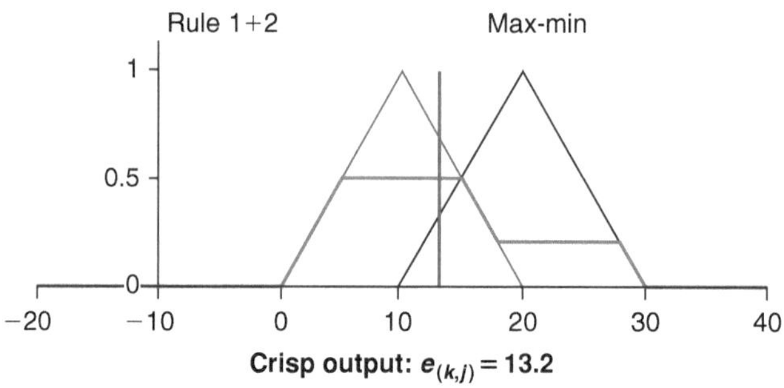

Figure 8.39 Crisp value using the centroid defuzzification method.

(center) value using the "defuzzification" process. (Refer to Figure 8.38.) The centroid defuzzification method was used to convert the fuzzy output to a crisp one. The method is defined by the algebraic expression in equation (8.31):

$$z = \frac{\int z\mu_{\underline{A}}(z)dz}{\int \mu_{\underline{A}}(z)dz} \quad \text{for continuous } \mu_{\underline{A}} \tag{8.33a}$$

$$z = \frac{\sum_{i=1}^{N} z_i \, \mu_{\underline{A}}(z_i)}{\sum_{i=1}^{N} \mu_{\underline{A}}(z_i)} \quad \text{for discrete } \mu_{\underline{A}} \tag{8.33b}$$

where N is the number of discrete points of the membership function. As an illustration, the crisp value of the fuzzy output in Figure 8.22 is the centroid (center of area) of the graph. Using equation (8.33a), we compute the centroid to be $e_{k,j} = 13.2$, which is illustrated in Figure 8.23.

8.10 Model Validation and Results

To measure the effectiveness of load prediction, we use the mean absolute percentage error with respect to the actual load to measure the error over the 24 hours of a day, and we use its standard deviation to measure its variation. They are given by the equations

$$\text{MAPE} = \left(\frac{|d_1|}{y_{a1}} + \cdots + \frac{|d_{24}|}{y_{a24}}\right)\left(\frac{100}{24}\right) \tag{8.34}$$

$$\text{SDAPE} = \left[\left(\frac{100|d_1|}{y_{a1}} - \text{MAPE}\right)^2 + \cdots + \left(\frac{100|d_1|}{y_{a1}} - \text{MAPE}\right)^2\right]\left(\frac{1}{24}\right) \tag{8.35}$$

where $d_k = y_{pk} - y_{ak}$, $k = 1, \ldots, 24$; and y_{pk} and y_{ak} are the predicted and actual loads at time instant k, respectively.

8.10.1 One-Day Parameter Estimation and Load Prediction

The forecast technique is tested on the 1994 summer data (May to August) for one of the largest power utilities in Canada. Special consideration is taken in treating weekdays differently than weekends. Thus, two sets of data are generated: a weekdays set and a weekend set. The estimation algorithm is executed on each set separately. The fuzzy model of equation (8.28) is used to model the load demand. The load and weather data for four months (May to August 1994) are used to estimate the load model coefficients. Moreover, three extra points between every two actual values of load as well as weather are generated using cubic interpolation to boost the Kalman filter iterations. For each hour of the day, a set of 12 parameters is estimated (6 for the center and 6 for the spread). The estimated parameters are used to predict the load and its spread for the first hour of the first day of September 1994. The one-hour prediction is repeated for 24 hours of that day, producing 24 12-coefficient sets and resulting in 24 predicted values of the load and its spread.

As a sample of the estimated parameters, Figure 8.40 presents the estimated model coefficients of the Kalman filter iterations for hour 0 and hour 1. Similar results are obtained for the rest of the 24 hours. As illustrated, all estimated parameters converge to their steady-state values after some transient fluctuations. Table 8.9 presents a sample of the 24 sets of steady-state values of estimated coefficients for the 24 hours of the day; each set consists of 12 values. The variations of these coefficients over the 24 hours are shown in Figure 8.41.

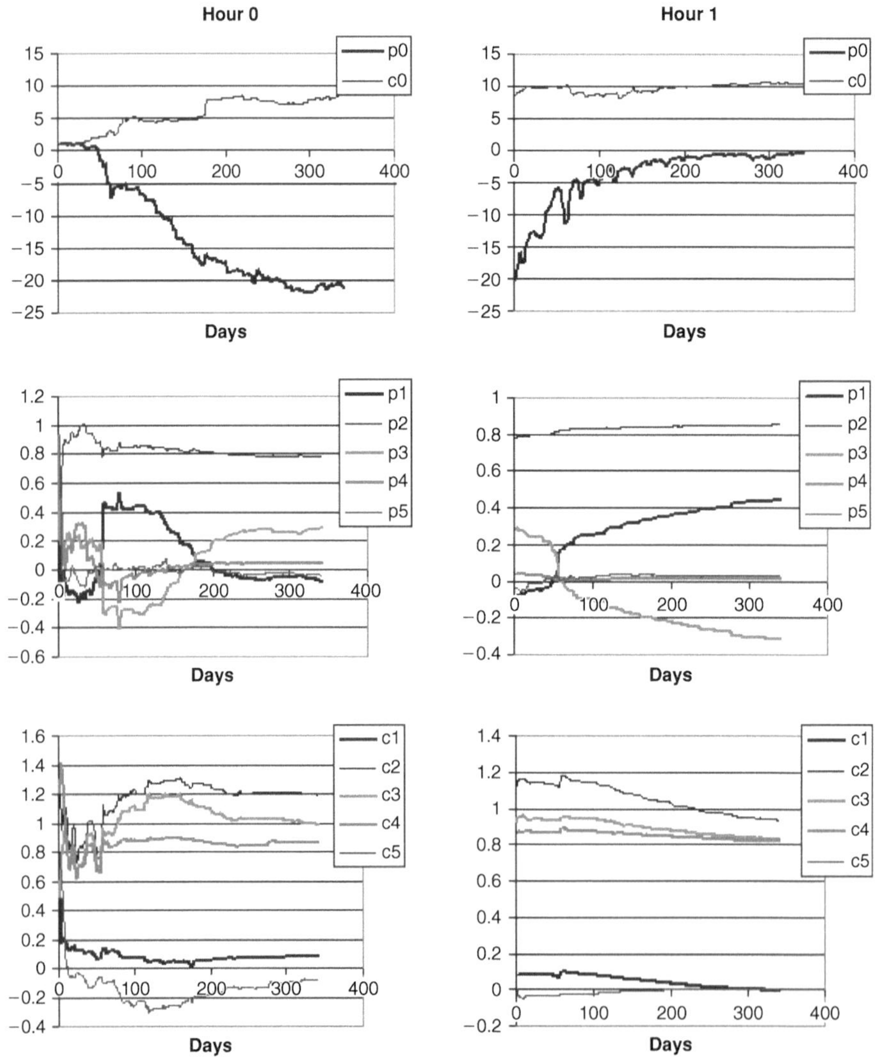

Figure 8.40 Estimation results: convergence of the coefficient's centers and spreads.

Table 8.9 Sample of the 24 Sets of Steady-State Centers and Spreads of Fuzzy Model Coefficients

	Center Values of Model Fuzzy Coefficients						Spread Values of Model Fuzzy Coefficients					
Hour	p_0	p_1	p_2	p_3	p_4	p_5	c_0	c_1	c_2	c_3	c_4	c_5
0	–20.708	–0.079	0.782	0.289	0.043	–0.035	8.994	0.082	1.197	0.992	0.861	–0.085
1	–0.437	0.441	0.853	–0.317	0.015	0.030	10.398	–0.007	0.938	0.830	0.818	0.006
2	8.098	0.560	0.906	–0.491	0.004	0.028	3.934	–0.058	0.852	0.819	0.743	–0.012
10	1.595	0.852	0.994	–0.846	0.002	–0.018	0.949	–0.253	0.336	0.570	0.563	–0.017
11	–0.016	0.847	0.994	–0.841	0.001	–0.004	0.824	–0.263	0.310	0.554	0.547	–0.018
20	0.771	0.867	0.991	–0.860	0.001	–0.004	0.128	–0.262	0.242	0.479	0.473	–0.019

The estimated parameters are used to predict demand load for one day ahead. The results are shown in Table 8.10, which presents two sets of prediction results: one for September 1, 1994, and the other for September 1, 1995. In each set, the actual and predicted load (center and spread) for the following 24 hours are shown. The last two columns of each set show the prediction error and its percentage, respectively. To compare the two sets of results, Figures 8.41, 8.43, 8.44, and 8.45 display both the 1994 as well as 1995 results. Figure 8.44 presents two sketches of the load prediction error over 24 hours. A comparison between the predicted load resulting from the estimated parameters and the actual load is shown in Figure 8.41. The results show how closely the prediction model matches the actual load. The load's MAPE for the predicted day (September 1, 1994) is 0.606%, and the standard deviation of the absolute percentage error (SDAPE) is 0.837% (the two numbers are the mean and standard deviation of the absolute values of the last column, respectively). We should note that although the resulting load's MAPE error is 0.606%, it is dependent on the data; different data may result in a different value of MAPE. We have obtained higher and lower MAPE errors for different days. The results shown in Tables 8.9 and 8.10 are samples to illustrate the effectiveness of the method. Figure 8.44 illustrates the relation between the predicted spread values and the predicted error, and Figure 8.45 displays the actual load and the lower and upper envelopes of the predicted spreads.

8.10.2 Up to 60 Days of Load Prediction

A sample of the Kalman filter parameter estimation iterations for hour 10 is presented in Table 8.11. This table illustrates that the load-estimated coefficients are almost constant for high iterations. The same results are obtained for the other hours of the day. To predict the load for a few days ahead, we will use only one set of the estimated coefficients for each hour. For example, the set at iteration 390 in Table 8.11 can be used to predict the load for the following days without the need to estimate the coefficients for those days.

Tables 8.12 and 8.13 (shortened to five hours) present the predicted load absolute percentage error for the last 10 days of August 1994. Table 8.12 uses a newly estimated set of coefficients for each predicted day, and Table 8.13 uses only *one set* of estimated parameters. The last two rows of the two tables display, respectively, the MAPE and SDAPE for each day (24 hours). The averages of the 10 days' MAPEs are 0.217 and 0.218 for Tables 8.12 and 8.13, respectively; and their standard deviations, SDAPEs, are 0.219 MW and 0.222 MW, respectively. The two tables display very close prediction results. Figure 8.46 demonstrates the prediction results of day 30 when predicting 30 days ahead.

Next, the prediction technique is used to forecast the load for 60 days ahead using one set of estimated coefficients. The coefficients are estimated using load and weather working days' data from May 1, 1994, to October 7, 1994, and are used to predict the load of the working days from October 10, 1994, to December 30, 1994. The averages of MAPEs and SDAPEs for the number of days are summarized in Table 8.14.

Table 8.10 Prediction Results (September 1, 1994, and September 1, 1995)

	1994					1995				
	Actual	Predicted		Prediction Error		Actual	Predicted		Prediction Error	
Hour	Load	Center	Spread	Error	%	Load	Center	Spread	Error	%
0	672.6	683.6	−14.18	−10.99	−1.63	955.4	949.3	3.82	6.10	0.64
1	607.6	609.8	−4.11	−2.17	−0.36	853.6	846.7	10.94	6.92	0.81
2	572.0	573.6	−2.98	−1.63	−0.29	796.3	790.4	10.61	5.95	0.75
3	560.5	559.9	−0.61	0.55	0.10	775.9	769.6	10.18	6.24	0.80
4	552.0	552.1	−1.12	−0.08	−0.02	771.1	764.5	9.59	6.57	0.85
5	557.5	557.4	−1.69	0.05	0.01	775.8	770.7	7.80	5.12	0.66
6	584.8	583.5	−0.39	1.23	0.21	801.8	796.9	8.08	4.97	0.62
7	664.9	658.5	4.06	6.47	0.97	872.8	870.5	7.55	2.36	0.27
8	790.0	788.0	3.77	1.94	0.25	986.8	986.9	4.84	−0.15	−0.02
9	902.4	902.1	3.32	0.35	0.04	1094.8	1095.3	2.86	−0.54	−0.05
10	943.3	947.6	−0.94	−4.25	−0.45	1185.8	1181.2	4.59	4.58	0.39
11	1115.7	1077.1	22.19	38.56	3.46	1222.0	1220.7	3.66	1.35	0.11
12	1018.0	1044.5	3.84	−26.49	−2.60	1234.7	1234.3	2.97	0.34	0.03
13	1008.9	1007.3	4.21	1.63	0.16	1212.9	1212.3	3.12	0.60	0.05
14	998.9	994.6	6.18	4.30	0.43	1182.1	1181.3	3.28	0.80	0.07
15	981.1	976.6	7.59	4.42	0.45	1150.8	1151.9	2.26	−1.07	−0.09
16	978.6	976.7	7.38	1.94	0.20	1152.5	1153.9	1.21	−1.39	−0.12
17	1003.1	996.5	10.31	6.59	0.66	1232.5	1233.5	0.21	−1.06	−0.09
18	975.8	975.0	8.09	0.84	0.09	1343.5	1342.3	0.27	1.16	0.09
19	915.0	920.0	4.04	−4.97	−0.54	1329.5	1329.2	−0.01	0.30	0.02
20	890.2	888.3	4.89	1.88	0.21	1289.8	1290.8	−0.79	−0.94	−0.07
21	916.8	915.3	5.32	1.54	0.17	1251.0	1250.8	−0.50	0.20	0.02
22	908.9	902.2	7.16	6.67	0.73	1186.0	1188.1	−1.58	−2.08	−0.18
23	812.7	817.0	2.88	−4.31	−0.53	1085.9	1085.6	−1.45	0.30	0.03

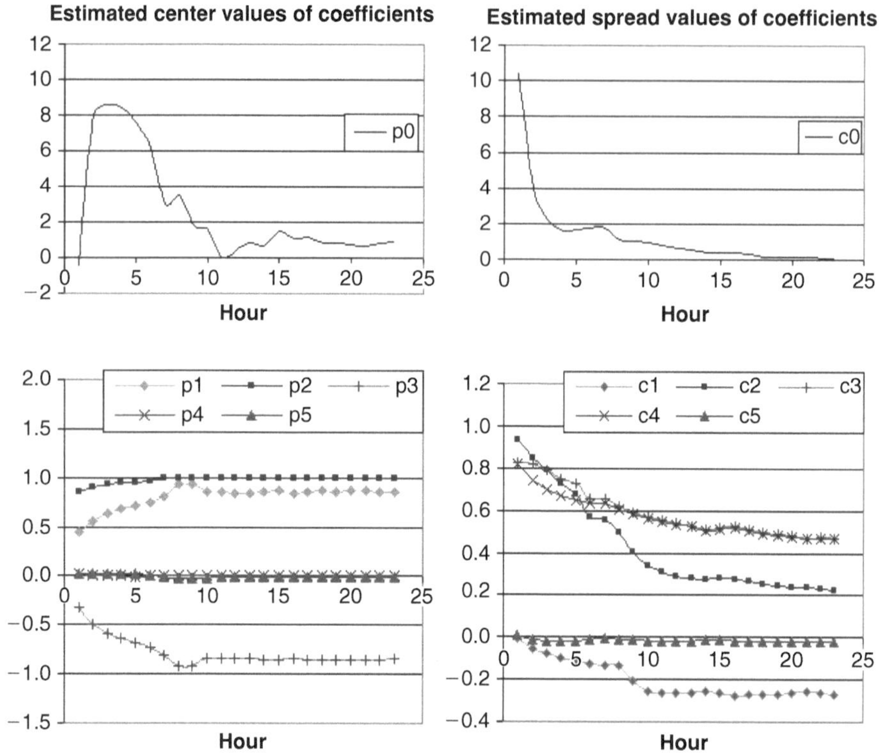

Figure 8.41 Steady state of estimated centers and spreads of fuzzy coefficients.

Finally, to show the effectiveness of the technique, Table 8.15 presents a comparison of various forecasting methods. It illustrates that the presented fuzzy Kalman filter approach has a lower mean absolute percentage error compared to the other techniques. We should mention that the numbers in Table 8.15 are data dependent; they reflect the general performance of the various forecasting approaches rather than their preciseness.

8.10.3 Conclusions

In this section we presented a load prediction methodology for the short-term load-forecasting problem to estimate hourly load. The prediction was carried out using a linear fuzzy model of the previous year's loads and weather. The Kalman filtering technique with a fuzzy rule-based inference was employed to estimate the model fuzzy coefficients. The rule-based inference was used to estimate load error utilizing our knowledge of the daily and seasonal load profiles. The method produced both crisp and spread values of the forecast load. The results showed that the prediction has 0.7% MAPE and 0.84% SDAPE over 24 hours. Moreover, the predicted

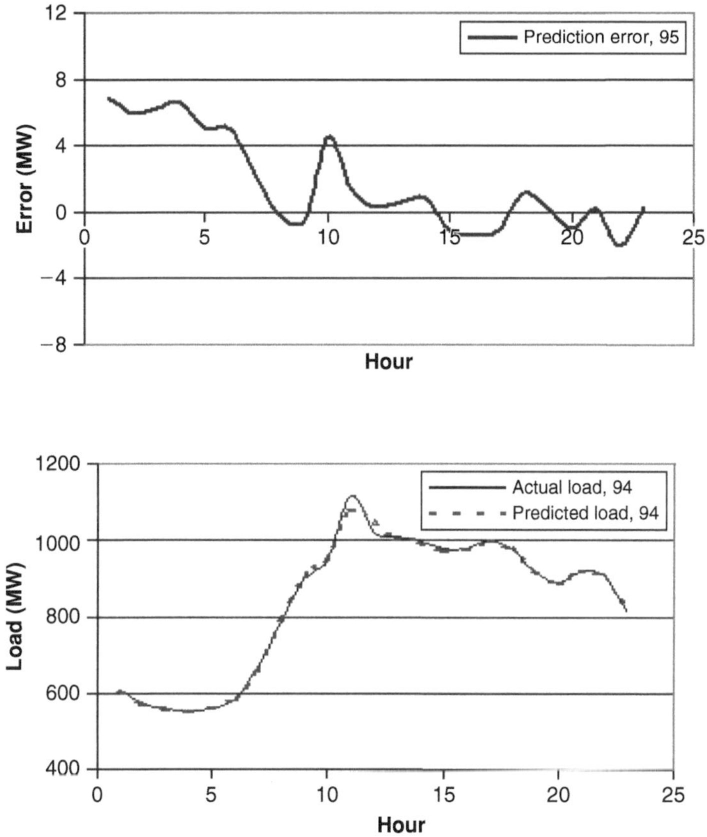

Figure 8.42 Load prediction error.

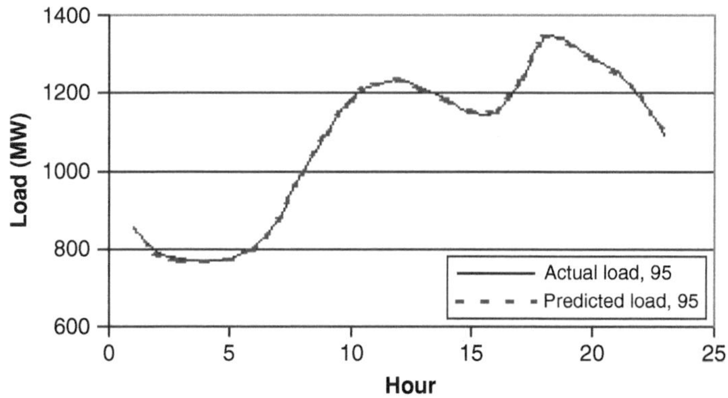

Figure 8.43 Comparing actual and predicted loads.

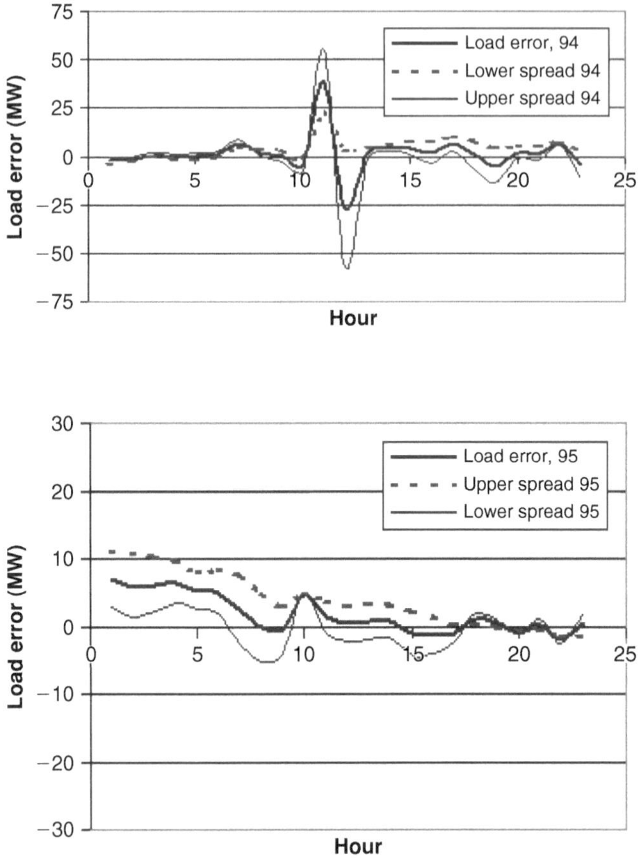

Figure 8.44 Comparing load predicted error and predicted spread.

spread value is useful for determining the extremes the load may reach at each hour.

8.11 Recursive Least Error Squares

In this section, we present the recursive least error squares (RLES) algorithm to predict the short-term electric load. The three models—model A, model B, and model C—explained in Chapter 3 are used. Model A is a regression model of the weather factors, and model B is a harmonic model of the time horizon only. Furthermore, model C is a hybrid model of models A and B. The parameters of these three models are estimated using the proposed algorithm. These coefficients are used to predict the load hour by hour for a summer day and a winter day. The results obtained for a practical system in operation are compared with the past least

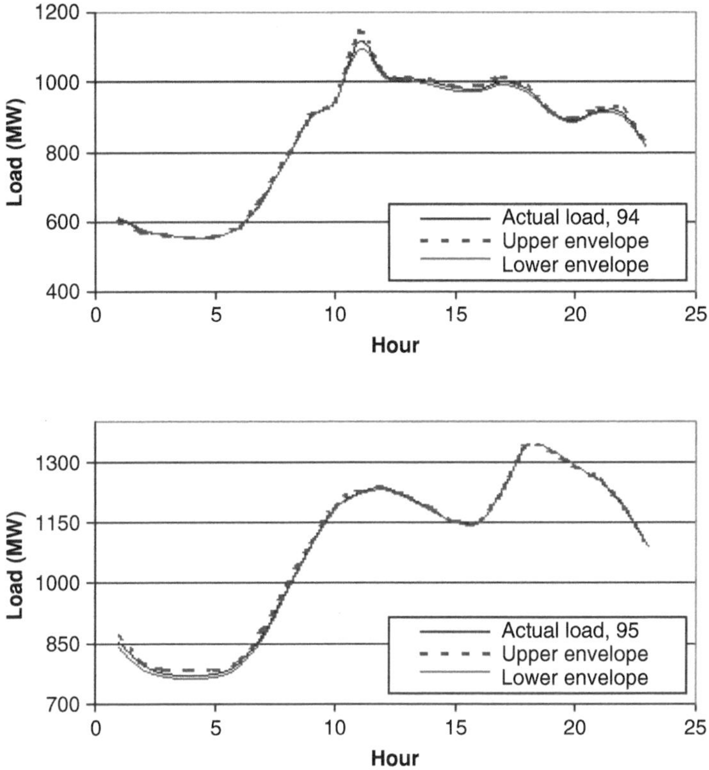

Figure 8.45 Actual load and spread envelopes.

error squares and the least absolute value (static estimation). It has been shown that the proposed algorithm produces good results for the load model.

8.11.1 Testing the Algorithm

In this section we use the proposed algorithm to predict the load power one hour ahead, where we use the load models developed. The parameters of these models are estimated using the past history data for summer weekdays and weekend days as well as for winter weekdays and weekend days. Then these models are used to predict the load power for 24 hours ahead, in both summer and winter seasons. The results are given in the form of figures for the estimated and predicted loads for the summer season.

8.11.1.1 Model A

The proposed algorithm is used to estimate the parameters of model A for a summer weekday. These parameters are used to reconstruct the load. Figure 8.47 gives the estimated load for a summer weekday using the proposed algorithm, where we compare with the results obtained using the past least error squares (LES) and least

Table 8.11 Estimated Parameters for a Few Kalman Filter Iterations for Hour 10

	Center Values of Model Fuzzy Coefficients						Spread Values of Model Fuzzy Coefficients					
Iter	p_0	p_1	p_2	p_3	p_4	p_5	c_0	c_1	c_2	c_3	c_4	c_5
390	4.0722	0.9322	0.9940	–0.9274	–0.0021	–0.0347	1.3086	–0.1651	0.3066	0.4415	0.4348	–0.0267
391	4.0737	0.9321	0.9940	–0.9274	–0.0021	–0.0347	1.3077	–0.1651	0.3066	0.4415	0.4348	–0.0267
392	4.0741	0.9321	0.9940	–0.9274	–0.0021	–0.0346	1.3068	–0.1651	0.3066	0.4415	0.4348	–0.0267
393	4.0740	0.9321	0.9940	–0.9274	–0.0021	–0.0346	1.3062	–0.1651	0.3065	0.4415	0.4348	–0.0266
394	4.0734	0.9322	0.9940	–0.9274	–0.0021	–0.0346	1.3057	–0.1651	0.3065	0.4415	0.4348	–0.0266

Table 8.12 Prediction Using New Estimated Parameters for Everyday

	Prediction Absolute Percent Error (Only 5 Hours Shown)									
Hour	**Day 1**	**Day 2**	**Day 3**	**Day 4**	**Day 5**	**Day 6**	**Day 7**	**Day 8**	**Day 9**	**Day 10**
0	0.091	0.406	0.807	0.744	0.172	0.656	0.850	0.082	0.291	0.388
1	0.049	0.448	0.491	0.011	0.181	0.428	0.117	0.180	0.348	0.159
2	0.010	0.044	0.516	0.652	0.075	0.077	0.183	0.141	0.137	0.066
3	0.078	0.216	0.289	0.315	0.037	0.097	0.004	0.218	0.097	0.046
4	0.093	0.071	0.334	0.132	0.061	0.060	0.034	0.120	0.019	0.085
MAPE	0.161	0.212	0.283	0.236	0.110	0.310	0.184	0.242	0.152	0.277
SDAPE	0.124	0.143	0.212	0.201	0.084	0.394	0.195	0.307	0.103	0.431

Table 8.13 Prediction Using One Set of Estimated Parameters

	Prediction Absolute Percent Error (Only 5 Hours Shown)									
Hour	**Day 1**	**Day 2**	**Day 3**	**Day 4**	**Day 5**	**Day 6**	**Day 7**	**Day 8**	**Day 9**	**Day 10**
0	0.091	0.410	0.828	0.778	0.228	0.675	0.882	0.053	0.322	0.456
1	0.049	0.450	0.510	0.000	0.175	0.437	0.137	0.197	0.375	0.140
2	0.010	0.044	0.512	0.660	0.063	0.086	0.194	0.153	0.151	0.053
3	0.078	0.216	0.283	0.318	0.038	0.102	0.001	0.220	0.103	0.042
4	0.093	0.071	0.335	0.136	0.059	0.063	0.037	0.123	0.015	0.082
MAPE	0.161	0.213	0.284	0.238	0.112	0.312	0.186	0.244	0.155	0.280
SDAPE	0.124	0.143	0.215	0.206	0.086	0.394	0.198	0.309	0.108	0.435

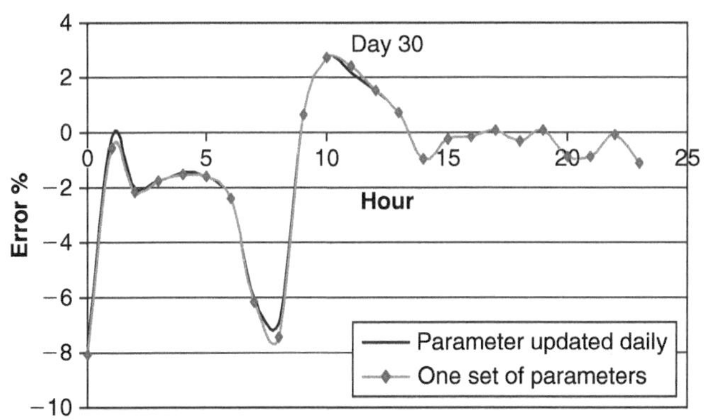

Figure 8.46 Comparison of prediction error for 30 days ahead using two different sets of estimated parameters.

Table 8.14 Prediction Errors for Up to 60 Days' Forecasts (Only 20 Shown)

No. of Days	Avg (MAPE)	Avg (SDAPE)	No. of Days	Avg (MAPE)	Avg (SDAPE)
1	1.256	1.702	51	0.665	0.650
2	0.979	1.290	52	0.667	0.649
3	0.784	0.946	53	0.668	0.651
4	0.659	0.763	54	0.667	0.650
5	0.611	0.683	55	0.670	0.652
6	0.581	0.647	56	0.668	0.647
7	0.560	0.607	57	0.668	0.643
8	0.552	0.584	58	0.665	0.638
9	0.541	0.559	59	0.675	0.646
10	0.557	0.561	60	0.686	0.659

Table 8.15 Comparison of MAPE Forecast Errors

Fuzzy Forecast Using Kalman Filter, Presented	Parallel Neural Network Fuzzy Expert System	Kohonen Neural Network And Wavelet Transform	Artificial Neural Network and Fuzzy Expert System	Regression Model	Multiple Linear Regression Used by Electric Power Utilities
This chapter	Ref. [14]	Ref. [5]	Ref. [13]	Ref. [6]	Ref. [14]
0.70	0.83	0.98	1.07	1.37	2.20

absolute value (LAV) algorithms. Examining this curve reveals the following observations:

- The parameters estimated using the proposed algorithm are as good as those estimated using the past LES and LAV.
- The estimated load is close to the actual load, which means that the model assumed with the number of parameters, 10 parameters, is accurate enough to present the load.

The parameters estimated in this test are used to predict the load 24 hours ahead for a summer weekday. Figure 8.48 gives the results obtained, where we also compare the proposed algorithm with the past LES and LAV. Examining this curve reveals the following observations:

- The proposed algorithm produces a good prediction that is better than the past LES and LAV.
- The proposed algorithm produces exactly the actual load.

Figures 8.49 and 8.50 give the results obtained for a summer weekend day. In Figure 8.49 the parameters are estimated using the proposed technique, and the

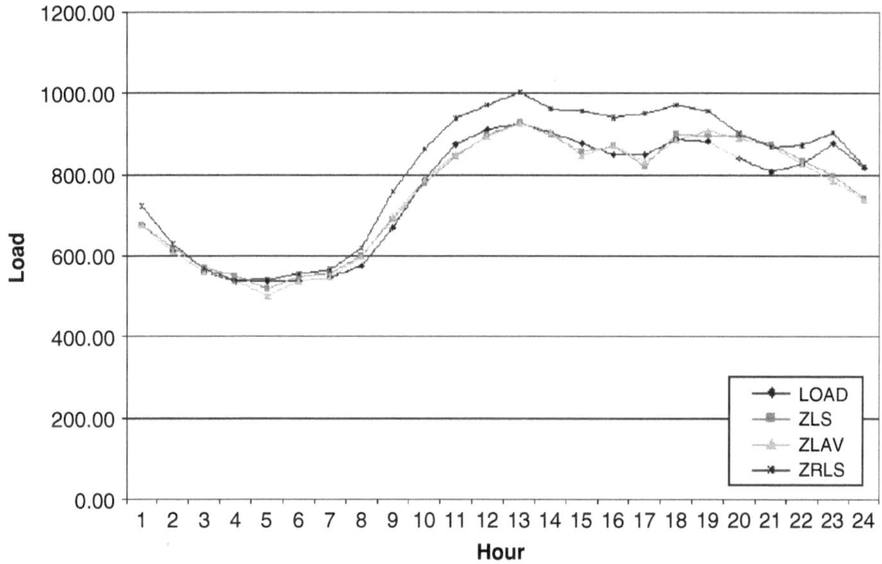

Figure 8.47 Summer workday estimated load curve: model A.

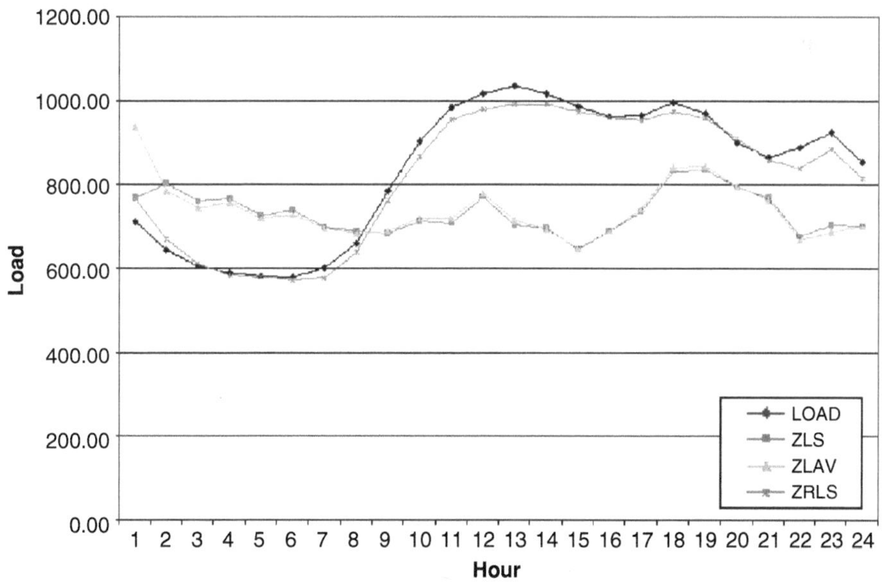

Figure 8.48 Summer workday predicted load curve: model A.

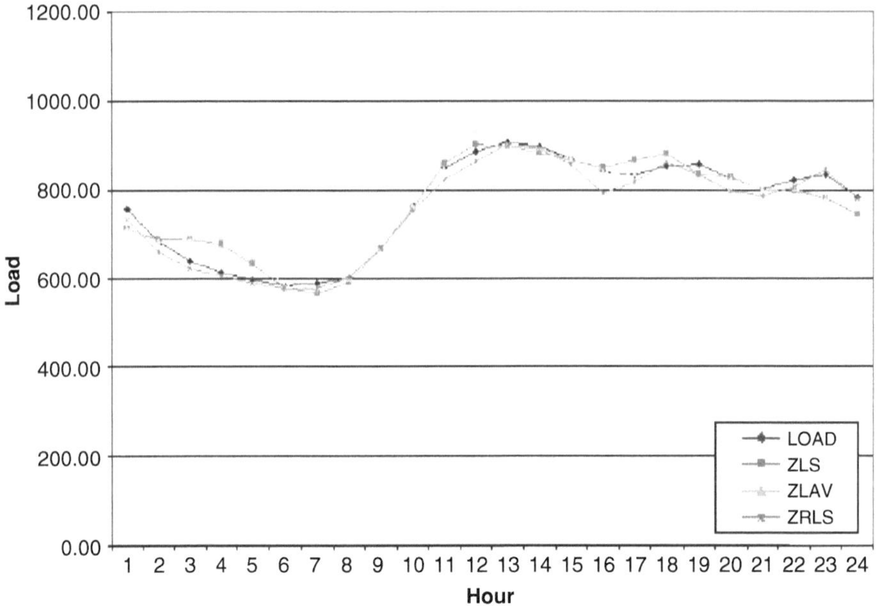

Figure 8.49 Summer weekend estimated load curve: model A.

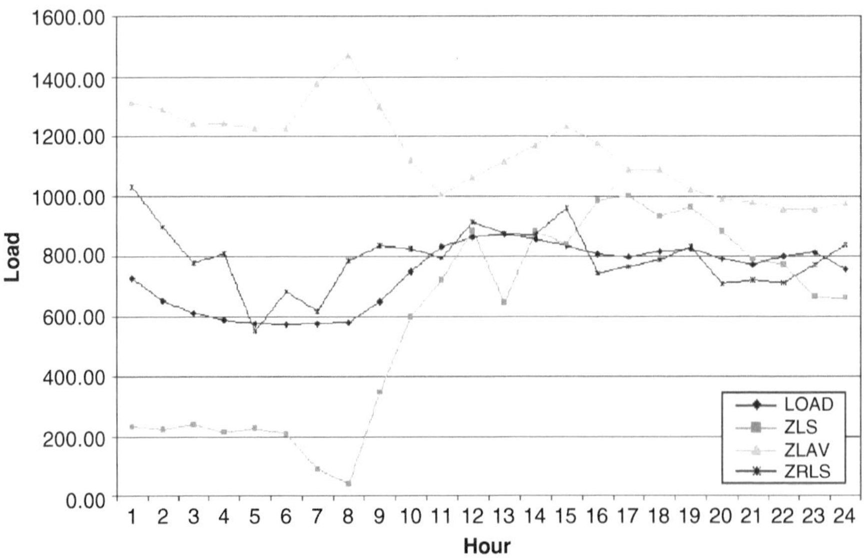

Figure 8.50 Summer weekend predicted load curve: model A.

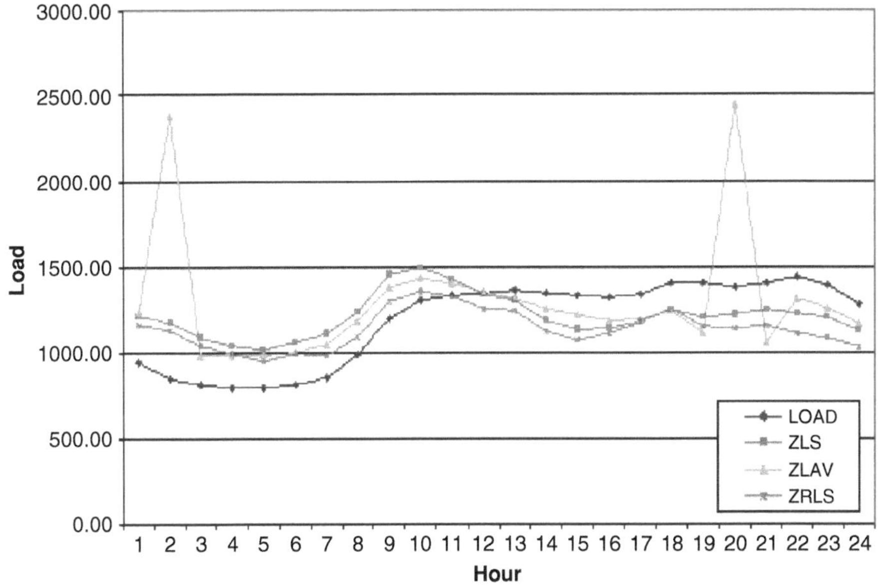

Figure 8.51 Winter workday predicted load curve: model B.

load power is reconstructed using these parameters. Examining this curve, we notice that the proposed algorithm estimates these parameters as good as the other past LES and LAV techniques. Figure 8.50 gives the predicted load using the estimated parameters. A good prediction is produced using the proposed algorithm.

8.11.1.2 Model B

As we explained in the preceding section, model B is a harmonics model and is a function only of the hour in question. Figure 8.51 gives the estimated and predicted load for such a model. Note that the matrix $f(t)$ in this model is constant independent of the weather parameters; it does mean that the coefficient for the summer weekdays or summer weekends is the same, as well as for the winter sessions. Examining Figure 8.51 reveals the following observations:

- The proposed algorithm produces results as good as the other two techniques.
- This model can be used for either the summer or winter sessions because it is independent of the weather factors.

8.11.1.3 Model C

The proposed algorithm is used to estimate the parameters of model C for a summer weekday. Figures 8.52 and 8.53 show the results obtained. Whereas in Figure 8.52 we reconstructed the load curve using the estimated parameters, Figure 8.53 gives the

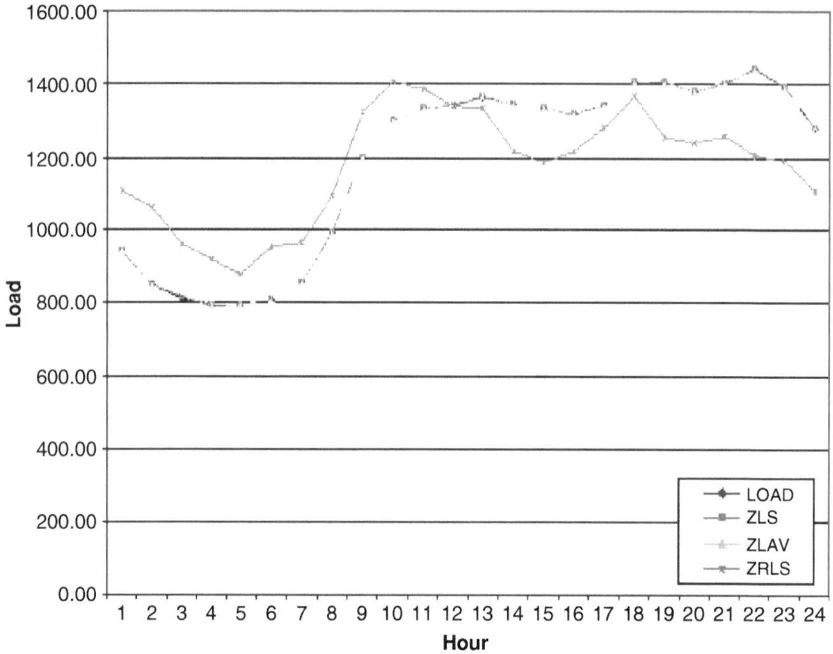

Figure 8.52 Winter workday estimated load curve: model C.

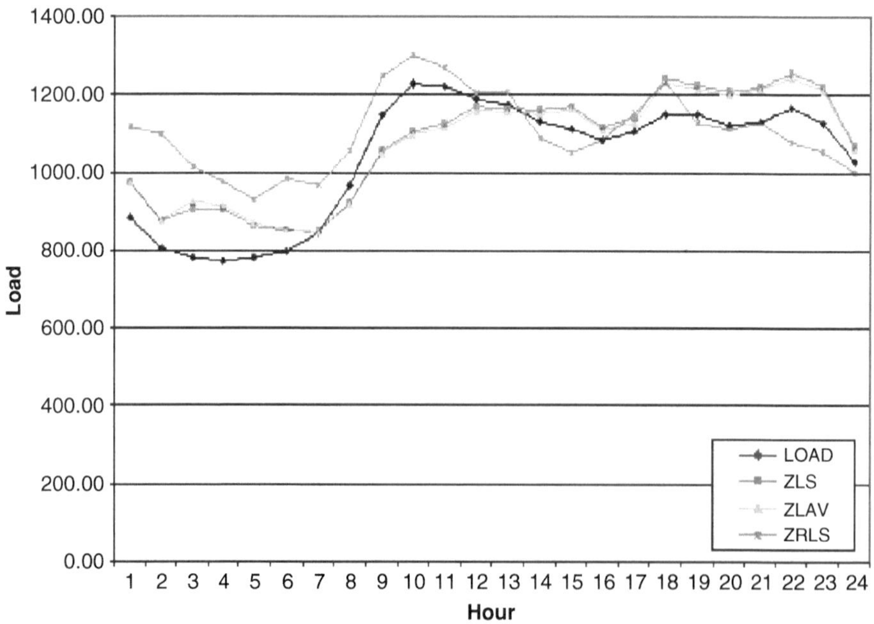

Figure 8.53 Winter workday estimated load curve: model C.

predicted load for 24 hours ahead using these parameters. Examining these figures, we note that

- The estimated load is of a good estimate type, but not better than those obtained using the past LES and LAV algorithms.
- The predicted load is better than the predicted load using the past LES and LAV algorithms.

8.11.2 Conclusions

In this section we presented the recursive least error squares algorithm for short-term electric load forecasting. The estimation algorithm was used to estimate the parameters of three proposed models: model A, model B, and model C. We have shown that, for summer workdays, the proposed algorithm produces acceptable estimation and prediction results compared to past static LES and LAV algorithms.

References

[1] H.M. Al-Hamadi, S.A. Soliman, Short-term electric load forecasting based on Kalman filtering algorithm with moving window weather and load models, Electric Power Syst. Res. 68 (2004) 47–59.
[2] R.E. Kalman, A new approach to linear filtering and prediction problems, Trans. ASME-J. Basic Eng. 82 (1960) 35–45.
[3] R.G. Brown, Introduction to Random Signal Analysis and Kalman Filtering, John Wiley and Sons, New York, 1983.
[4] G.F. Franklin, J.D. Powel, M.L. Workman, Digital Control of Dynamic System, second ed., Addison Wesley, Boston, MA, 1990.
[5] C. Kim, I. Yu, Y.H. Song, Kohonen neural network and wavelet transform based approach to short-term load forecasting, Electric Power Syst. Res. 63 (2002) 169–176.
[6] H.L. Willis, L.A. Finley, M.J. Buri, Forecasting electric demand of distribution system in rural and sparsely populated regions, IEEE Trans. Power Syst. 10 (4) (1995) 2008–2013.
[7] P.H. Henault, R.B. Eastvedt, J. Peschon, L.P. Hajdu, Power system long term planning in the presence of uncertainty, IEEE Trans. Power Apparatus Syst. PAS-89 (1970) 156–164.
[8] H.C. Wu, C. Lu, Automatic fuzzy model identification for short term load forecast, IEE Proc. Gener. Transm. Distrib.-C 146 (5) (1999) 477–482.
[9] W. Charytoniuk, M.S. Chen, Very short-term load forecasting using artificial neural networks, IEEE Trans. Power Syst. 15 (1) (2000) 263–268.
[10] K.H. Kim, H.S. Youn, Y.C. Kang, Short-term forecasting for special days in anomalous load conditions using neural networks and fuzzy inference method, IEEE Trans. Power Syst. 15 (2) (2000) 559–565.
[11] P.A. da Silva, L.S. Moulin, Confidence intervals for neural network based short-term load forecasting, IEEE Trans. Power Syst. 15 (4) (2000) 1191–1196.
[12] S.A. Villalba, C.A. Bel, Hybrid demand model for load estimation and short-term load forecasting in distribution electric systems, IEEE Trans. Power Syst. 15 (2) (2000) 764–769.
[13] R.H. Liang, C.C. Cheng, Combined regression-fuzzy approach for short-term load forecasting, IEE Proc. Gener. Transm. Distrib. 147 (4) (2000) 261–266.
[14] H.S. Hippert, C.E. Pedreira, R.C. Souza, Neural networks for short-term load forecasting: a review and evaluation, IEEE Trans. Power Syst. 16 (1) (2001) 44–55.

[15] M. Huang, H.T. Yang, Evolving wavelet-based networks for short-term load forecasting, IEE Proc. Gener. Transm. Distrib. 148 (3) (2001) 222–228.
[16] H. Mori, A. Yuihara, Deterministic annealing clustering for ANN-based short-term load forecasting, IEEE Trans. Power Syst. 16 (3) (2001) 545–551.
[17] N. Amjady, Short-term hourly load forecasting using time-series modeling with peak load estimation capability, IEEE Trans. Power Syst. 16 (3) (2001) 498–505.
[18] M.A. Abu-El-Maged, N.K. Sinha, Short-term load demand modeling and forecasting: a review, IEEE Trans. Syst. Man Cybern. SMC-12 (3) (1982) 370–382.
[19] I. Moghram, S. Rahman, Analysis and evaluation of five short-term load forecasting techniques, IEEE Trans. Power Syst. 4 (1989) 1484–1491.
[20] A. Papalexopoulos, T. Hesterburg, A regression based approach to short-term load forecasting, IEEE Trans. Power Syst. 5 (4) (1990) 1535–1547.
[21] K. Liu, et al., Comparison of short-term load forecasting techniques, in: Presented at IEEE PES'95 SM, Portland, 95 SM 547-0 PWRS (1995).
[22] M.T. Hagan, S.M. Behr, The time series approach to short-term forecasting, IEEE Trans. Power Syst. 2 (3) (1987) 785–791.
[23] W.R. Christiaanse, Short-term load forecasting using general exponential smoothing, IEEE Trans. Power Apparatus Syst. PAS-90 (2) (1971) 900–910.
[24] R. Campo, P. Ruiz, Adaptive weather sensitive short-term load forecasting, IEEE Trans. Power Syst. 2 (3) (1987) 592–600.
[25] M.E. El-Hawary, G.A. Mbamalu, Short-term power system load forecasting using the iteratively reweighted least squares algorithm, Electric Power Syst. Res. 19 (1990) 11–22.
[26] K. Srinivasan, R. Pronovost, Short-term load forecasting using multiple correlation models, IEEE Trans. Power Apparatus Syst. PAS-94 (5) (1997) 1854–1858.
[27] S.A. Soliman, S. Persaud, K. El-Nagar, M. El-Hawary, Application of least absolute value parameter estimation based on linear programming to short-term load forecasting, Electr. Power Energy Syst. 19 (3) (1997) 209–216.
[28] J. Toyoda, M. Chen, Y. Inouye, An application of the state estimation to the short-term load forecasting, part I and part II, IEEE Trans. Power Apparatus Syst. PAS-89 (7) (1970) 1678–1688.
[29] P.C. Gupta, K. Yamada, Adaptive short term forecasting of hourly loads using weather information, IEEE Trans. Power Apparatus Syst. PAS-91 (1970) 2085–2094.
[30] G. Singh, K.K. Biswas, A.K. Mahalanabis, Power system load forecasting using smoothing techniques, Int. J. Syst. Sci. 9 (4) (1978) 363–368.
[31] F.D. Galiana, F.C. Schweppes, A weather-dependent probabilistic model for short-term load forecasting, in: IEEE Winter Power Meeting, paper C72 171–2 (1972).
[32] F.D. Galiana, E. Haschin, A.R. Fiechter, Identification of stochastic electric load model from physical data, IEEE Trans. Automat. Contr. AC-19 (6) (1974) (887–893).

9 Electric Load Modeling for Long-Term Forecasting

9.1 Introduction

Long-term electric peak-load forecasting is an important issue in effective and efficient planning. Over- or underestimation can greatly affect the revenue of the electric utility industry. Overestimation of the future load may lead to spending more money in building new power stations to supply this load. Moreover, underestimation of load may cause troubles in supplying this load from the available electric supplies and produce a shortage in the spinning reserve of the system that may lead to an insecure and unreliable system. Therefore, an accurate method is needed to forecast loads, as is an accurate model that takes into account the factors that affect the growth of the load over a number of years. Furthermore, an accurate algorithm is needed to estimate the parameters of such models.

The growth in electricity consumption in many developing countries has outstripped existing projections, and accordingly, the uncertainties of forecasting have increased [1]. Variables such as economic growth, population, and efficiency standards, coupled with other factors inherent in the mathematical development of forecasting models, make accurate projection difficult [1, 2]. Unfortunately, an accurate forecast depends on the judgment of the forecaster, and it is impossible to rely strictly on analytical procedures to obtain an accurate forecast.

The objective of the forecasting task is to provide energy and peak-load predictions that meet planning requirements in a consistent and credible manner. A wide variety of techniques for short-term load forecasting (hour-by-hour forecasting) are available in the literature [6–18]; they include the autoregressive moving average (ARMA); Kalman filtering algorithm; artificial neural networks (ANNs) [6–8]; expert system (ES); fuzzy system (FS) [14], etc. A few of them have been applied to long-term annual load forecasting. These techniques range from the simplest approach, such as use of the most recent observation as the forecast, to highly complex approaches, such as an econometric system of simultaneous equations.

The methods used for forecasting electrical peak load and energy for long-term planning fall within two main categories—namely, the econometric and extrapolation methods [3].

Reference [4] applies the least absolute value (LAV) estimation algorithm to estimate the parameters of the annual peak-load model. The model used is a function of the time only (one year is equivalent to one time step). Different orders for the peak-load models are developed. However, all of them are linear in the parameters to be estimated.

DOI: 10.1016/B978-0-12-381543-9.00009-9

In this chapter we introduce different techniques used to estimate the annual peak load, where different models are used. In the first part, the LAV and least error squares (LES) static state estimation algorithms are used to estimate the parameters of the load model, and a comparative study is performed between the two techniques.

9.2 Peak-Load-Demand Model

A model for peak-load demand should take into account the following factors or a part of them, depending on the country in which this model is going to be implemented. There is no unique model that can be applied for utility companies.

These factors are

- The gross domestic product (GDP)
- The population (POP)
- The GDP per capita (GDP/CAP)
- The multiplication of electricity consumption by population (EP)
- The power system losses (LOSS)
- The load factor (LF)
- The cost of one kilowatt-hour (the average rate per unit sale; R/S) (mill/kWh).

The first four factors depend on the behavior of the public; thus, they may vary from country to country, whereas the last three factors depend on the electric power system and the load itself, as well as the consumption of power generated.

Let us begin by putting aside the last three factors for a while, and focus on the first four factors. We call these the country dependency factors. The peak-load demand in this case can be written as

$$P_L = f(\text{GDP}) + g(\text{POP}) + h(\text{EP}) + k(\text{GDP/CAP}) \tag{9.1}$$

where f, g, h, and k are functions of the variable stated between parentheses. They may be linear and/or nonlinear functions. We assume, for simplicity's sake, linear relations between the peak-load demand and write these factors as

$$P_L = a_0 + a_1\text{GDP} + a_2\text{POP} + a_3\text{EP} + a_4(\text{GDP/CAP}) \tag{9.2}$$

where a_0, a_1, a_2, a_3, and a_4 are the regression parameters to be determined by the LES and LAV algorithms. The problem now is to determine these parameters using the past data available:

$$P_{L_i} = \begin{bmatrix} 1 & \text{GDP} & \text{POP} & \text{EP} & \dfrac{\text{GDP}}{\text{CAP}} \end{bmatrix}_i \begin{bmatrix} a_0 \\ a_1 \\ a_2 \\ a_3 \\ a_4 \end{bmatrix}; \quad i = 1, m \tag{9.3}$$

for $i = 1, \ldots, m$; m is the number of year observations available from past data history; $m \geq 4$. In vector form, equation (9.3) can be rewritten as

$$Z = HX + \xi \tag{9.4}$$

where Z is the $m \times 1$ measurement vector of peak-load demand, H is an $m \times n$ observation matrix containing the factors that affect the peak load, X is the 5×1 column vector of the load parameters to be estimated, and ξ is the $m \times 1$ error vector to be minimized. At least the past five years' data should be given to determine the peak-load-demand parameter X.

The solution to equation (9.4) based on the least error squares algorithm is

$$X^* = \left[H^T H\right]^{-1} H^T Z \tag{9.5}$$

Furthermore, the least absolute value algorithm stated in references [4] and [5] is also implemented to compute the best estimate based on LAV minimization criteria. The steps behind the LAV algorithm are explained in reference [3] and Chapter 3.

Having identified the peak-load-demand parameters, we can predict the load for a specified year, using equation (9.1), provided that the other variables in this equation are known in advance for this year.

9.2.1 Example

The model proposed in the preceding section is tested using the data for a big utility company [4]; the data are given for the years 1981 to 1988 and listed in Table 9.1. The load model parameters are estimated using only the data of the year 1981 to the year 1988 ($m = 8$ observations). Table 9.2 gives these parameters using the LES and LAV algorithms. Table 9.3 gives the predicted peak-load demand for the years 1989 to 1996 and the percentage error in this prediction using the two estimation algorithms.

The absolute error for both techniques (residual vector) resulting from these parameters for the eight years is given as

$$\zeta_{\text{LES}} = \begin{bmatrix} -20.6 \\ 85.72 \\ -141.21 \\ 71.60 \\ 68.69 \\ -71.88 \\ -9.69 \\ 17.38 \end{bmatrix} \quad \text{For the LAV is } \zeta_{\text{LAV}} = \begin{bmatrix} 0.00 \\ 37.99 \\ -216.87 \\ 0.0 \\ 0.0 \\ -128.54 \\ 0.0 \\ 0.0 \end{bmatrix}$$

Note that due to the interpolation property of the LAV, the algorithm fits five data points. The estimated parameters in Table 9.2 are used to predict the peak load for the years from 1989 to 1996.

Table 9.1 Data for a Big Utility Company (Egyptian Unified Network, or EUN)

Year	Peak Load (GW)	GDP (Million EP)	POP (Million)	EP	GDP/ CAP	System Losses (MW)	Load Factor (%)	Cost of Energy (Mill/ kWh)
1981	3179	18985	42.11	30.11	450.85	4288.1	71.35	142.173
1982	3694	20628	43.33	33.58	476.07	4563.5	67.66	131.499
1983	3981	22450	44.50	35.67	504.49	4977.5	70.37	123.205
1984	4672	24042	45.77	37.06	525.26	5592.7	67.78	110.147
1985	5158	25691	46.99	38.60	546.73	6478.7	66.69	94.4155
1986	5361	26842	48.32	40.20	555.50	6159.0	68.66	101.59
1987	5803	27912	50.50	41.20	552.71	6862.6	69.25	86.200
1988	6152	29172	51.51	43.90	566.34	7479.1	70.22	70.2631
1989	6279	30417	52.54	46.02	578.93	7473.9	71.96	67.2093
1990	6664	31726	53.59	47.96	592.00	7369.8	71.34	65.6462
1991	7004	32799	54.66	50.58	600.02	7411.8	70.86	63.8323
1992	7215	33448	55.76	52.74	599.90	8124.7	71.96	58.9473
1993	7503	34282	56.87	54.58	602.80	8456.0	71.65	57.4074
1994	7657	35624	58.01	56.48	614.12	8415.3	72.46	58.33
1995	8149	37298	59.17	56.48	630.37	8555.8	71.90	61.2318
1996	8491	39161	60.35	58.54	648.87	8787.4	72.23	71.8849

Table 9.2 Estimated Parameters for the Peak-Load-Demand Model

Parameters	LES Algorithm	LAV Algorithm
a_0	−2479.2	−4081.65
a_1	0.329	0.2314
a_2	28.5	39.39
a_3	−37.86	−10.55
a_4	−1.379	3.39

The predicted loads as well as the errors in this prediction, using LES and LAV techniques, are given in Table 9.3. Examining this table reveals that both techniques produce fairly good estimates for such type of forecast and such type of peak-load model.

9.2.2 A More Detailed Model

Using four variables in the first model may not be adequate; thus, we need an accurate model that takes into account all the factors stated previously. We may assume this model to be

Table 9.3 Predicted Peak Load and the Percentage Error in This Prediction

	Actual Load	LES Estimates		LAV Estimates	
Year	**MW**	**Peak-Load Power**	**% Error**	**Peak-Load Power**	**% Error**
1989	6279	6484.72	−3.26	6803.46	−3.57
1990	6664	6853.84	−2.85	6871.56	−3.11
1991	7004	7127.10	−1.78	7161.55	−5.25
1992	7215	7290.35	−1.04	7331.86	−1.62
1993	7503	7522.71	−0.26	7558.99	−0.75
1994	7657	7909.18	−3.29	7932.76	−3.6
1995	8149	8470.57	−3.95	8420.90	−3.34
1996	8491	9013.63	−6.15	8939.46	−5.28

Table 9.4 Estimated Parameters for a Detailed Peak-Load-Demand Model

Parameters	a_0	a_1	a_2	a_3	a_4	a_5	a_6	a_7
LES	4713.214	0.4192	−13.38	−27.997	−10.389	0.1535	−61.865	2.394
LAV	6398.75	0.4932	−33.835	−28.328	−13.8612	0.1256	−73.128	4.3864

$$P_L = a_0 + a_1(\text{GDP}) + a_2(\text{POP}) + a_3(\text{EP}) + a_4(\text{GDP/CAP}) + a_5(\text{system losses}) + a_6(\text{LF}) + a_7(\text{cost of kWh}). \tag{9.6}$$

Equation (9.6) is a linear equation in the parameters to be estimated, a_0 to a_7. Thus, equation (9.6) can be rewritten in the form of equation (9.4) as

$$Z = HX + \xi. \tag{9.7}$$

In equation (9.6), the following vectors and matrices are defined as follows:

Z is an $m \times 1$ measurement vector of the past history of the peak-load demand;
H is an $m \times 8$ measurement matrix of which the elements contain the seven factors stated in equation (9.6);
X is the 8×1 load parameters a_0 to a_7;
ξ is an $m \times 1$ error vector associated with each measurement to be minimized.

Therefore, we have eight parameters to be estimated, and at least eight measurements should be available to estimate these parameters. Using only eight measurements may produce a poor estimate because we force the errors vector to be zero (because the number of equations equals the number of unknowns). Here, we use 12 measurements to estimate the eight parameters using LES and LAV techniques. The solution to equation (9.7) is similar to that given in equation (9.5). Table 9.4 gives the estimated parameters using both techniques.

The validity of the proposed model and the accuracy of the estimated parameters are checked by implementing the model to predict the peak-load power for the years

Table 9.5 Predicted Peak-Load Power with the Percentage Errors

	Actual Load	LES Estimates		LAV Estimates	
Year	**MW**	**Peak-Load Power**	**% Error**	**Peak-Load Power**	**% Error**
1993	7503	7535.9	−0.438	7626.93	−1.652
1994	7657	7857.9	−2.624	7979.76	−4.215
1995	8149	8438.5	−3.552	8611.65	−5.678
1996	8491	8994.42	−5.929	9227.78	−8.68

1993 to 1996, using the factors given in Table 9.1 for the same years. Table 9.5 gives the estimated peak load and the percentage error in these estimates.

Examining Table 9.5 reveals the following observations:

- The predicted load using both techniques is accurate enough for such long-term forecasting.
- The predicted load for this estimation period using eight parameters is almost the same as those using the five parameters stated in Table 9.2.
- The maximum predicted error for LES is about 6%, whereas it is about 9% for LAV, both for the year 1996. These are fairly good estimates for such long-term forecasting.

9.2.3 A Time-Dependent Model

If the year under consideration is taken into account (the time horizon), then the peak-load power demand can be written as

$$P_L = a_0 + a_1(\text{GDP}) + a_2(\text{POP}) + a_3(\text{EP}) + a_4(\text{GDP/CAP}) \\ + a_5(\text{system losses}) + a_6(\text{LF}) + a_7(\text{cost of kWh}) + a_8(\text{time}) \quad (9.8)$$

In equation (9.8) the *time* takes values $0, 1, \ldots, T_f$, where 0 is the starting year, 1 is the next year, and so on. Furthermore, T_f is the number of years minus one used in this study. Equation (9.8) can be rewritten in vector form as

$$Z = HX + \xi \quad (9.9)$$

In equation (9.9), the vectors and matrices are defined as follows:

Z is an $m \times 1$ measurement vector of the past history of the peak-load demand;
H is an $m \times 9$ measurement matrix of which the elements contain the eight factors stated in equation (9.8);
X is the 9×1 load parameters a_0 to a_8;
ξ is an $m \times 1$ error vector associated with each measurement to be minimized.

Therefore, we have nine parameters to be estimated, and at least nine measurements should be available to estimate these parameters. Using nine measurements may produce a poor estimate because we force the errors vector to be zero. Here, we use 12 measurements to estimate the nine parameters using LES and LAV

Table 9.6 Estimated Parameters for a Time-Dependent Model

Parameters	a_0	a_1	a_2	a_3	a_4	a_5	a_6	a_7	a_8
LES	5613.5	−0.0816	81.49	−74.81	6.67	−0.01915	−62.593	−4.925	397.62
LAV	5459.73	1.133	−182.43	14.473	−34.8	0.3293	−60.76	14.563	−455.8

Table 9.7 Predicted Peak-Load Power with the Percentage Errors

	Actual Load	**LES Estimates**		**LAV Estimates**	
Year	**MW**	**Peak-Load Power**	**% Error**	**Peak-Load Power**	**% Error**
1993	7503	7553.86	−0.68	7557.13	−0.455
1994	7657	7812.18	−2.027	7977.57	−4.187
1995	8149	8299.56	−1.847	8763.84	−7.545
1996	8491	8541.92	−0.600	9800.82	−15.428

techniques. The solution to equation (9.9) is similar to that given in equation (9.5). Table 9.6 gives the estimated parameters using both techniques.

The estimated parameters in Table 9.6 are used to predict the peak-load-demand power for the past four years. Table 9.7 gives the predicted load and percentage error in this prediction.

Examining Table 9.7, by using the time horizon, we note that

- The LES algorithm produces an accurate prediction for the peak-load power, whereas the LAV produces a fairly accurate prediction.
- The results obtained using this model, especially for the LES estimation, are better than those mentioned in Table 9.5.
- Examining Tables 9.5 and 9.7, we can conclude that the time horizon has little effect on the prediction of the peak-load power.

9.3 Time-Series Analysis

In a time-series analysis model, a time series is constructed that takes into account the effect of load for the previous years on the load for the year in question. The order of this time difference series depends on the accuracy of the prediction needed as well as the data available from the past history. The general form for this time series can be formulated as

$$P_L(k) = a_1 P_L(k-1) + a_2 P_L(k-2) + a_3 P_L(k-3) + \cdots + \cdots + a_n P_L(k-n) \tag{9.10}$$

where $k = K, K-1, K-2, \ldots, 1$, K is the year in question, and n is the degree of the time series. In this model, we use $n = 4$. Equation (9.10) in this case becomes

$$P_L(k) = \begin{bmatrix} P_L(k-1) & P_L(k-2) & P_L(k-3) & P_L(k-4) \end{bmatrix} \begin{bmatrix} a_1 \\ a_2 \\ a_3 \\ a_4 \end{bmatrix} \tag{9.11}$$

Equation (9.11) can be rewritten in vector form as

$$P_L = BX + \delta \tag{9.12}$$

where P_L is a $K \times 1$ peak-load power, B is a $K \times 4$ measurement matrix that contains the elements of the previous peak-load power, X is a 4×1 series parameters vector to be estimated, and δ is a $K \times 1$ error vector to be minimized.

The estimation problem formulated in equation (9.12) can be solved using the two proposed algorithms, LES and LAV, explained in the previous sections. Having identified the series parameters, we can then predict the peak-load power for the forthcoming year.

9.3.1 Example for the Time-Series Model

The time-series model explained in this section is used to predict the load for the utility mentioned in the previous example. First, the series model parameters are estimated using the LES and LAV algorithms. Table 9.8 gives these parameters.

The accuracy of these parameters is tested by predicting the annual peak power from the years 1985 to 1996.

Table 9.9 gives the predicated annual peak-load power and the percentage errors in this prediction using the proposed two algorithms. Examining this table reveals the following:

- The model used in this section is an adequate model.
- Both the LES and LAV techniques produce accurate estimates, but the LES estimates are better than the LAV estimates.
- The results obtained for this model are much better than those obtained in the other proposed models.
- The model in this section is independent of the system variables, but it depends on the history peak-load power available.

9.3.2 Remarks

In this section, we discussed the following points:

1. Different models are developed and tested for long-term peak-load power forecasting.
2. We studied the effects of GDP, POP, EP, GDP/CAP, etc. on the performance of each developed model.

Table 9.8 Estimated Parameters for a Time-Series Model

Parameters	a_1	a_2	a_3	a_4
LES	1.14735	−0.29612	0.78316	−0.61930
LAV	1.0020	−0.081380	0.61208	−0.504513

Table 9.9 Predicted Load Using the Time-Series Model

	Actual Load	LES Estimates		LAV Estimates	
Year	**MW**	**Peak-Load Power**	**% Error**	**Peak-Load Power**	**% Error**
1985	5158	5106	1.00	5014	2.8
1986	5361	5305	1.10	5217	2.6
1987	5803	5768	0.60	5671	2.28
1988	6152	6152	0.00	5969	2.97
1989	6279	6343	−1.0	6183	−1.02
1990	6664	6688	−0.40	6548	1.74
1991	7004	7041	−0.50	6851	2.2
1992	7215	7256	−0.57	7104	−0.57
1993	7503	7549	−0.60	7450	−0.61
1994	7657	7885	−3.00	7776	−2.98
1995	8149	8134	0.20	8077	0.88
1996	8491	8416	0.90	8436	0.65

3. We studied the applications of two parameter estimation algorithms, the LES and LAV algorithms, on the prediction of the annual peak load.
4. In the time-dependent model, the LES algorithm produces better-predicted results than the LAV algorithm.
5. The time-series model is the best one for such systems because it has the lowest error rate among the other developed models.

It is worthwhile to state here that every power system has its own model; the one suitable for a particular system may not be suitable for another system.

9.4 Kalman Filtering Algorithm

Long-term forecasting is characterized by its high uncertainty owing to its high dependence on socioeconomic factors; for this reason, an error level up to 10% is acceptable [28]. These results are highly dependent on uncertain parameters such as electric utility, region, country, economic growth, population growth, and population habits. Moreover, the data on which the long-term forecasting technique is tested play an important factor in determining the level of the forecast error. An algorithm that gives a low average forecast error for a certain electric utility in a certain country may not give the same level of error for a different utility in a different country. Therefore, any attempt for comparing different forecasting techniques should utilize the same testing data.

The technique used in this section combines regression estimation with a time-series load model suited for the Kalman filtering approach. Historic load data over a certain period of time, say one year, are arranged in a two-dimensional (2D) (24 hours × 52 weeks) layout. It is worth mentioning that a time period of one year is highly suggested not only because this period provides a reasonable amount of data, but also because it entirely exploits the underlying daily and seasonal load variations. The technique used in this section employs the following primary features of the long-term forecasting problem:

- Seasonal and daily load-demand behavior: The cyclic behavior of the load demand in response to seasonal and daily variations is modeled using *short-term* linear regression techniques over a specific period of time (one year). The short-term forecasting accuracy is high due to the high correlation of the load time series. Therefore, the resulting model is *reasonably* accurate and establishes the basis for future (next year) trends of the load demand.
- Annual load-demand growth: The overall load demand of a system continually increases due to population and industrial growth as well as increases in industrial consumption. A third-order regression model is used to develop the annual growth in load demand as a function of time. The annual growth provides an approximate correction factor for the load-demand behavior for the next year.

Any long-term forecast is always inaccurate due to the complexity of the load-affecting factors. For example, peak demand is very much dependent on temperature. The fluctuation in temperature is extremely hard to forecast for a long period of time. Therefore, the main objective of long-term load forecasting is to increase forecast accuracy. The load time-series behavior is developed as a linear time-varying mathematical model relating the load at time instant k as a function of the load at time instances less than k. The load model is then used to form a time-varying discrete dynamic system suited for the Kalman filter, which is employed to estimate and predict the next year's load demand. The Kalman filter is fed with the estimated load augmented with the annual load growth obtained from the previous two steps as measurement values.

9.4.1 Estimating Multiple Regression Models

The electric load depends on a number of complex factors that have nonlinear characteristics, and good results may not be obtained using a single linear model. The approach taken in this section is the decomposition of the problem into multiple simple (first-order) linear regression models to capture the global nonlinear behavior of the load. Each of the linear regression models extracts the short-term correlation of a certain set of data. One year's data are arranged into a two-dimensional layout with 24 columns representing 24 hours of a day and 52 rows representing 52 weeks of the year. Figure 9.1 illustrates the 2D layout of the load data.

Special consideration is taken for the load variation during the weekends. Accordingly, weekends are treated separately but in exactly the same manner as the working days. The $L(i, k)$ cell in Figure 9.1 is the average load of the working days of the ith week at the kth hour. With this setup of the load data, obvious great intrinsic correlations exist between successive columns as well as between successive rows, as illustrated in Figures 9.2 and 9.3, respectively. These two figures, and all subsequent results, are based on the load demand of one of the largest electric power utilities in Canada for the years 1994 and 1995.

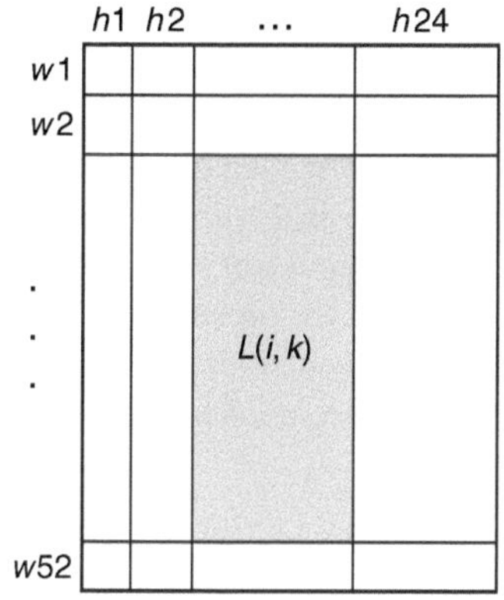

Figure 9.1 Two-dimensional layout of the load data.

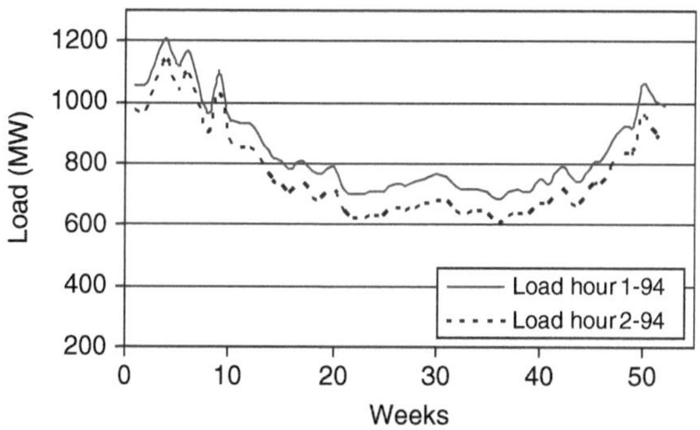

Figure 9.2 Comparing weekly average load of hour 1 and hour 2, 1994.

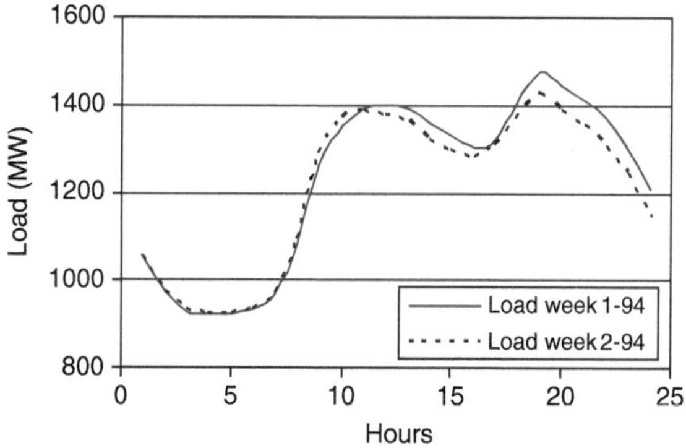

Figure 9.3 Comparing weekly average load of weeks 1 and 2, 1994.

Figure 9.2 shows the load correlation between hour 1 AM and hour 2 AM throughout the whole year 1994. The correlation factor was calculated as 0.997. Similarly, Figure 9.3 shows the load correlation of week 1 and week 2 of year 1994, with a correlation factor of 0.985. The strong correlation is maintained over the entire year for all 24 hours of the day, as illustrated in Figures 9.4 and 9.5.

The persistent correlation of the prevalent load patterns suggests the use of short-term simple linear regression models for successive hours (see equation (9.13a)) and another set for successive weeks (see equation (9.13b)). This results in 24 × 52 simple linear regression models, which are used to draw the shape of the 2D load behavior contour for one year.

$$L(i,k) = a(k)\ L(i,k-1) + b(k) \qquad k = 1,\ldots,24 \tag{9.13a}$$

$$L(i,k) = c(i)\ L(i-1,k) + d(i) \qquad i = 1,\ldots,52 \tag{9.13b}$$

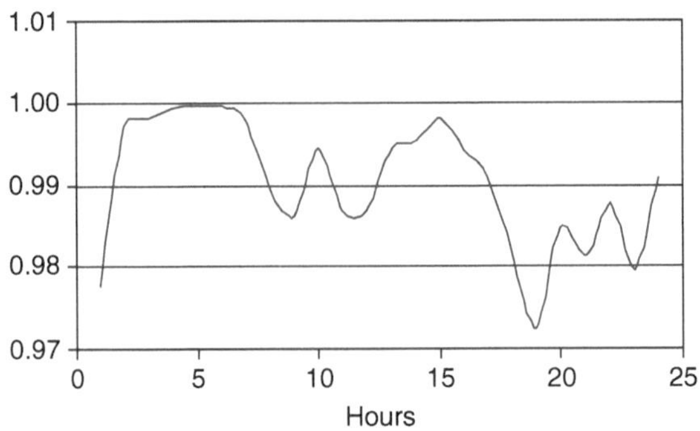

Figure 9.4 Correlation factor for successive hours of 52 weeks of 1994.

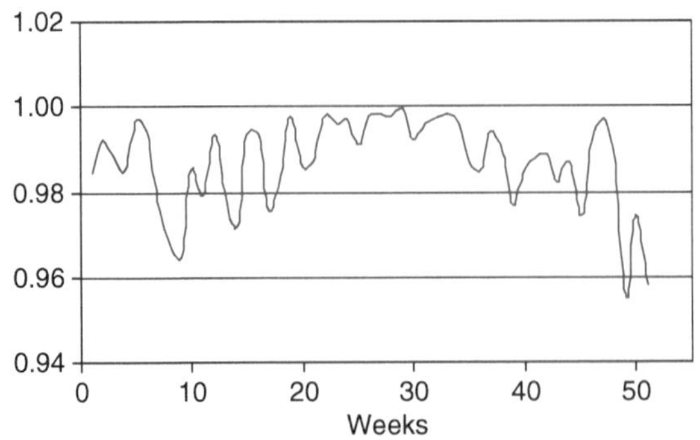

Figure 9.5 Correlation factor for successive weeks over 24 hours of 1994.

where

$a(k)$ and $b(k)$ are regression parameters at the kth hour; $k = 1, 2, \ldots, 24$, which are determined using the load pairs $[L(i, k), L(i, k-1)]$ for all $i = 1, 2, \ldots, 52$, by the least squares method;
$L(i, k)$ and $L(i, k-1)$ are the weekly average load at hours k and $k-1$, respectively, for all weeks $i = 1, \ldots, 52$, with the initial condition $L(i, 0) = L(i-1, 24)$;
$c(i)$ and $d(i)$ are regression parameters of the ith week, $i = 1, 2, \ldots, 52$, which are determined using the load pairs $[L(i, k), L(i-1, k)]$ for all $k = 1, 2, \ldots, 24$, by the least squares method;
$L(i, k)$ and $L(i-1, k)$ are the weekly average load in the ith and $(i-1)$th weeks, respectively, for all hours $k = 1, \ldots, 24$, with the initial condition $L(0, k) = [L(52, k)$ of the previous year].

9.4.2 Estimating the Next Year's Load Contour

The preceding regression models are used to project the load trends for the next year. Figures 9.6 and 9.7 demonstrate the fact that successive years have nearly identical load behavior contours.

A recursive procedure used to estimate next year's load contour utilizing regression models of the previous year is as follows:

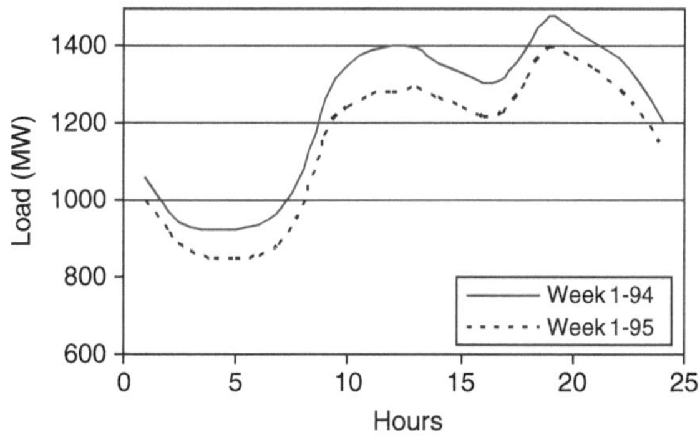

Figure 9.6 Comparing average weekly loads for various weeks of 1994 and 1995.

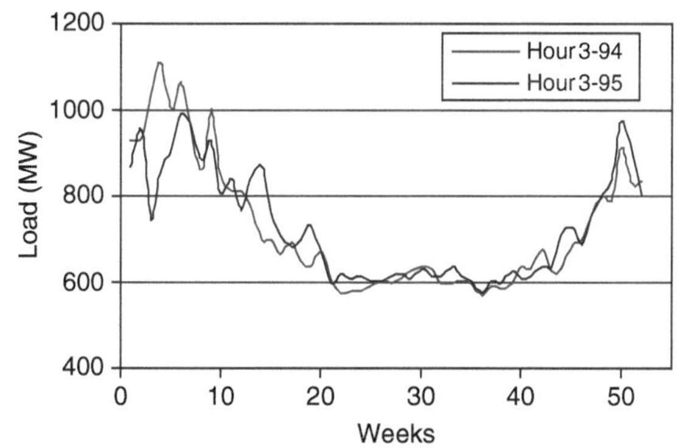

Figure 9.7 Comparing average weekly loads for various hours of 1994 and 1995.

1. Estimate for the first week the weekly average load: This process corresponds to estimating the first row of the next year's load; refer to Figure 9.8(a). Using equation (9.13b), we calculate $\hat{L}(1,k)$

$$\hat{L}(1,k) = c(k)\hat{L}(0,k) + d(k) \quad k = 1,2,\ldots,24 \tag{9.14}$$

where $\hat{L}(1,k)$ is the estimated weekly average load of the first week at the kth hour; $\hat{L}(0,k)$ is taken as $L(52,k)_{\text{last-year}}$, which is the weekly average load of last year's 52nd week; and $[c(k),\ d(k)]$ is a pair of regression coefficients of the kth hour, obtained from equation (9.13b) using last year's data.
2. Estimating for the first hour the weekly average load: This process corresponds to estimating the first column of the next year's load; refer to Figure 9.8(b). Using equation (9.13a), we calculate $\hat{L}(i,1)$

$$\hat{L}(i,1) = a(i)\hat{L}(i,0) + b(i) \qquad i = 2,3,\ldots,52 \tag{9.15}$$

where $\hat{L}(i,1)$ is the estimated weekly average load of the first hour in the ith week; $\hat{L}(i,0)$ is taken as $L(i-1,\ 24)$, which is the weekly average load of the 24th hour of the previous week; and the $[a(i),\ b(i)]$ is a pair of regression coefficients of the ith week, obtained from equation (9.13a) using last year's data.
3. Estimating for the second week the weekly average load: This process corresponds to estimating the second row of the next year's load; refer to Figure 9.8(c). Using equation (9.13b), we calculate $\hat{L}(2,k)$

$$\hat{L}(2,k) = c(k)\hat{L}(1,k) + d(k) \qquad k = 2,3,\ldots,24 \tag{9.16}$$

where $\hat{L}(2,k)$ is the estimated weekly average load of the second week at the kth hour; and $\hat{L}(1,k)$ is obtained using equation (9.14).
4. Estimating for the second hour the weekly average load: This process corresponds to estimating the second column of the next year's load; refer to Figure 9.8(d). Using equation (9.13a), we calculate $\hat{L}(i,2)$

$$\hat{L}(i,2) = a(i)\hat{L}(i,1) + b(i) \qquad i = 3,4,\ldots,52 \tag{9.17}$$

where $\hat{L}(i,2)$ is the estimated weekly average load of the second hour in the ith week; and $\hat{L}(i,1)$ is obtained using equation (9.15).
5. The recursive iterations are repeated until $i=52$ and $k=24$.
6. Steps 1 through 5 are repeated for forecasting more years.

The preceding procedure produces a two-dimensional contour of the load behavior for one year based on regression coefficients of the previous year. The load contour will then be augmented by the annual load growth to account for the load change between successive years.

9.5 Annual Load Growth

To maximize the accuracy of next year's load-demand estimation, we estimate and employ annual load growth as an adjusting factor. It is evident that there is a very strong

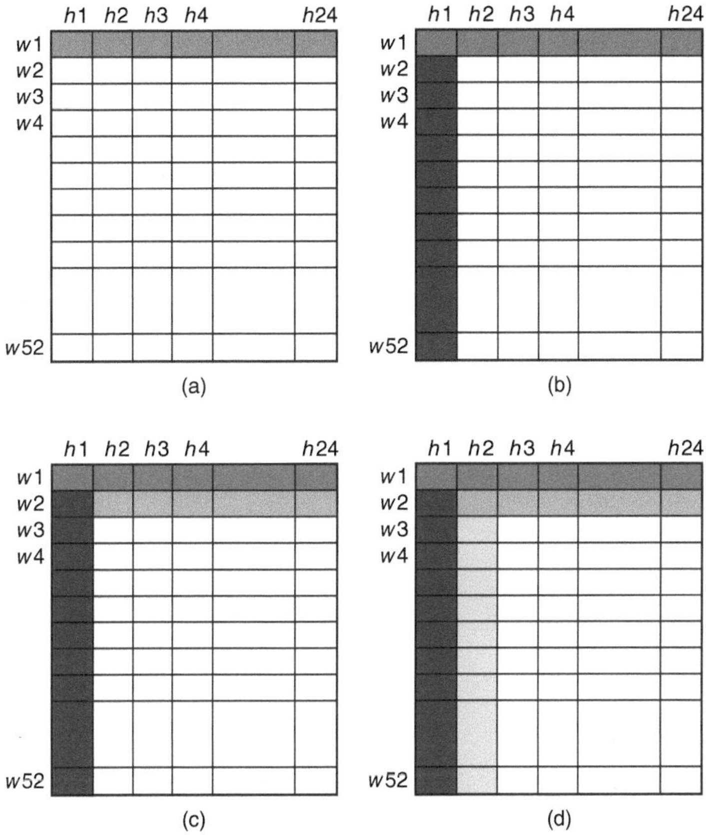

Figure 9.8 (a) First week (row) load estimation resulting from the first iteration. (b) First hour (column) load estimation resulting from the second iteration. (c) Second week (row) load estimation resulting from the third iteration. (d) Second hour (column) load estimation resulting from the fourth iteration.

dependence of the load demand on time. Typical load profiles of successive years reveal very strong correlation at certain periodic time intervals. For example, refer to Figure 9.7; the two load curves at a certain hour over the whole year for two successive years retain the same shape. Moreover, there is, on average, a clear load increase over the previous year. This increase amounts to an annual load growth at that hour as a function of time (weeks) throughout the whole year. The load growth is modeled as the difference between the load curves of two successive years as a function of time.

A third-order polynomial is utilized to model the load as a function of time at the kth hour as a function of the load of the previous hour. The regression model is as follows:

$$L(i,k) = \beta_0(k) + \beta_1(k)\,L(i,k-1) + \beta_2(k)\,L^2(i,k-1) + \beta_3(k)\,L^3(i,k-1) \tag{9.18}$$

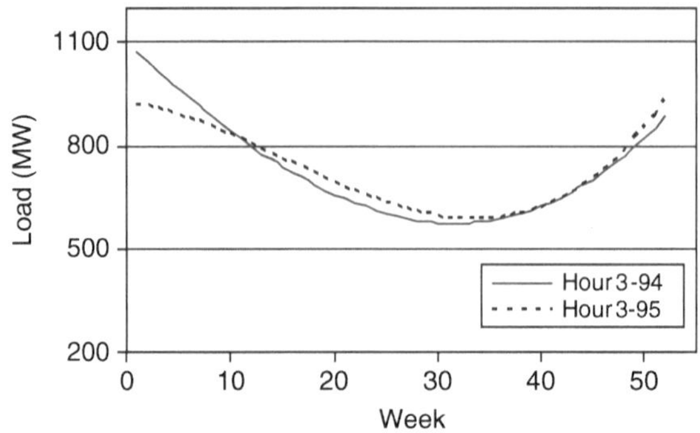

Figure 9.9 Approximate curves of load of hour 3 of 1994 and 1995.

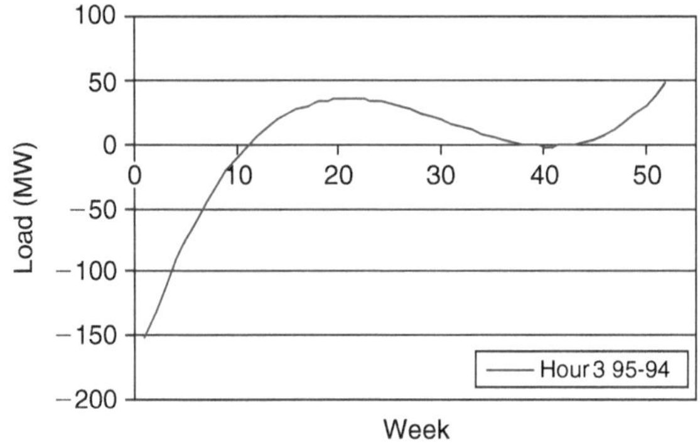

Figure 9.10 Annual load growth variations during 52 weeks of a year.

where $\beta_j(k)$, $j = 0, 1, 2, 3$ are regression variables at the kth hour, and $k = 1, 2, \ldots, 24$, which are determined using the load pairs [$L(i, k)$, $L(i, k-1)$ for all $i = 1, 2, \ldots, 52$] by the least squares method. The initial values $L(i, 0)$ are set to $L(i-1, 24)$. The two curves that approximate the relationship between $L(i, k)$ and $L(i, k-1)$ corresponding to the load behavior of the two years in Figure 9.7 are shown in Figure 9.9.

Next, the procedure for evaluating the annual load growth is as follows; we assume that the annual load growth is calculated between 1993 and 1994 to be used for predicting the 1995 load:

1. Using equation (9.18), we determine the regression coefficients (24 sets) for 24 hours for the year 1993. The coefficients define 24 approximate curves of the weekly average load, one curve per hour.
2. We repeat the calculations of the previous step to the 1994 data.
3. We define the annual load growth as the difference of the approximate load curves of 1994 and 1993:

$$\text{Annual Load Growth } (i) = L(i,k)_{(95)} - L(i,k)_{(94)} \quad k = 1, 2, \ldots, 24, \quad i = 1, 2, \ldots, 52 \tag{9.19}$$

The annual load growth curve is obtained by subtracting the approximate curve of 1994 from the approximate curve of 1993, as shown in Figure 9.10.

After estimating the shape of the load contour and augmenting it with the annual load growth, we use the Kalman filtering algorithm to predict the next year's load demand. First, we express the dynamic variation of load with respect to load values at previous hours as a time-varying linear model. Second, we construct a dynamic time-varying state space model and adapt it for the Kalman filtering technique. Last, we use the estimated regression and the annual load growth results as measurement inputs for the Kalman filtering algorithm.

9.5.1 Load Modeling for the Kalman Filtering Algorithm

Generally, the load at any discrete time instant $k = 1, 2, \ldots, 24$, corresponding to 24 hours of one day, can be expressed as a fourth-order time-varying linear model as follows:

$$L(i,k) = \alpha_0(k) + \alpha_1(k)\,L(i,k-1) + \alpha_2(k)\,L(i-1,k) + \alpha_3(k)\,L(i-1,k-1) \tag{9.20}$$

where

$L(i,k)$ = weekly average load at time instant: ith week and kth hour;
$\alpha_0(k)$ = base load at time instant k;
$\alpha_j(k) = j = 1, 2, 3$, load coefficients at the kth hour.

The model assumes that the load coefficients are constant over each discrete time instant $k = 1, \ldots, 24$, of the 24 hours of the day. Parameter estimation is carried out for each of the 24 discrete instances in a day. Accordingly, 24 sets of coefficients are required to be estimated for one day. The estimated coefficients can be plugged into the model to predict hourly loads for the next day.

9.5.2 Kalman Filter Parameter Estimation Algorithm

In this section we address only the necessary equation for the development of the basic recursive discrete Kalman filter. Given the discrete state equations

$$\begin{aligned} x(k+1) &= A(k)x(k) + w(k) \\ z(k) &= C(k)x(k) + v(k) \end{aligned} \tag{9.21}$$

where

$x(k)$ is $n \times 1$ system states;
$A(k)$ is an $n \times n$ time-varying state transition matrix;
$z(k)$ is an $m \times 1$ measurement vector;
$C(k)$ is an $m \times n$ time-varying output matrix;
$w(k)$ is an $n \times 1$ system error;
$v(k)$ is an $m \times 1$ measurement error.

The noise vectors $w(k)$ and $v(k)$ are uncorrelated white noises. The basic discrete-time Kalman filter algorithm recursive equations appropriate for forecasting problems were discussed earlier. The load model is used to form a time-varying discrete dynamic system relevant to the Kalman filter. The dynamic system of equation (9.21) is used with the following definitions:

1. The state transition matrix, $A(k)$, is a constant 4×4 identity matrix.
2. The error covariance matrices are chosen to be identity matrices for this simulation; they would be assigned better values if more knowledge were obtained on the sensor accuracy and process error.
3. The state vector, $x(k)$, consists of four parameters: $[\alpha_0(k), \alpha_1(k), \alpha_2(k), \alpha_3(k)]^T$.
4. $C(k)$ is a four-element time-varying row vector, which relates the measured load data to the state vector. (Refer to equation (9.21).)
5. The observation vector, $z(k)$, for this application is a scalar representing the load at time instant k. (Refer to equation (9.21).)

The observation equation $z(k) = C(k)\, x(k)$ has the form

$$z(k) = \hat{L}(i,k) = [1 \quad \hat{L}(i,k-1) \quad \hat{L}(i-1,k) \quad \hat{L}(i-1,k-1)] \begin{bmatrix} \alpha_0(k) \\ \alpha_1(k) \\ \alpha_2(k) \\ \alpha_3(k) \end{bmatrix} \tag{9.22}$$

where the parameters and load values are defined in equation (9.20), with k representing the time instant of the 24 discrete hours of the day, $k = 1, \ldots, 24$. $\hat{L}(i,k)$ is the estimated weekly average load using regression parameters and annual load growth. For any time instant k, the Kalman filter iterates over all available load data, $\hat{L}(i,k)$ for all weeks, $i = 1, \ldots, 52$, with additional interpolation load points to estimate the set of parameters $[\alpha_0(k), \alpha_1(k), \alpha_2(k), \alpha_3(k)]$. Interpolation of load points is required to accelerate Kalman filter convergence by increasing its input data.

9.6 Computer Exercises

To verify the effectiveness of the proposed load-demand forecasting technique, we used load data for one of the largest utility companies in Canada for the years 1994 and 1995. Regression models are obtained from 1994 data and used to project load demand for 1995. Kalman filtering is used to increase the estimation accuracy of the year 1995, and then the forecasted results are compared with the actual data of 1995.

9.6.1 Multiple Regression Models Results

Using equation (9.13a), we calculate 24 sets of regression coefficients. Table 9.10 shows the first seven of these sets as a sample. This table also lists the correlation factors of successive hours (columns) of the 1994 load data. Similarly, using equation (9.13b), we calculate 52 sets of regression coefficients. Table 9.11 shows the first seven of these sets as a sample, together with the correlation factors of successive weeks (rows) of the 1994 load data.

Table 9.10 Correlation Factors and Regression Coefficients for Seven Hours of 1994

1994	Hour 1	Hour 2	Hour 3	Hour 4	Hour 5	Hour 6	Hour 7
k = hour of the day	1	2	3	4	5	6	7
Correlation Factor	0.978	0.997	0.998	0.999	0.999	1.000	0.998
$a(k)$	0.973	0.994	1.014	1.022	1.025	1.024	1.049
$b(k)$	−89.311	−76.835	−49.053	−31.009	−21.580	−11.659	−6.003

Table 9.11 Correlation Factors and Regression Coefficients for Seven Weeks of 1994

1994	i = Week Number	Correlation Factor	$c_{(i)}$	$d_{(i)}$
Week 1	1	0.985	0.918	80.911
Week 2	2	0.993	0.964	137.674
Week 3	3	0.987	0.953	123.455
Week 4	4	0.985	0.983	−86.209
Week 5	5	0.997	1.025	43.987
Week 6	6	0.994	0.909	−5.718
Week 7	7	0.976	1.161	−252.143

9.6.2 Estimating the 1995 Load Contour

The mean absolute percentage error (MAPE) with respect to the actual load is used to measure the effectiveness of the estimated results. For n estimated load values, the MAPE error is given by the equation

$$\text{MAPE} = \frac{100}{n}\sum_{i=1}^{n}\frac{\left|\hat{L}_{est,i} - L_{act,i}\right|}{L_{act,i}} \tag{9.23}$$

where $\hat{L}_{est,i}$ and $L_{act,i}$ are the estimated and actual ith load values, respectively.

The recursive procedure outlined in Section 9.4.2 is used to project the shape of the 1995 load contour. The regression coefficients determined earlier—namely [$c(i)$, $d(i)$] and [$a(k)$, $b(k)$]—are alternatively used to estimate a row and a column, respectively, of the 1995 contour described in Figure 9.1. The procedure is carried out for 24 iterations converging to the actual 1995 load. Figure 9.11 shows a sample of the MAPE error convergence for each hour over the 24 iterations. As shown, the error for each hour converges to its minimum. The overall MAPE error for the whole year was found to be 5.12%.

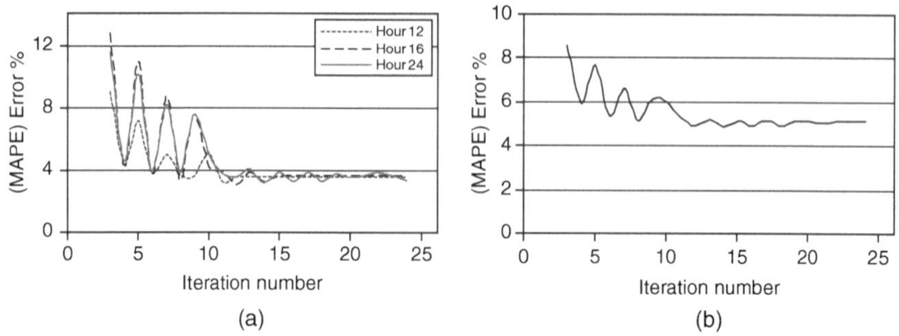

Figure 9.11 (a) Regression estimation (MAPE) error over 52 weeks of 1995; (b) overall regression estimation (MAPE) error over 52 weeks of 1995.

9.6.3 Kalman Filter Prediction Results

To prepare the input (measurement) data for Kalman filtering, we produce the annual load growth and use it to augment the estimated load contours determined in Section 9.7.2. The third-order polynomial load models described in equation (9.18) are used to calculate the annual load growth for each hour of the day. Figure 9.9 shows the approximate fitted curves for hour 3 of 1994 and 1995, and Figure 9.10 shows the annual load growth for that hour. The annual load growth curves for all hours follow almost the same shape with very minimal variations, as illustrated by Figure 9.12. During almost the first 10 weeks, the annual load growth is negative. This accounts for the unexpectedly low load demand during these weeks in 1995, as noticed in Figure 9.7. The low power consumption in these weeks of 1995 was mainly due to the above-normal high temperatures. The model naturally responds to the given data. It will react differently to different data from different utilities. To reduce the dependency of the annual load growth on uncontrollable short-term weather variations, we can calculate the average of the annual growth over several years. Furthermore, second-order models will not be sufficient to pick up such annual load growth variations. Third-order models or higher must be used. Models with orders 3, 4, 5, and 6 were tested. It was found that models with orders higher than third order were very sensitive to round-off errors and produce "very" incorrect results.

The fourth-order dynamic time-varying state space model for the Kalman filter described in Section 9.5.2 is employed to implement the following steps:

Step 1 The initial condition of the parameter vector is fixed arbitrarily to ones.

Step 2 Run the Kalman filter for the first hour of the day (the first column of Figure 9.1) using the actual load values of 1994 in the observation equation, equation (9.22), of the Kalman filter model. We used Cubic-Hermit interpolation to generate five extra points between each pair of load values to boost up the Kalman filter convergence. Save the four load-model estimated parameters for prediction. Set $i=1$ (i represents the week number of 1995).

Step 3 Predict the load value of the ith week of 1995 using the saved load-model estimated parameters:

Set $i=i+1$; if i is greater than 52 weeks, go to step 5.

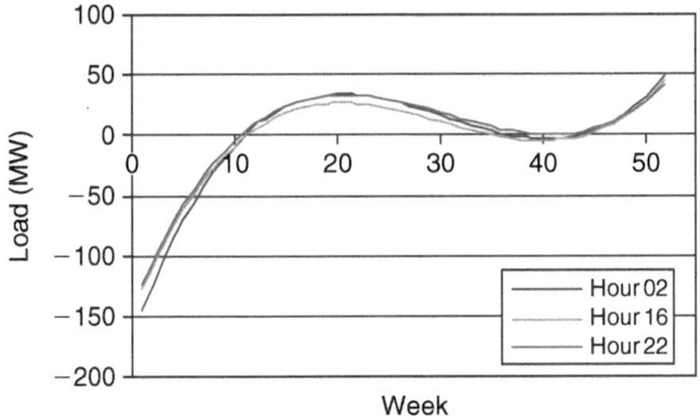

Figure 9.12 Annual load growth throughout 52 weeks of the year.

Step 4 Use estimated regression load values of the *i*th week of 1995 data (from Section 5.2) in the measurement equation, equation (9.22), to estimate the next set of load-model parameters using the Kalman filter. Save the load-model parameters. Go to step 3.

Step 5 Use the estimated parameters of the previous hour as the initial condition for estimating the next hour's coefficient using the Kalman filter. Repeat steps 3 and 4 for all 24 hours of the day.

The five steps of the preceding algorithm are illustrated using the flow diagram in Figure 9.13.

Figure 9.14 presents the estimated Kalman filter load-model parameters. As illustrated, all estimated parameters converge to their steady-state value after some transient fluctuations. Table 9.12 presents only a sample (10 weeks) of the estimated load using the Kalman filter together with the actual load demand of 1995. The large MAPE error in some weeks, especially in the second to fourth weeks, is attributed to the sudden, unexpected, and aberrant load condition in either year, which could not be explained by any model variables. The prediction method used captures the general behavior of the load over the year based on the previous year's load data rather than its short-term fluctuations. To reduce such aberrant effects, we could base the load prediction on the average of several previous years instead of only one year's data. Figure 9.15 illustrates the improvement in accuracy of the Kalman filtering algorithm by reducing the error compared to that obtained by the regression technique. A comparison between loads resulting from the estimated parameters and the actual load is shown in Figures 9.16 and 9.17. The results show how closely the estimated model matches with the actual load. Figure 9.16 also displays the MAPE error between the estimated and actual loads. The overall MAPE error for the whole year 1995 was calculated to be 2.24%, and the overall standard deviation was found to be 4.6 MW.

9.6.4 Remarks

This part of the chapter presented a composite technique for long-term load forecasting using multiple linear regression models and the Kalman filtering algorithm. Simple linear regression models, which capture 2D load behavior over one year, are utilized

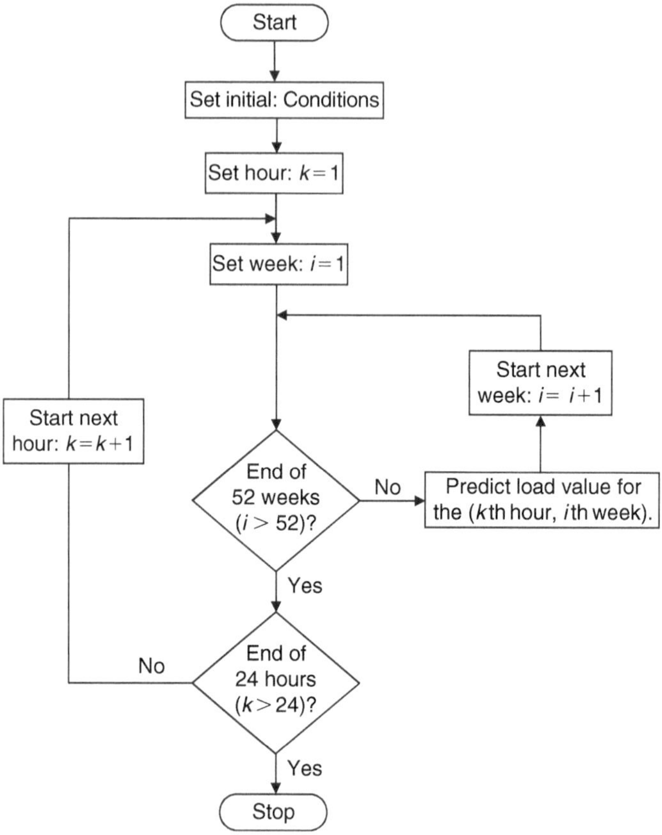

Figure 9.13 Kalman filtering load prediction flow diagram.

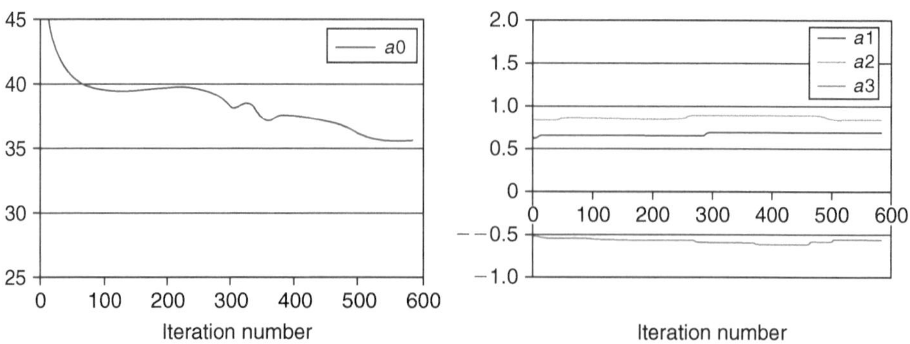

Figure 9.14 Kalman filter load-model parameters' convergence, hour 9.

Table 9.12 Sample of the Predicted Weekly Average Load Results for 1995 Using the Kalman Filter

	Hour 1			Hour 7			Hour 17			Hour 22		
Week	**Actual**	**Predicted**	**MAPE**	**Actual**	**Predicted**	**MAPE**	**Actual**	**Predicted**	**MAPE**	**Actual**	**Predicted**	**MAPE**
1	998	993	0.57	887	871	1.76	1228	1212	1.36	1289	1277	0.89
2	1078	1000	7.25	1002	893	10.82	1293	1206	6.75	1339	1274	4.87
3	871	1100	26.40	767	1010	31.69	1136	1286	13.27	1145	1368	19.51
4	967	1151	19.09	880	1056	20.01	1244	1325	6.45	1302	1399	7.43
5	1024	1056	3.10	955	982	2.84	1238	1265	2.15	1304	1342	2.87
6	1101	1115	1.22	1055	1025	2.82	1290	1313	1.80	1375	1385	0.71
7	1082	1007	6.93	1017	928	8.72	1200	1219	1.59	1298	1287	0.81
8	999	939	5.93	924	872	5.60	1213	1190	1.91	1265	1268	0.27
9	1040	1043	0.26	976	961	1.51	1253	1244	0.70	1307	1322	1.21
10	924	914	1.13	843	835	0.95	1179	1160	1.61	1222	1237	1.18

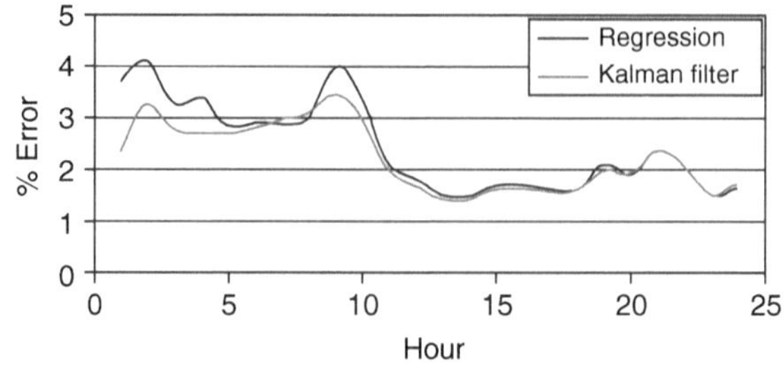

Figure 9.15 Comparing regression and Kalman filter MAPE errors.

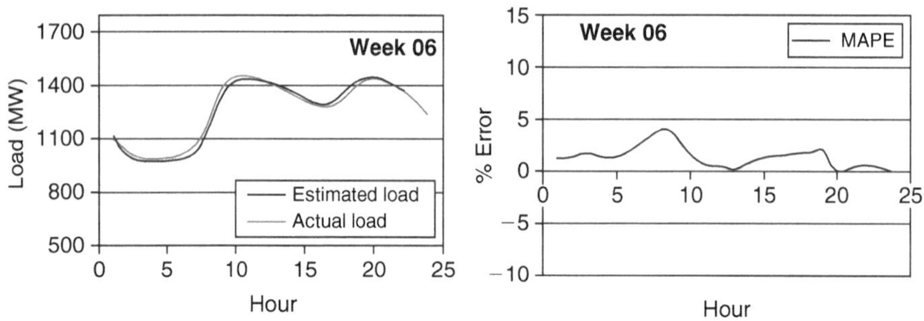

Figure 9.16 Estimated and actual load for 1995 during 24 hours.

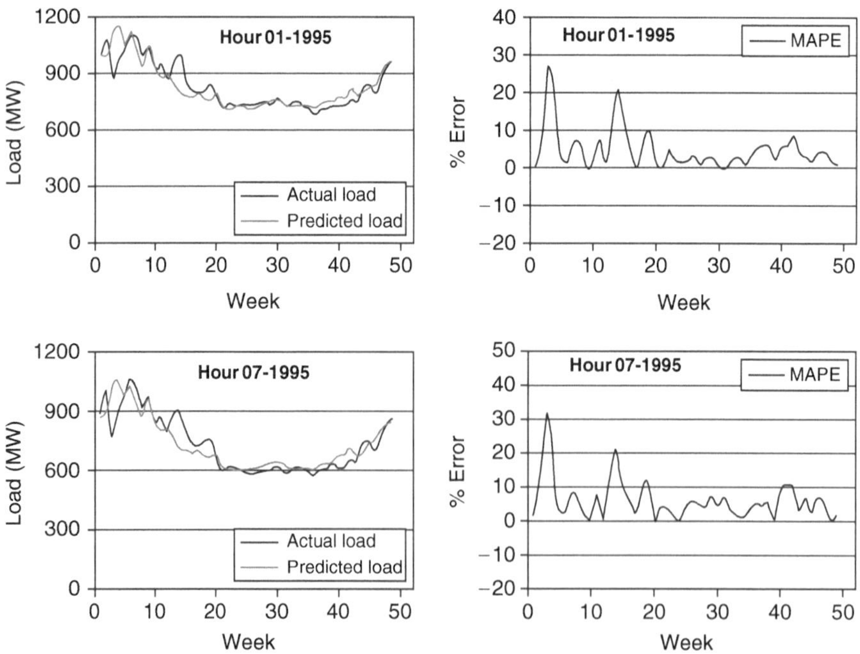

Figure 9.17 Estimated and actual loads for 1995 throughout 52 weeks.

recursively to project the load behavior of the next years. The Kalman filtering algorithm exploits the annual load growth to effectively improve the forecasting accuracy. The results indicate that the mean absolute percentage error of the predicted daily load does not exceed 2.3% of the actual load over a whole year period. With the produced results, the proposed composite technique provides a significant advantage compared to those typically seen in the literature in increasing the forecast accuracy.

9.7 Long-Term/Mid term Forecasting (Short-Term Correlation and Annual Growth)

The great importance of long-term and mid term load forecasting for electric power utility planning and its economic consequences is encouraging the development of forecasting approaches in electric power research to improve its accuracy [20–36]. Since the 1980s, many techniques have been developed to improve long-term and mid term forecasting accuracy. Regression models utilize the strong correlation of load with load-affecting factors such as weather. A method of mathematical modeling for global forecasting based on regression analysis was used to forecast load demand up to 2000. Long-term forecasting based on linear and linear-log regression models of six predetermined sectors has been developed. The time-series models—autoregressive (AR), moving average (MA), and autoregressive moving average (ARMA)—are popular and widely accepted by power utilities at present. They require a massive amount of historical data to produce optimal models. Gray system theory is successfully used to develop dynamic load-forecasting models.

By nature, long-term electric load forecasting is a complex problem. Among other factors, its accuracy is extremely influenced by the weather as well as social behavior of the community of that load. These factors are difficult to predict for the long-term load-forecasting time horizon. Conversely, short-term forecasting, though affected by weather and daily social habits, is small enough to predict load with high accuracy. Some short-term forecasting algorithms report to have results with a mean absolute error of less than 1%. Consequently, short-term correlation of daily (24 hours) and yearly (52 weeks) load demand of a previous year is utilized to construct a one-year load-demand behavior. The load trends obtained thus far are adjusted with the annual load growth (ALG) to project load demand for the next year. Daily and yearly correlations are modeled as simple linear regressions on weekly average load (WAL) for the 24 hours and 52 weeks resulting in (24×52) simple linear regression equations. Daily regression is used to depict the relation between the loads at each hour with the hour prior to it, and weekly regression relates the average weekly load with the week prior to it.

9.7.1 *Load Regression Models*

The mid term and long-term electric load demand as a function of time has a complex nonlinear behavior. It depends on a number of complex factors such as daily and seasonal weather, national economic growth, and social habits. All these factors depend on time in a complex way. Therefore, a single mid term and long-term electric load-demand model that accommodates most of these factors will have high nonlinear characteristics and may not give accurate prediction results. The approach taken in this section is the decomposition of the problem into multiple simple (first-order) linear regression models to capture the global nonlinear behavior of the load. Each of the linear regression models extracts the *short-term* correlation of a certain set of data. Then a recursive iterative algorithm is used to tie up the short-term results to capture the global load prediction.

One year's worth of data is arranged into a two-dimensional layout with 24 columns representing 24 hours of a day and 52 rows representing 52 weeks of the year. Figure 9.18 illustrates the 2D layout of the load data.

Special consideration is taken for the load variation during the weekends. Accordingly, weekends are treated separately but in the same manner exactly as the working days. The $L(i, k)$ cell in Figure 9.18 is the average load of the working days of the ith week at the kth hour. With this setup of the load data, obvious great intrinsic correlations exist between successive columns as well as between successive rows, as illustrated in Figures 9.19 and 9.20, respectively. These two figures and all subsequent results are based on the load demand of one of the largest electric power utilities in Canada for the years 1994 and 1995.

Figure 9.19 shows the load correlation between hour 1 AM and hour 2 AM throughout the whole year 1994. The correlation factor was calculated as 0.997. Similarly, Figure 9.20 shows the load correlation of week 1 and week 2 of year 1994, with a correlation factor of 0.985. The strong correlation is maintained over the entire year for all 24 hours of the day, as illustrated in Figures 9.21 and 9.22.

The persistent correlation of the prevalent load patterns suggests the use of short-term simple linear regression models for successive hours (see equation (9.24a)) and another set for successive weeks (see equation (9.24b)). This results in 24 × 52 simple linear regression models, which are used to draw the shape of the 2D load behavior contour for one year.

$$L(i,k) = a(k)L(i,k-1) + b(k) \qquad k = 1,\ldots,24 \tag{9.24a}$$

$$L(i,k) = c(i)L(i-1,k) + d(i) \qquad i = 1,\ldots,52 \tag{9.24b}$$

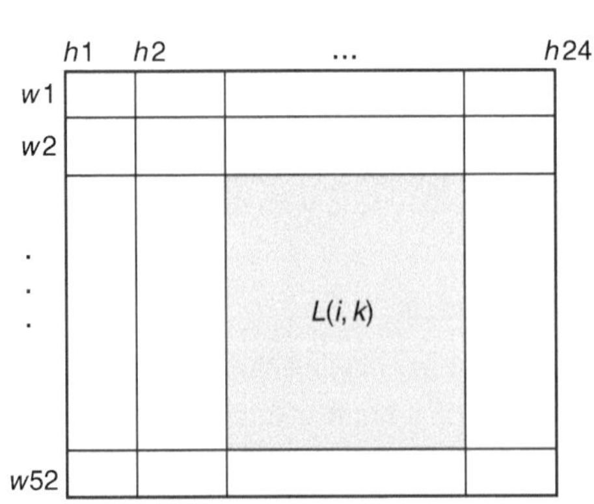

Figure 9.18 Two-dimensional layout of the load data.

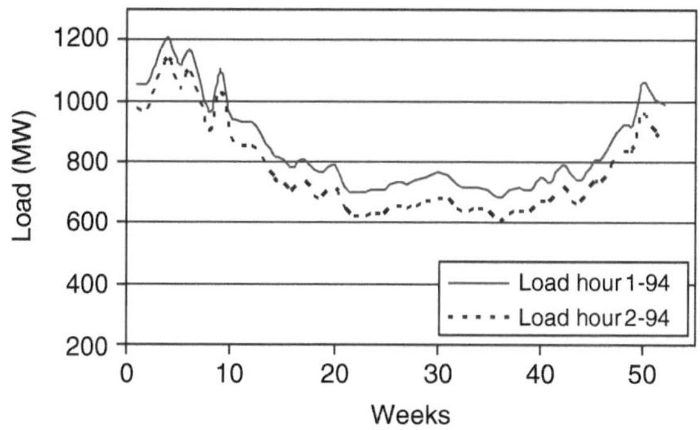

Figure 9.19 Comparing weekly average load of hours 1 and 2, 1994.

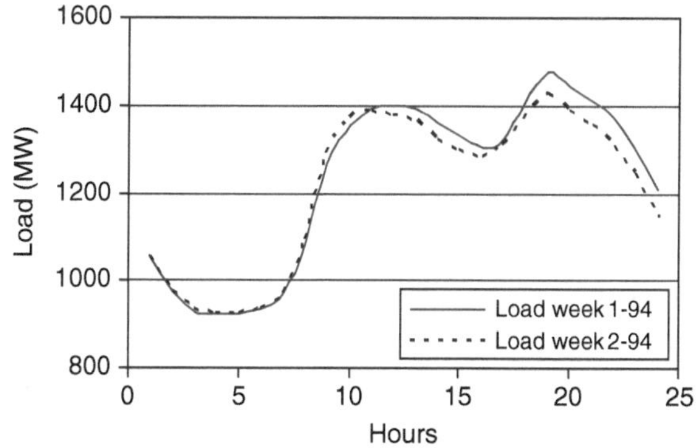

Figure 9.20 Comparing weekly average load of weeks 1 and 2, 1994.

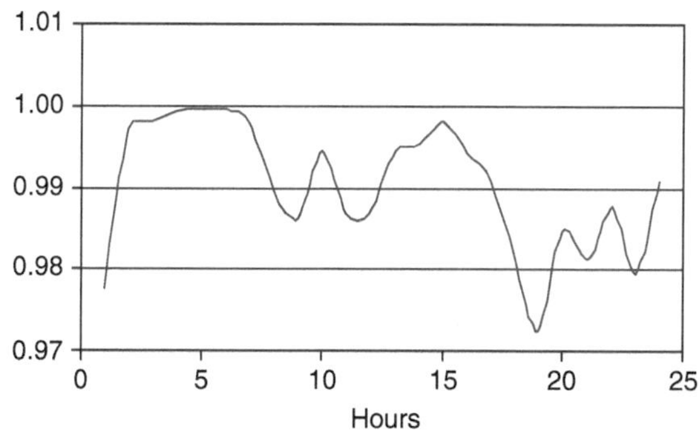

Figure 9.21 Correlation factor for successive hours over 52 weeks of 1994.

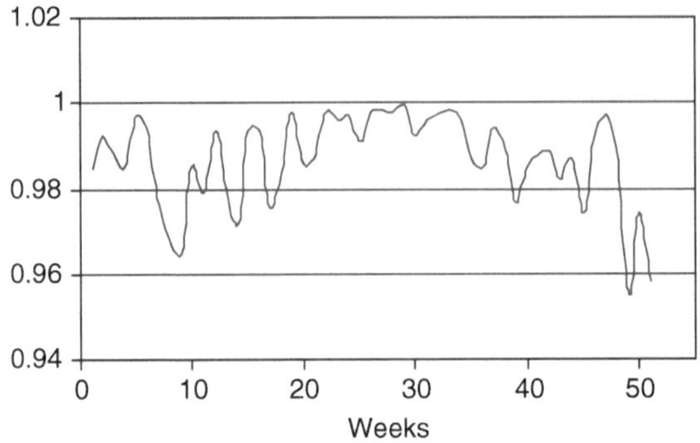

Figure 9.22 Correlation factor of successive weeks over 24 hours of 1994.

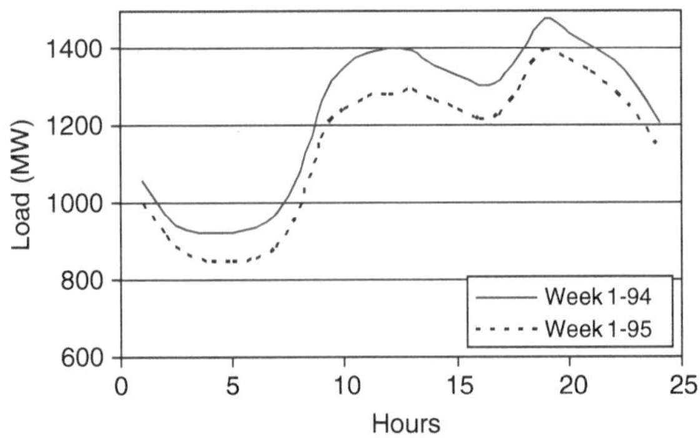

Figure 9.23 Comparing weekly average load of the first weeks of 1994 and 1995.

where

$a(k)$ and $b(k)$ are regression parameters at the kth hour; $k = 1, 2, \ldots, 24$, which are estimated using the load pairs $[L(i, k), L(i, k-1)]$ for all $i = 1, 2, \ldots, 52$, by the least squares method; $L(i, k)$ and $L(i, k-1)$ are the weekly average load at hours k and $k-1$, respectively, for all weeks $i = 1, \ldots, 52$, with the initial condition $L(i, 0) = L(i-1, 24)$;

$c(i)$ and $d(i)$ are regression parameters of the ith week, $i = 1, 2, \ldots, 52$, which are estimated using the load pairs $[L(i, k), L(i-1, \mathrm{k})]$ for all $k = 1, 2, \ldots, 24$, by the least squares method; $L(i, k)$ and $L(i-1, k)$ are the weekly average load in ith and $(i-1)$th weeks, respectively, for all hours $k = 1, \ldots, 24$, with the initial condition $L(0, k) =$ [$L(52, k)$ of the previous year].

9.7.2 Estimating the Next Year's Load Contour

The first-order regression models developed in the preceding section are used to project the load trends for the next year. Figures 9.23 and 9.24 demonstrate the fact that successive years have nearly identical load behavior contours. The load contours of the previous year (1994) coupled with the annual load growth are utilized to predict the next year's load (1995). Each regression model depicts a local relation of the load contours of the two years. The 24 linear regression models of equation (9.24a) relate

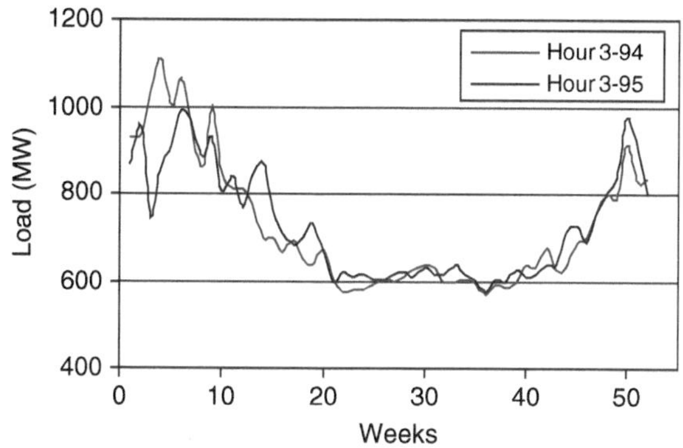

Figure 9.24 Comparing weekly average load of hour 3 of 1994 and 1995.

the load demand of successive hours of a day. They model the daily behavior of the load. The seasonal behavior of the load is modeled by the 52 linear regression models of equation (9.24b).

A recursive procedure used to estimate the next year's load contour utilizing regression models of the previous year is as follows:

1. Estimating for the first week the weekly average load: This process corresponds to estimating the first row of the next year's load; refer to Figure 9.25(a). Using equation (9.24b), we calculate $\hat{L}(1,k)$:

$$\hat{L}(1,k) = c(k)\ \hat{L}(0,k) + d(k) \qquad k = 1,2,\ldots,24 \tag{9.25}$$

where $\hat{L}(1,k)$ is the estimated weekly average load of the first week at the kth hour; $\hat{L}(1,k)$ is set to $L(52, k)_{\text{last-year}}$, which is the weekly average load of last year's 52nd week; and $[c(k), d(k)]$ is a pair of regression coefficients of the kth hour, obtained from equation (9.24b) using last year's data.

2. Estimating for the first hour the weekly average load: This process corresponds to estimating the first column of the next year's load; refer to Figure 9.25(b). Using equation (9.24a), we calculate $\hat{L}(i,1)$:

$$\hat{L}(i,1) = a(i)\ \hat{L}(i,0) + b(i) \qquad i = 2,3,\ldots,52 \tag{9.26}$$

where $\hat{L}(i,1)$ is the estimated weekly average load of the first hour in the ith week; $\hat{L}(i,0)$ is set to $L(i-1, 24)$, which is the weekly average load of the 24th hour of the previous week; and $[a(i), b(i)]$ is a pair of regression coefficients of the ith week, obtained from equation (9.24a) using last year's data.

3. Estimating for the second week the weekly average load: This process corresponds to estimating the second row of the next year's load; refer to Figure 9.25(c). Using equation (9.24b), we calculate $\hat{L}(2,k)$:

$$\hat{L}(2,k) = c(k)\ \hat{L}(1,k) + d(k) \qquad k = 2,3,\ldots,24 \tag{9.27}$$

where $\hat{L}(2,k)$ is the estimated weekly average load of the second week at the kth hour, and $\hat{L}(2,k)$ is obtained using equation (9.25).

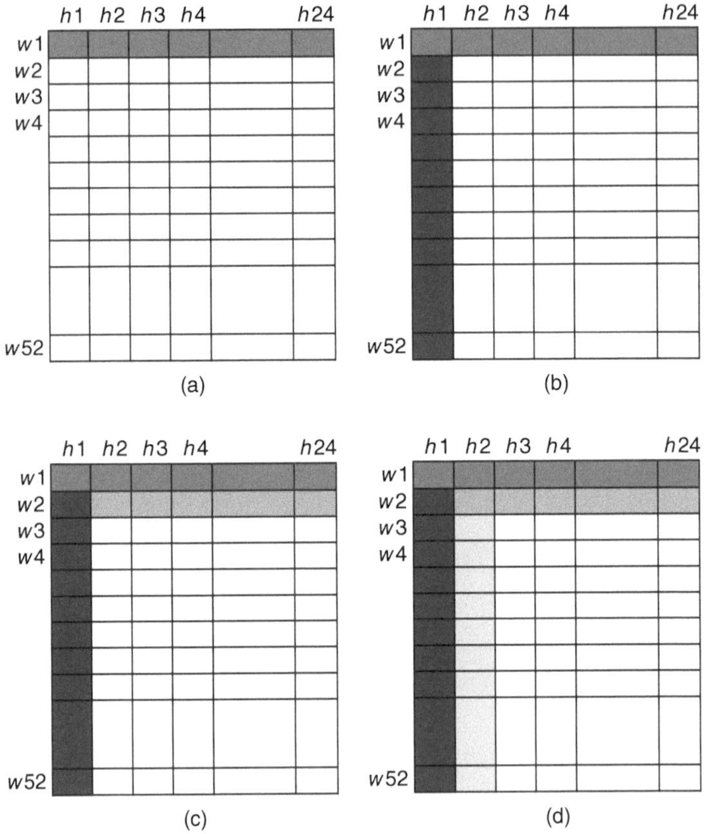

Figure 9.25 (a) First week (row) load estimation resulting from the first iteration. (b) First hour (column) load estimation resulting from the second iteration. (c) Second week (row) load estimation resulting from the third iteration. (d) Second hour (column) load estimation resulting from the fourth iteration.

4. Estimating for the second hour the weekly average load: This process corresponds to estimating the second column of the next year's load; refer to Figure 9.25(d). Using equation (9.24a), we calculate $\hat{L}(i,2)$:

$$\hat{L}(i,2) = a(i)\ \hat{L}(i,1) + b(i) \qquad i = 3,4,\ldots,52 \tag{9.28}$$

   where $\hat{L}(i,2)$ is the estimated weekly average load of the second hour in the ith week and is obtained using equation (9.26).
5. The recursive iterations are repeated until $i = k = 24$.

The preceding procedure produces a two-dimensional contour of the load behavior for one year based on regression coefficients of the previous year. The load contour will then be augmented by the annual load growth to account for the load change between successive years.

9.7.3 Annual Load Growth

To maximize the accuracy of next year's load-demand estimation, we estimate and employ annual load growth as an adjusting factor. It is evident that load demand has a very strong dependence on time. Typical load profiles of successive years reveal very strong correlation at certain periodic time intervals. For example, refer to Figure 9.24; the two load curves at a certain hour over the whole year for two successive years retain the same shape. Moreover, there is, on average, a clear load increase over the previous year. This increase amounts to an annual load growth at that hour as a function of time (weeks) throughout the whole year. The load growth is modeled as the difference between the load curves of two successive years as a function of time.

Practical load profiles show that second-order models will not be sufficient to pick up the annual load growth variations. Third-order models or higher must be used. Models with orders 3, 4, 5, and 6 were tested to best fit load profiles. It was found that models with orders higher than third order were very sensitive to round-off errors and produce "very" incorrect results. A third-order polynomial is utilized to model the load as a function of time at the kth hour as a function of the load of the previous hour. The regression model is as follows:

$$L(i,k) = \beta_0(k) + \beta_1(k)\, L(i,k-1) + \beta_2(k)\, L^2(i,k-1) + \beta_3(k)\, L^3(i,k-1) \tag{9.29}$$

where $\beta_j(k)$, $j=0, 1, 2, 3$, are regression variables at the kth hour, and $k=1,2,\ldots,24$, which are determined using the load pairs [$L(i, k)$, $L(i, k-1)$, for all $i=1,2,\ldots,52$] by the least squares method. The initial values $L(i, 0)$ are set to $L(i-1, 24)$. The two curves that approximate the relationship between $L(i, k)$ and $L(i, k-1)$ corresponding to the load behavior of the two years in Figure 9.24 are shown in Figure 9.26. The annual load growth curve is obtained by subtracting the approximate curve of 1995 (estimated data using regression models) from the approximate curve of 1994 (actual data), as shown in Figure 9.27.

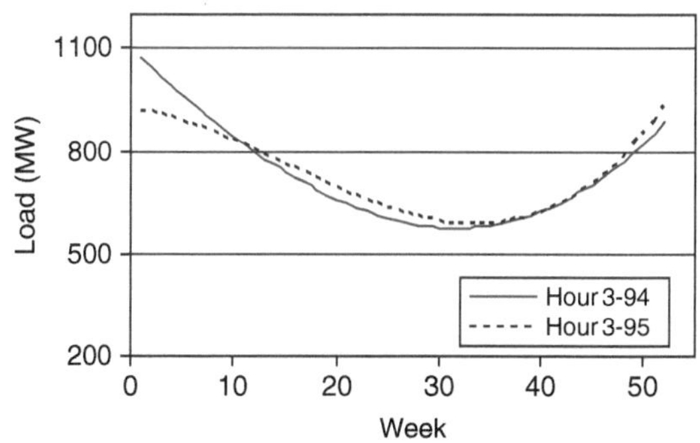

Figure 9.26 Approximate curves of load of third hour of 1994 and 1995.

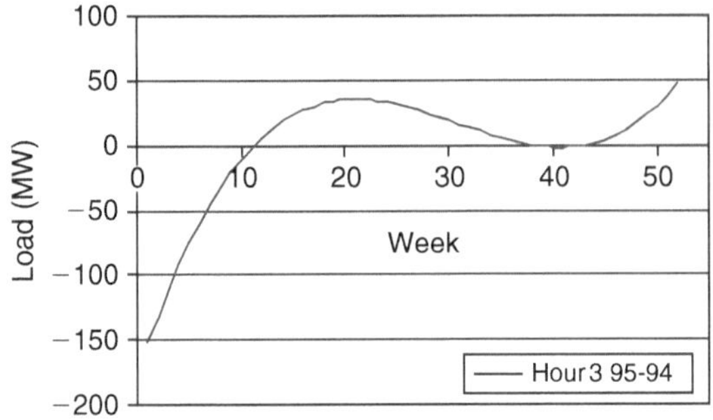

Figure 9.27 Annual load growth variations during 52 weeks of the year.

Next, the procedure for evaluating the annual load growth is as follows; we assume that the annual load growth is calculated between 1994 and 1995:

1. Using equation (9.29), we determine the regression coefficients (24 sets) for 24 hours for the actual date from year 1994. The coefficients define 24 approximate curves of the weekly average load, one curve per hour.
2. We repeat the calculations of the preceding step to the 1995 estimated data obtained using the regression models.
3. We define the annual load growth as the difference of the approximate load curves of 1995 and 1994 of steps 2 and 1, respectively:

$$\text{Annual Load Growth } (i) = L(i,k)_{(95)} - L(i,k)_{(94)} \quad k = 1,2,\ldots,24, \quad i = 1,2,\ldots,52 \tag{9.30}$$

For each hour, the annual load growth is added to the 1995 estimated data obtained using the regression models to produce the final prediction results.

9.8 Examples of Long-Term/Mid Term Forecasting

To verify the effectiveness of the proposed load-demand forecasting technique, we used load data for one of the largest utility companies in Canada for the years 1994 and 1995. Regression models are obtained from 1994 data and used to project load demand for 1995.

9.8.1 Multiple Regression Model Results

Using equation (9.24a), we calculate 24 sets of regression coefficients. Table 9.13 shows the first seven of these sets as a sample. This table also lists the correlation factors of successive hours (columns) of the 1994 load data. Similarly, using equation (9.24b), we calculate 52 sets of regression coefficients. Table 9.14 shows the first seven of these sets as a sample, together with the correlation factors of successive weeks (rows) of the 1994 load data.

Table 9.13 Correlation Factors and Regression Coefficients for Seven Hours of 1994

1994	Hour 1	Hour 2	Hour 3	Hour 4	Hour 5	Hour 6	Hour 7
k = hour of the day	**1**	**2**	**3**	**4**	**5**	**6**	**7**
Correlation Factor	0.978	0.997	0.998	0.999	0.999	1.000	0.998
$a(k)$	0.973	0.994	1.014	1.022	1.025	1.024	1.049
$b(k)$	−89.311	−76.835	−49.053	−31.009	−21.580	−11.659	−6.003

Table 9.14 Correlation Factors and Regression Coefficients of Seven Weeks of 1994

1994	i = Week Number	Correlation Factor	$c_{(i)}$	$d_{(i)}$
Week 1	1	0.985	0.918	80.911
Week 2	2	0.993	0.964	137.674
Week 3	3	0.987	0.953	123.455
Week 4	4	0.985	0.983	−86.209
Week 5	5	0.997	1.025	43.987
Week 6	6	0.994	0.909	−5.718
Week 7	7	0.976	1.161	−252.143

9.8.2 Estimating the 1995 Load Contour

The MAPE with respect to the actual load is used to measure the effectiveness of the estimated results. For n estimated load values, the MAPE error is given by the equation

$$\text{MAPE} = \frac{100}{n}\sum_{i=1}^{n}\frac{|\hat{L}_{est,\,i} - L_{act,\,i}|}{L_{act,\,i}} \tag{9.31}$$

where $\hat{L}_{est,\,i}$ and $L_{act,i}$ are the estimated and actual ith load values, respectively. The recursive procedure outlined in Section 9.7.2 is used to project the shape of the 1995 load contour. The regression coefficients—namely [$c(i)$, $d(i)$] and [$a(k)$, $b(k)$]—are alternatively used to estimate a row and a column, respectively, of the 1995 contour described in Figure 9.18. The procedure is carried out for 24 iterations converging to the actual 1995 load. Figure 9.28(a) shows a sample of the MAPE error convergence for each hour over the 24 iterations. As shown, the error for each hour converges to its minimum. Figure 9.28(b) shows the convergence of the overall MAPE error for the whole year, which was found to be 5.12%.

9.8.3 Annual Load Growth Results

The annual load growth is evaluated and used to augment the estimated load contours determined in Section 9.8.2. The third-order polynomial load models described in

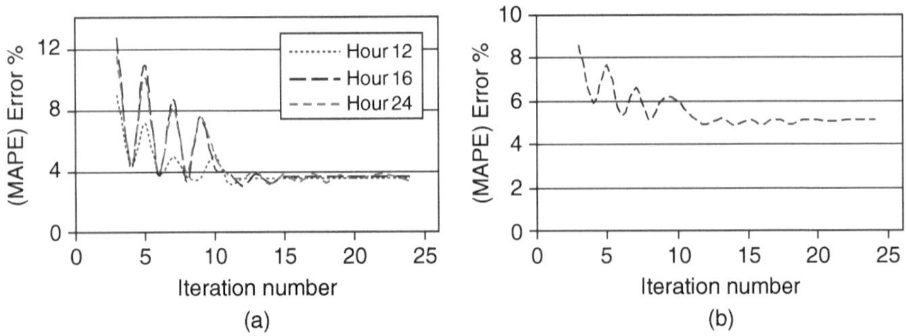

Figure 9.28 (a) Regression estimation (MAPE) error over 52 weeks of 1995. (b) Overall regression estimation (MAPE) error over 52 weeks of 1995.

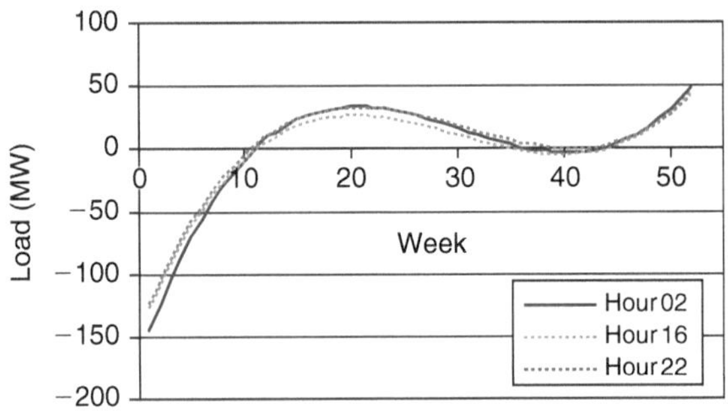

Figure 9.29 Annual load growth throughout 52 weeks of the year.

equation (9.28) are used to calculate the annual load growth for each hour of the day. Figure 9.26 shows the approximate fitted curves for hour 3 of 1994 and 1995, and Figure 9.27 shows the annual load growth for that hour. The annual load growth curves for all hours follow almost the same shape with very minimal variations, as illustrated by Figures 9.29 and 9.30. During almost the first 10 weeks, the annual load growth is negative. This accounts for the unexpectedly low load demand during these weeks in 1995, as shown in Figure 9.27. The low power consumption in these weeks of 1995 was mainly due to the above-normal high temperatures. The model naturally responds to the given data. It will react differently to different data from different utilities. To reduce the dependency of the annual load growth on uncontrollable short-term weather variations, we can calculate the average of the annual growth over several years.

Figure 9.31 shows a sample of the estimated weekly average load curves for some weeks together with MAPE error over 24 hours of the day. Similarly, Figure 9.32 presents a sample of the weekly average load for some hours varying over 52 weeks of the year. Introducing annual load growth improved the estimation results obtained in Section 9.8.2. The resulting overall MAPE is 3.8 with a standard deviation of 4.14.

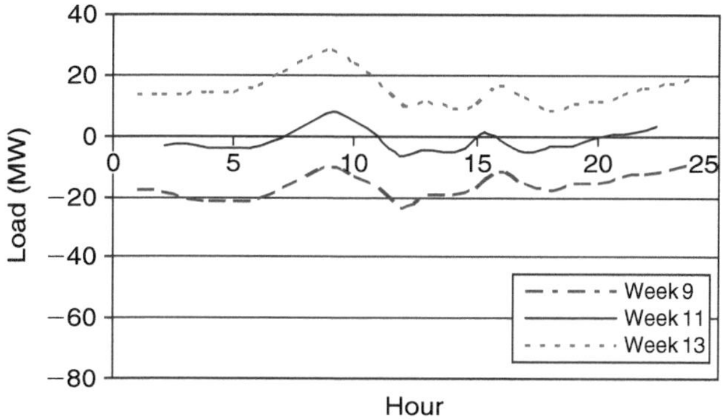

Figure 9.30 Annual growth variation during hours of a day.

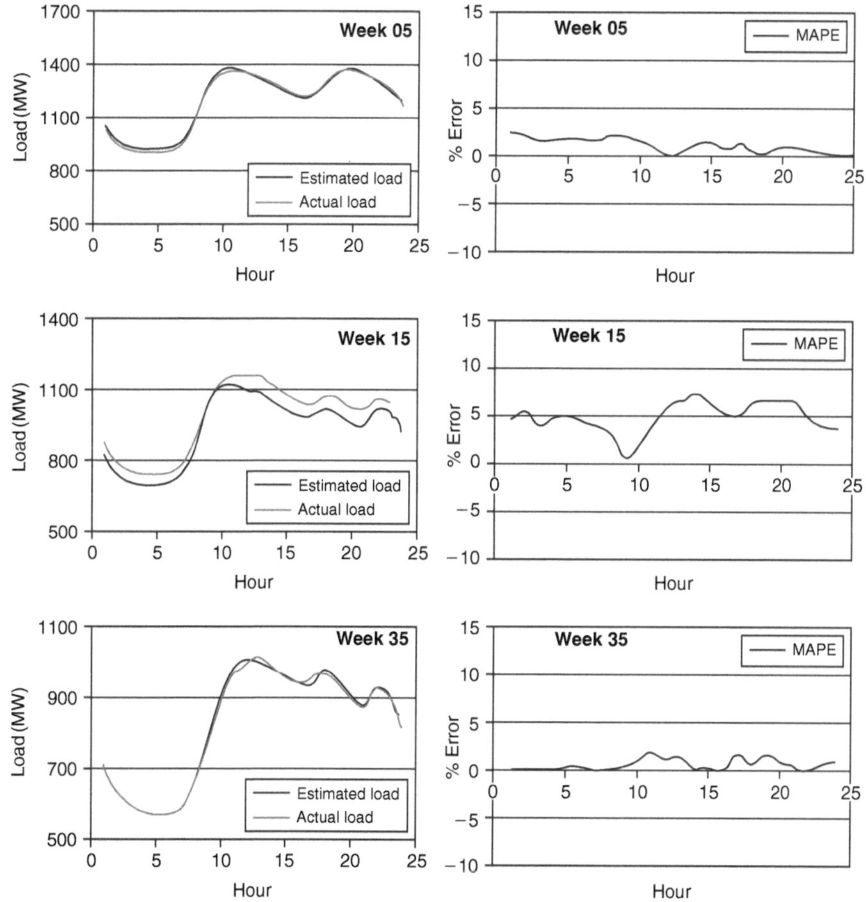

Figure 9.31 Comparison of a sample of estimated and actual load for 1995 during 24 hours.

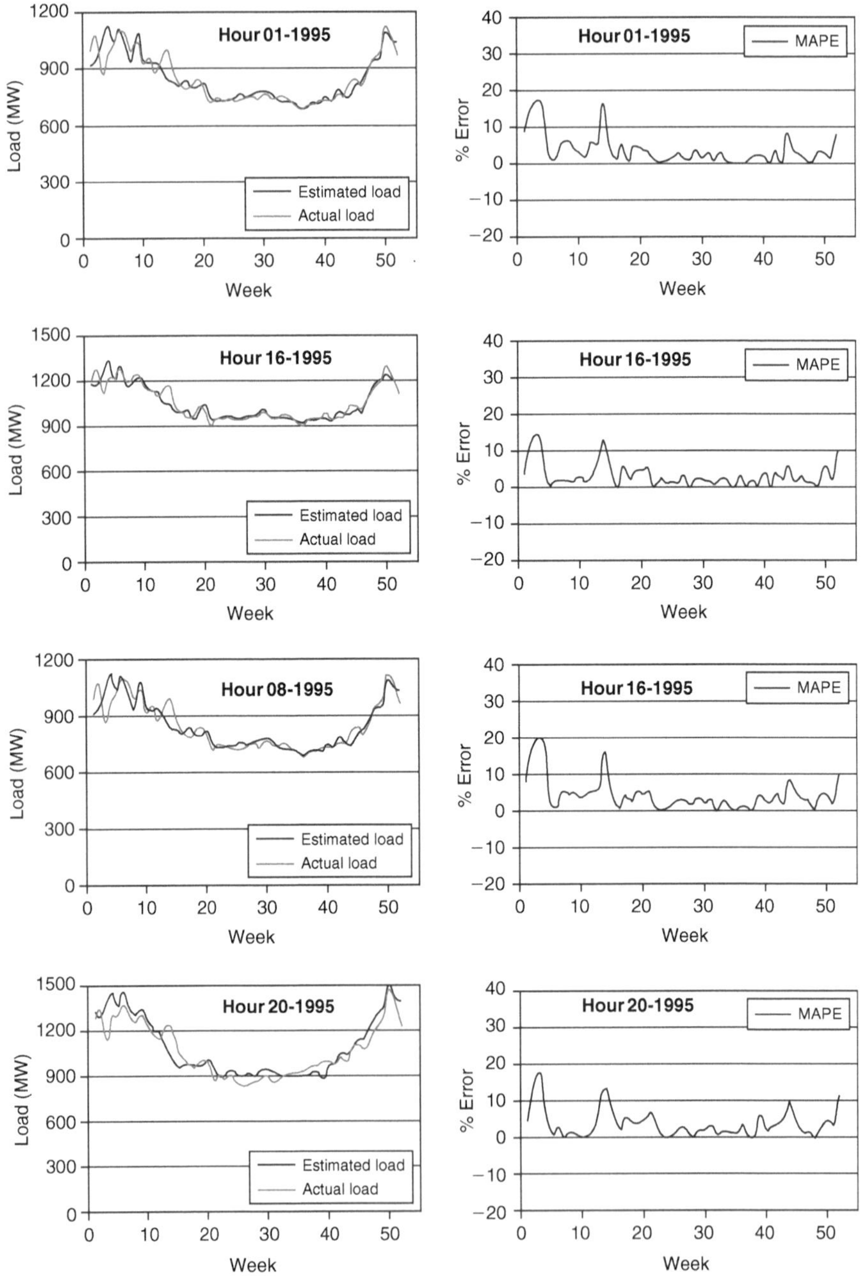

Figure 9.32 Comparison of a sample of estimated and actual loads for 1995 throughout 52 weeks.

9.8.4 Remarks

This section demonstrated a long-term and mid term electric load-forecasting technique for forecasting hourly daily load demand for a lead time of several weeks to a few years. It was achieved utilizing short-term correlation of load behavior together with its annual growth. First, using historic data over a specific period of time (one year), we obtained the hourly daily load shape using multiple simple linear regression parametric load models. Second, we employed the parametric models obtained using alternating hourly and weekly load estimations to determine the shape of the load behavior for the next year. Last, we added annual growth load to correct the shape of the next year's load. The results indicated that the mean absolute error of the predicted weekly average daily load did not exceed 3.8% of the actual load over a whole year period. With the produced results, the proposed model and forecast technique used provide a significant advantage compared to those typically seen in the literature for reducing the average absolute error between the forecasted and actual loads over a forecast period of one year ahead.

9.9 Fuzzy Regression for Peak-Load Forecasting

In power system planning, a utility establishes goals and objectives, seeks to predict environmental factors, and then selects actions that result in the accomplishment of these goals and objectives [37–52]. The need for electric load forecasting is increasing as power system planning attempts to decrease its dependence on chance and becomes realistic in dealing with its environment.

Frequently, there is a time lag between awareness of an impending event or need and the occurrence of that event. This time lag is the main reason for power system planning and electric load forecasting. If the time lag is long and the outcome of the final event is conditional upon identifiable factors, power system planning can play an important role. In such situations, electric load forecasting is needed to determine when a need will arise so that the appropriate action can be taken.

The load growth of a geographical area served by a utility company is the most important factor influencing the expansion of a power system. Therefore, the forecasting of an increasing load and power system reaction to such load growth is essential to the planning process. Electric load forecasting can be regarded as answering this question: What amount of electricity should be arranged to supply a specific number and type of customer over a specific period of time? Forecasting can be achieved by performing analysis of past and/or present data, identifying trends and patterns that exist in the data that are then used to project load into the future.

This section presents the application of a fuzzy regression technique to long-term annual peak-load forecasting. The proposed technique takes into account the uncertainties in the nature of the peak load. Different factors are taken into account on modeling the peak load. These factors include the gross domestic product (GDP), population (POP), GDP/POP, the multiplication of the consumption of electricity and population (EP), the system losses (PL), and the rate of sale of electricity (RS; the price). Finally, we consider the time in question. Different fuzzy models are developed that relate these variables with the peak load. This section offers an example for estimating the peak load for the Egyptian Unified Network (EUN) to explain the main features of the proposed algorithm.

9.9.1 Modeling of Electric Annual Peak Load

Annual peak-load demand mainly depends on the community and the nation within this community. The main factors that greatly affect the growth of the load on a power system are different from one nation to another. For the Egyptian Unified Network, the following factors are to be considered when modeling the annual peak load:

- The gross domestic product (GDP)
- The population (POP)
- The gross domestic product per population (GDP/POP)
- The electric population (EP)
- The system losses (SL)
- The load factor (LF)
- The rate of sale (RS) measured in mill/kWh
- The time horizon (the year in question; T)

The annual peak-load demand is a function of these variables. The technique developed in reference [9] uses some of these factors to estimate the annual peak-load demand of Japan. In this section, we consider all these variables to obtain a fuzzy model for the annual peak load.

9.9.2 A Nonfuzzy Peak Load with Fuzzy Parameters

In this section, we assume that the peak load is nonfuzzy, whereas the parameters of the load are fuzzy parameters with a symmetrical triangular membership function. In this case, the annual load model can be written as

$$\begin{aligned} P_L = {} & A_0 + A_1(\text{GDP}) + A_2(\text{POP}) + A_3(\text{EP}) + A_4(\text{GDP/POP}) \\ & + A_5(\text{SL}) + A_6(\text{LF}) + A_7(\text{RS}) + A_8(\text{T}) \end{aligned} \tag{9.32}$$

where $\underline{A}_0, \underline{A}_2, \ldots, \underline{A}_8$ are the model fuzzy parameters to be estimated, and each parameter has a certain middle p and a certain spread c. Equation (9.32) can be rewritten as

$$\begin{aligned} P_L = {} & (p_0, c_0) + (p_1, c_1)(\text{GDP}) + (p_2, c_2)(\text{POP}) + (p_3, c_3)(\text{EP}) \\ & + (p_4, c_4)(\text{GDP/POP}) + (p_5, c_5)(\text{SL}) + (p_6, c_6)(\text{LF}) \\ & + (p_7, c_7)(\text{RS}) + (p_8, c_8)(\text{T}) \end{aligned} \tag{9.33}$$

In fuzzy regression, we seek to find the fuzzy coefficients that minimize the spread of fuzzy output for all the data sets. In mathematical form, the objective function to be minimized is

$$\begin{aligned} O = \min \Bigg\{ & \sum_{j=1}^{m} [c_0 + c_1(\text{GDP})_j + c_2(\text{POP})_j + c_3(\text{EP})_j + c_4(\text{GDP/CAP})_j \\ & + c_5(\text{SL})_j + c_6(\text{LF})_j + c_7(\text{RS})_j + c_8(\text{T})_j] \Bigg\} \end{aligned} \tag{9.34}$$

This is subject to satisfying two constraints on each annual peak-load demand as

$$\begin{aligned}(P_L)_j \geq \{(p_0 + p_1(\text{GDP})_j + p_2(\text{POP})_j + p_3(\text{EP})_j + p_4(\text{GDP/CAP})_j + p_5(\text{SL})_j \\ + p_6(\text{LP})_j + p_7(\text{RS})_j + p_8(\text{T})_j\} - (1-\lambda)\{(c_0 + c_1(\text{GDP})_j + c_2(\text{POP})_j \\ + c_3(\text{EP})_j + c_4(\text{GDP/CAP})_j + c_5(\text{SL})_j + c_6(\text{LP})_j + c_7(\text{RS})_j + c_8(\text{T})_j\}; \\ j = 1, \ldots, m\end{aligned} \tag{9.35}$$

$$\begin{aligned}(P_L)_j \leq \{(p_0 + p_1(\text{GDP})_j + p_2(\text{POP})_j + p_3(\text{EP})_j + p_4(\text{GDP/CAP})_j + p_5(\text{SL})_j \\ + p_6(\text{LP})_j + p_7(\text{RS})_j + p_8(\text{T})_j\} + (1-\lambda)\{(c_0 + c_1(\text{GDP})_j + c_2(\text{POP})_j \\ + c_3(\text{EP})_j + c_4(\text{GDP/CAP})_j + c_5(\text{SL})_j + c_6(\text{LP})_j + c_7(\text{RS})_j + c_8(\text{T})_j)\}, \\ j = 1, \ldots, m\end{aligned} \tag{9.36}$$

where λ is the degree of fuzziness.

The problem formulated in equations (9.34) to (9.36) is a standard linear programming problem and can be solved using linear programming based on the simplex method available in the IMSL/STAT library.

Having identified the fuzzy parameters of the model, we could easily forecast the annual peak-load demand for any year, providing that the factors mentioned in Section 9.9.1 are available.

9.9.3 A Fuzzy Peak-Load Demand

Due to the uncertainties in the annual peak-load-demand forecasting, we assume that this load is a fuzzy load having a certain power m_j with a spread α_j, $j = 1, \ldots, m$. In this case equation (9.32) can be rewritten as

$$\begin{aligned}(m_j, \alpha_j) = (p_0, c_0) + (p_1, c_1)(\text{GDP})_j + (p_2, c_2)(\text{POP})_j + (p_3, c_3)(\text{EP})_j \\ + (p_4, c_4)(\text{GDP/POP}) + (p_5, c_5)(\text{SL})_j + (p_6, c_6)(\text{LF})_j \\ + (p_7, c_7)(\text{RS})_j + (p_8, c_8)(\text{T})_j\end{aligned} \tag{9.37}$$

Two fuzzy numbers are equal if and only if their middles and spreads are equal—that is

$$\begin{aligned}m_j = p_0 + p_1(\text{GDP})_j + p_2(\text{POP})_j + p_3(\text{EP})_j + p_4(\text{GDP/POP})_j \\ + p_5(\text{SL})_j + p_6(\text{LF})_j + p_7(\text{RS})_j + p_8(\text{T})_j; \qquad j = 1, \ldots, m\end{aligned} \tag{9.38}$$

and

$$\begin{aligned}\alpha_j = c_0 + c_1(\text{GDP})_j + c_2(\text{POP})_j + c_3(\text{EP})_j + c_4(\text{GDP/POP})_j \\ + c_5(\text{SL})_j + c_6(\text{LF})_j + c_7(\text{RS})_j + c_8(\text{T})_j; \qquad j = 1, \ldots, m\end{aligned} \tag{9.39}$$

The problem now turns out to be: Given the previous history of the fuzzy annual peak load in the form of (m_j, α_j), we need to find the fuzzy parameters $\underline{A}_0, \ldots, \underline{A}_8$ that minimize the cost function given by

$$O = \min\left\{\sum_{j=1}^{m}[c_0 + c_1(\text{GDP})_j + c_2(\text{POP})_j + c_3(\text{EP})_j + c_4(\text{GDP/CAP})_j + c_5(\text{SL})_j + c_6(\text{LF})_j + c_7(\text{RS})_j + c_8(\text{T})_j\right\} \tag{9.40}$$

This is subject to satisfying the following two constraints in each load power

$$\begin{aligned} m_j - (1-h)\,\alpha_j \geq \{&p_0 + p_1(\text{GDP})_j + p_2(\text{POP})_j + p_3(\text{EP})_j + p_4(\text{GDP/POP})_j \\ &+ p_5(\text{SL})_j + p_6(\text{LF})_j + p_7(\text{RS})_j + p_8(\text{T})_j\} - (1-\lambda)\{c_0 \\ &+ c_1(\text{GDP})_j + c_2(\text{POP})_j + c_3(\text{EP})_j + c_4(\text{GDP/POP})_j \\ &+ c_5(\text{SL})_j + c_6(\text{LF})_j + c_7(\text{RS})_j + c_8(\text{T})_j\}; \quad j = 1, \ldots, m \end{aligned} \tag{9.41}$$

and

$$\begin{aligned} m_j + (1-h)\,\alpha_j \leq \{&p_0 + p_1(\text{GDP})_j + p_2(\text{POP})_j + p_3(\text{EP})_j + p_4(\text{GDP/POP})j \\ &+ p_5(\text{SL})_j + p_6(\text{LF})_j + p_7(\text{RS})_j + p_8(\text{T})_j\} + (1-\lambda)\{c_0 \\ &+ c_1(\text{GDP})_j + c_2(\text{POP})_j + c_3(\text{EP})_j + c_4(\text{GDP/POP})_j \\ &+ c_5(\text{SL})_j + c_6(\text{LF})_j + c_7(\text{RS})_j + c_8(\text{T})_j\}, \quad j = 1, \ldots, m \end{aligned} \tag{9.42}$$

Again, the problem formulated in this section is a linear programming problem that can be solved using the simplex method.

Having identified the model fuzzy parameters, we can estimate the peak annual load for the forthcoming years.

9.10 Testing the Algorithm

9.10.1 Nonfuzzy Annual Peak Load

In this section we test the proposed algorithm for the data of the EUN [37–52]. The data are given in Table 9.15. The data from year 1981 to year 1992, $T = 0$ to $T = 11$, are used to estimate the fuzzy parameter of the model given in equation (9.37). The unseen data for the rest of the years are used to evaluate the accuracy of the estimated parameters. The linear programming available in the IMSL/STAT library is used to solve the linear optimization problem. The fuzzy coefficients obtained are given as

$\underline{A}_0 = (0.0, 222.382)$ $\underline{A}_5 = (0.2652, 0.0)$
$\underline{A}_1 = (0.075, 0.0)$ $\underline{A}_6 = (0.0, 0.0)$
$\underline{A}_2 = (0.0, 0.0)$ $\underline{A}_7 = (0.0, 0.0)$
$\underline{A}_3 = (0.0, 0.0)$ $\underline{A}_8 = (154.135, 0.0)$
$\underline{A}_4 = (1.561, 0.0)$

Note that $\underline{A}_0$ is the only fuzzy parameter. These estimated parameters are used to estimate the annual peak load for the unseen data. Table 9.16 gives the results obtained.

Table 9.15 Actual, Estimated Annual Peak Load

Year	Actual Load	Estimated Load	Error (MW)	% Error
1993	7503	7603.61	−100.61	−1.34
1994	7657	7866.61	−209.61	−2.78
1995	8149	8213.92	−64.92	−0.80
1996	8491	8591.76	−100.76	−1.19

Table 9.16 Estimated Parameters at Different Degrees of Fuzziness

Parameter	$\lambda = 0.25$	$\lambda = 0.5$	$\lambda = 0.75$
$\underline{A}_0$	(0, 745)	(0, 805)	(0, 1013)
$\underline{A}_1$	(0.165, 0)	(0.1337, 0)	(0.088, 0)
$\underline{A}_2$	(0, 0)	(7.985, 0)	(21.84, 0)
$\underline{A}_3$	(0, 0)	(0, 0)	(0, 0)
$\underline{A}_4$	(0, 0)	(0, 0)	(0, 0)
$\underline{A}_5$	(0.067, 0)	(0.11482, 0)	(0.179, 0)
$\underline{A}_6$	(0, 0)	(0, 0)	(0, 0)
$\underline{A}_7$	(0, 0)	(0, 0)	(0, 0)
$\underline{A}_8$	(108, 0)	(127.82, 0)	(153.5, 0)

Examining Table 9.16 reveals that the proposed algorithm estimates the annual peak load very accurately, and the errors in the estimates are small compared to the other techniques described in the literature.

We examined the effects of the degree of fuzziness on the estimated parameters in this test, where we changed λ from a small value, 0.0, to a large value, 1.0. It has been found that the degree of fuzziness has no effect on the middle of the fuzzy coefficients, but as the degree of fuzziness increases, the spread of the output increases to satisfy the increased measure of goodness of fit.

9.10.2 Fuzzy Annual Peak Load

In this section we assume that the annual peak load is fuzzy and the spread of each measurement is 0.1 from the actual peak load given in Table 9.15. The problem formulated in equations (9.37), (9.38), and (9.39) is solved using the simplex method based on linear programming. Table 9.16 gives the estimated parameters at different degrees of fuzziness, for the first 12 measurements of Table 9.15. Tables 9.17 and 9.18 give the estimated load for the rest of the data of Table 9.15, unseen data, as well as the error in the estimated value.

Examining these tables reveals the following observations:

- All the parameters are nonfuzzy parameters except the first one.
- As the degree of fuzziness increases, the spread of $\underline{A}_0$ increases.
- The estimated load is very close to the actual load, even the spread of the load, and still the actual load moves between the boundaries of the triangular membership function we assumed.

Table 9.17 Estimated Peak Load at $\lambda = 0.25$

Year	Actual P_L	Estimated P_l	% Error
1993	(7503, 750)	(7498, 745)	0.060
1994	(7657, 766)	(7824, 745)	−2.200
1995	(8149, 815)	(8217, 745)	−0.083
1996	(8491, 849)	(8647, 745)	−9.943

Table 9.18 Estimated Peak Load at $\lambda = 0.5$

Year	Actual P_L	Estimated P_l	% Error in the Estimates
1993	(7503, 750)	(7543, 805)	−0.5
1994	(7657, 766)	(7855, 805)	−2.6
1995	(8149, 815)	(8232, 805)	−1.02
1996	(8491, 849)	(8645, 805)	−1.8

- As λ changes from 0.25 to 0.5, the annual peak-load demand is changed, and having the same form as λ changes from 0.5 to 0.75. The degree of fuzziness has a great effect on the behavior of the model.

9.10.3 Remarks

In this section, we did the following:

- We developed a new fuzzy model for the annual peak-load demand for long-term planning.
- We developed models to solve the problem of uncertainties of the annual peak demand.
- We developed models to treat the long-term planning variables. Some of these variables depend on the nation of the community under investigation, whereas the others depend on the electric system itself.
- More investigation should be carried out to estimate the annual peak load for 15 or 20 years ahead. We were not able to do this due to the shortage of the data available to us.

9.11 Time-Series Models

A major aim of an electric power utility is to accurately forecast load requirements. In broad terms, power system load forecasting can be categorized into long-term and short-term functions [53–60]. Long-term load forecasting usually covers from 1 to 10 years ahead (monthly and yearly values) and is explicitly intended for applications in capacity expansion and long-term capital investment return studies.

In this section we mainly focus on the long-term load forecasting with mathematical methods. First, we introduce some basic foundations used with this forecasting.

9.11.1 Time Series

A time series can be defined as a sequential set of data measured over time, such as hourly, daily, or weekly peak load. The basic idea of forecasting is to first build a pattern matching available data as accurately as possible and then obtain the forecasted value with respect to time using the established model.

Generally, series are often described as having the characteristic

$$X(t) = T(t) + S(t) + R(t) \qquad t = \ldots -1, 0, 1, 2, \ldots \tag{9.43}$$

where $T(t)$ is the trend term, $S(t)$ is the seasonal term, and $R(t)$ is the irregular or random component. At this point, we do not consider the cyclic terms because these fluctuations can have a duration from 2 to 10 years or even longer; therefore, they are not applicable to short-term load forecasting.

We assume the following to make this example a bit easier:

1. The trend is a constant level.
2. The seasonal effect has a period s; that is, it repeats after s time periods. Or the sum of the seasonal components over a complete cycle or period is zero.

$$\sum_{j=1}^{s} S(t+j) = 0 \tag{9.44}$$

9.11.2 Forecasting Methods

To this point, we have used forecasting methods that are classified into two basic types: *qualitative* and *quantitative* methods.

Qualitative forecasting methods generally use the opinions of experts to predict future load subjectively. Such methods are useful when historical data are not available or are scarce. These methods include subjective curve fitting, the Delphi method, and technological comparisons.

Quantitative methods include regression analysis, decomposition methods, exponential smoothing, and the Box-Jenkins methodology.

9.11.3 Forecasting Errors

Unfortunately, all forecasting situations involve some degree of uncertainty, which makes errors unavoidable.

The forecast error for a particular forecast $\hat{X}_t$ with respect to actual value X_t is

$$e_t = X_t - \hat{X}_t \tag{9.45}$$

To avoid the offset of positive with negative errors, we need to use *absolute deviations*:

$$|e_t| = |X_t - \hat{X}_t| \tag{9.46}$$

Hence, we can define a measure known as the mean absolute deviation (MAD) as follows:

$$\mathrm{MAD} = \frac{\sum_{t=1}^{n} |e_t|}{n} = \frac{\sum_{t=1}^{n} |X_t - \hat{X}_t|}{n} \tag{9.47}$$

Another method is to use the mean square error (MSE) defined as follows:

$$\mathrm{MSE} = \frac{\sum_{t=1}^{n} e_t^2}{n} = \frac{\sum_{t=1}^{n} (X_t - \hat{X}_t)^2}{n} \tag{9.48}$$

9.12 Power System Load Forecasting

The power system load is assumed to be time dependent, evolving according to a probabilistic law [58]. It is common practice to employ a white noise sequence as input to a linear filter of which the output is the power system load. This is an adequate model for predicting the load time series. The noise input is assumed normally distributed, with zero mean and some variance σ^2. A number of classes of models exist for characterizing the linear filter.

9.12.1 A Simple Example of Power System Load Forecasting

Consider the data on fuel consumption given in Table 9.19.

We can average the seasonal values over the series and use these, minus the overall mean, as seasonal estimates shown here (overall mean is 761.65):

$$S(1) = 888.2 - 761.65 = 126.55 \qquad S(2) = 709.2 - 761.65 = -52.4$$
$$S(3) = 616.4 - 761.65 = -145.25 \qquad S(4) = 832.8 - 761.65 = 71.15$$

After subtraction of these values, the original series removes seasonal effects. It should be noted that this technique works well on series having linear trends with small slopes.

Table 9.19 Primary Energy Consumption in a Utility (Coal Equivalent)

Year	1	2	3	4
1965	874	679	616	816
1966	866	700	603	814
1967	843	719	594	819
1968	906	703	634	844
1969	952	745	635	871
Means	888.2	709.2	616.4	832.8

In addition, we can look at the averages for each complete seasonal cycle (the period) because the seasonal effect over an entire period is zero. To avoid losing too much data, we use a method called *moving average* (MA), which is simply the series of averages:

$$\frac{1}{s}\sum_{j=0}^{s-1} X_{t+j},\ \frac{1}{s}\sum_{j=1}^{s} X_{t+j},\ \frac{1}{s}\sum_{j=2}^{s+1} X_{t+j},\ \ldots$$

A problem is presented here if the period is even because the adjusted series values do not correspond to the original ones at time points. To overcome this problem, we use the *centered moving average* (CMA) to bring us back to the correct time points. This CMA is shown in Table 9.20.

Through looking at the differences between CMA and the original series, we can estimate the *k*th seasonal effect simply by average the *k*th quarter differences:

$$S(1) = 128.954 \qquad S(2) = -46.487 \qquad S(3) = -142 \qquad S(4) = 65$$

But the sum of these four values is 5.107. Recall that we assume the seasonal sum to be zero, so we need to add a correction factor of $-5.107/4 = -1.254$ to give

$$S(1) = 127.34 \qquad S(2) = -47.741 \qquad S(3) = -143.254 \qquad S(4) = 63.746$$

Table 9.20 Calculation of the Moving Average

Quarter	*X*(*t*)	MA	CMA(Order-2)	Difference
1	874			
2	679			
3	616	746.25	745.25	−129.25
4	816	744.25	746.875	69.125
5	866	749.5	747.875	118.125
6	700	746.25	746	−46
7	603	745.75	741.75	−138.75
8	814	737.75	740.125	73.875
9	834	742.5	741.375	92.625
10	719	740.25	740.875	−21.875
11	594	741.5	750.5	−156.5
12	819	759.5	757.5	61.5
13	906	755.5	760.5	145.5
14	703	765.5	768.625	−65.625
15	634	771.75	777.5	−143.5
16	844	783.25	788.5	55.5
17	952	793.75	793.875	158.125
18	745	794	797.375	−52.375
19	635	800.75		
20	871			

Table 9.21 Prediction of Energy Consumption

Period t	Trend T	Seasonal S	Predicted	Actual X
21	789.963	127.340	917.303	981
22	793.610	−47.741	747.123	759
23	797.257	−143.254	654.003	674
24	800.904	63.746	864.650	900

Now the irregular component can be easily calculated by subtracting both the CMA and the seasonal effects.

If we suppose the model (1) is appropriate, then we can use it to make predictions. To simplify, we omit the random data, so all we need to do is to predict the trend, say, a linear trend:

$$T(t) = a + bt$$

With the application to the CMA, we have

$$\hat{T}(t) = 713.376 + 3.647t$$

Hence, a prediction is shown in Table 9.21.

9.13 Linear Regression Method

The linear regression method is already used in short-term load forecasting and supposes that the load is affected by some factors such as high and low temperatures, weather condition, economic growth, etc. This relation can be expressed as

$$y = \beta_0 + \beta_1 x_1 + \beta_2 x_2 + \cdots + \beta_k x_k + \varepsilon \tag{9.49}$$

where y is the load, x_i is the affecting factors, β_i are regression parameters with respect to x_i, and ε is an error term.

For this model, we always assume that the error term ε has a mean value equal to zero and constant variance.

Since parameters β_i are unknown, they should be estimated from observations of y and x_i. Let b_i $(i = 0, 1, 2, \ldots, k)$ be the estimates in terms of β_i $(i = 0, 1, 2, \ldots, k)$. Recall that the error term has a 50% chance of being positive and negative, respectively, so we omit this term in calculating parameters, which means

$$\hat{y} = b_0 + b_1 x_1 + b_2 x_2 + \cdots + b_k x_k \tag{9.50}$$

Then, we use the least error squares estimate method, which minimizes the sum of squared residuals (SSE), to obtain the parameters b_i

$$\underline{B} = [b_0 \ b_1 \ b_2 \ldots b_k]^T = \left(\underline{X}^T \underline{X}\right)^{-1} \underline{X}^T \underline{Y} \tag{9.51}$$

where $\underline{Y}$ and $\underline{X}$ are the following column vector and matrix:

$$\underline{Y} = \begin{bmatrix} y_1 \\ y_2 \\ \vdots \\ y_n \end{bmatrix} \quad \text{and} \quad \underline{X} = \begin{bmatrix} 1 & x_{11} & x_{12} \dots x_{1k} \\ 1 & x_{21} & x_{22} \dots x_{2k} \\ \vdots & \vdots & \vdots & \vdots \\ 1 & x_{n1} & x_{n2} \dots x_{nk} \end{bmatrix} \tag{9.52}$$

After the parameters are calculated, this model can be used for prediction. It will be accurate in predicting y values if the standard error s is small:

$$s = \sqrt{\frac{SSE}{n-(k+1)}}, \qquad SSE = \sum_{i=1}^{n} (y_i - \hat{y}_i)^2, \qquad y_i : \text{observed}, \qquad \hat{y}_i : \text{estimated} \tag{9.53}$$

There are also some other ways to check the validity of a regression model [1].

9.14 Autoregressive (AR) Model

In the autoregressive model, the current value X_t of the time series is expressed linearly in terms of its previous values $X_{t-1}, X_{t-2}, \ldots$ and a white noise series $\{\varepsilon_t\}$ with zero mean and variance σ^2:

$$X_t = \phi_1 X_{t-1} + \phi_2 X_{t-2} + \cdots + \phi_p X_{t-p} + \varepsilon_t \tag{9.54}$$

By introducing the backshift operator B that defines $X_{t-1} = BX_t$, and consequently $X_{t-m} = B^m X_t$, we can rewrite equation (9.54) in the form

$$\phi(B) X_t = \varepsilon_t \tag{9.55}$$

where

$$\phi(B) = 1 - \phi_1 B - \phi_2 B^2 - \cdots - \phi_p B^p \tag{9.56}$$

Note that this model has a similar form to the multiple linear regression models. The difference is that in regression the variable of interest is regressed onto a linear function of other known (explanatory) variables, whereas here X_t is expressed as a linear function of its own past values—thus, the description "autoregressive." As the values of X_t at p previous times are involved in the model, it is said to be an AR (p) model.

Now we need to calculate the parameters φ_i for prediction. There are two such methods: least squares estimation and maximum likelihood estimation (MLE).

To calculate the least squares estimators, we need to minimize the expression (here, we let $p = 2$)

$$\sum_{t=1}^{N} (X_t - \phi_1 X_{t-1} - \phi_2 X_{t-2})^2 \tag{9.57}$$

with respect to φ_1 and φ_2. But because we do not have the information for $t = 1$ or $t = 2$, an assumption is made here that X_1 and X_2 are fixed, and excluding the first two terms from the sum of squares. That is, to minimize

$$\sum_{t=3}^{N} (X_t - \phi_1 X_{t-1} - \phi_2 X_{t-2})^2$$

then we use a similar approach to linear regression to obtain the parameters.

Maximum likelihood estimation is attractive because generally it is asymptotically unbiased and has minimum variance. Therefore, we introduce this method here.

Suppose that we have a sample of dependent observations X_t, $t = 1, \ldots, N$, each with $f(X_t)$. Then the joint density function is

$$f(X_1, X_2, \ldots, X_N) = \prod_{t=1}^{N} f(X_t | \underline{X}_{t-1}) \tag{9.58}$$

where $\underline{X}_t$ denotes all observations up to and including X_t. $f(X_t|\underline{X}_{t-1})$ is the conditional distribution of X_t given all observations prior to t.

We use the same model as before and suppose ε_t is normally distributed. So the mean of the conditional distribution is $\phi_1 X_{t-1} + \phi_2 X_{t-2}$, and the variance is σ^2. Therefore,

$$f(X_t|X_1, X_2, \ldots, X_N) = \frac{1}{\sqrt{(2\pi)}\sigma} \exp\left[\frac{-(X_t - \phi_1 X_{t-1} - \phi_2 X_{t-2})^2}{2\sigma^2}\right] \tag{9.59}$$

Similarly, we set X_1 and X_2 to be fixed and define the conditional likelihood as

$$L(\theta) = \prod_{t=3}^{N} f(X_t|X_1, X_2, \ldots, X_{t-1}) \tag{9.60}$$

By minimizing $L(\theta)$, we can obtain the parameters.

Consider a time series of the number of reported purse snatchings in a particular area 28 days apart, as shown in Figure 9.33.

If we use the MLE applied to the AR (2) model, the fitted model is

$$X_t = 0.0307841 X_{t-1} + 0.400178 X_{t-2} + \varepsilon_t var(\varepsilon_t) = 36.115343$$

Now, this model can be used to predict future data.

9.15 Moving Average (MA) Model

In the moving average process, the current value of the time series X_t is expressed linearly in terms of current and previous values of a white noise series $\varepsilon_t, \varepsilon_{t-1}, \ldots$. This noise series is constructed from the forecast errors or residuals when load observations become available. The order of this process depends on the oldest noise value

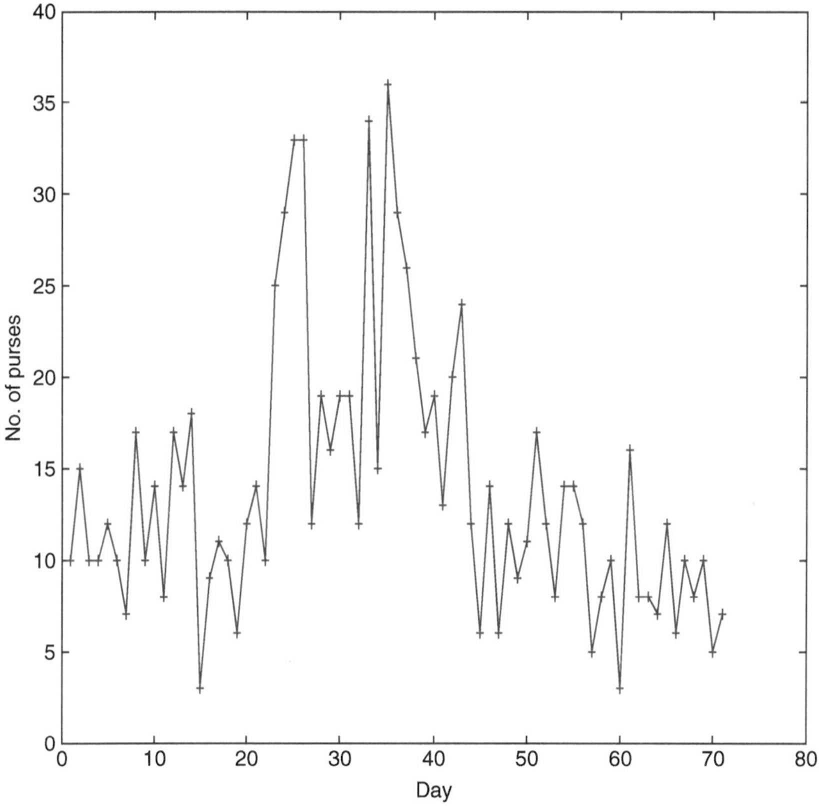

Figure 9.33 Reported purse snatching in an area.

at which X_t is regressed. For a moving average of order q (i.e., MA (q)), this model can be written as

$$X_t = \varepsilon_t - \theta_1\varepsilon_{t-1} - \theta_2\varepsilon_{t-2} - \cdots - \theta_q\varepsilon_{t-q} \tag{9.61}$$

A similar application of the backshift operator on the white noise series would allow equation (9.61) to be rewritten as

$$X_t = \theta(B)\varepsilon_t \tag{9.62}$$

where $\theta(B) = 1 - \theta_1 B - \theta_2 B^2 - \cdots - \theta_q B^q$.

9.16 Autoregressive Moving Average (ARMA, or Box-Jenkins) Model

If we combine the MA and AR models, we can present a broader class of model—that is the autoregressive moving average model—as

$$X_t = \phi_1 X_{t-1} + \phi_2 X_{t-2} + \cdots + \phi_p X_{t-p} + \varepsilon_t + \theta_1 \varepsilon_{t-1} + \theta_2 \varepsilon_{t-2} + \cdots + \theta_q \varepsilon_{t-q} \quad (9.63)$$

where ϕ_i and θ_j are called the *autoregressive* and *moving average* parameters, respectively. And in this case, this is an ARMA (p,q) model.

A methodology for ARMA models was developed largely by Box and Jenkins [60] so the models are often called Box-Jenkins models.

9.17 Autoregressive Integrated Moving Average (ARIMA) Model

The time series defined previously as an AR, MA, or ARMA process is called a *stationary process* [6]. This means that the mean of the series of any of these processes and the covariance among its observations do not change with time. Unfortunately, this is not often true in a power load. But previous knowledge is definitely useful in that the nonstationary series can be transformed into a stationary one with some tricks. This transformation can be achieved, for the time series that are nonstationary in the mean, by a differencing process. By introducing the ∇ operator, we can write a differenced time series of order one as

$$\nabla X_t = X_t - X_{t-1} = (1 - B)X_t \quad (9.64)$$

Consequently, an order d differenced time series is written as

$$\nabla^d X_t = (1 - B)^d X_t \quad (9.65)$$

The differenced stationary series can be modeled as AR, MA, or ARMA to yield an ARI, IMA, or ARIMA time-series process. For a series that needs to be differenced d times and has orders p and q for the AR and MA components (i.e., ARIMA (p, d, q)), the model is written as

$$\phi(B)\nabla^d X_t = \theta(B)\varepsilon_t \quad (9.66)$$

However, as a result of daily, weekly, yearly, or other periodicities, many time series exhibit periodic behaviors in response to one or more of these periodicities. Therefore, a seasonal ARIMA model is appropriate. It has been shown that the general multiplicative model $(p, d, q) * (P,D,Q)_s$ for a time-series model can be written in the form

$$\phi(B)\Phi\left(B^S\right)\nabla^d\nabla_S^D X_t = \theta(B)\Theta\left(B^S\right)\varepsilon_t \quad (9.67)$$

where definitions for $\Phi(B^S)$, ∇_S^D, $\Theta(B^S)$ are given in the following:

$$\nabla_S^D = (X_t - X_{t-S})^D = \left(1 - B^S\right)^D X_t \quad (9.68)$$

$$\Phi\left(B^S\right) = 1 - \Phi_1 B^S - \Phi_2 B^{2S} - \cdots - \Phi_p B^{pS} \quad (9.69)$$

$$\Theta\left(B^S\right) = 1 - \Theta_1 B^S - \Theta_2 B^{2S} - \cdots - \Theta_q B^{qS} \quad (9.70)$$

The model presented in equation (9.67) can obviously be extended to the case in which data for two seasons are accounted for. An example demonstrating seasonal time-series modeling is the model for hourly load data with a daily cycle. Such a model can be expressed using the model of equation (9.67) with $S = 24$.

To obtain this model, we use the parameters p, d, q, P, D, Q, and other coefficients. By studying the self-variance, covariance, and variance function of the order one or higher-order differentiation of variables we get the d and D. Then the model can be simplified into AR, MA, or ARMA models to calculate other values so that the model can be built.

9.18 ARMAX and ARIMAX Models

ARMA and ARIMA use only the time and load as input parameters. Because load generally depends on the weather and time of the day, exogenous variables sometimes can be included to give the ARMAX and ARIMAX models [7]. Other useful methods implementing evolutionary programming (EP) and fuzzy logic (FL) into conventional time-series models were also proposed. We will not consider these methods in detail here.

9.18.1 Remarks

No one method can be applicable to all situations. So a method should be chosen considering many factors, such as the time frame, pattern of data, cost of forecasting, desired accuracy, availability of data, and ease of operation and understanding. Therefore, more work still needs to be done to include all these factors.

References

[1] H.L. Willis, L.A. Finley, M.J. Buri, Forecasting electric demand of distribution system in rural and sparsely populated regions, IEEE Trans. Power Syst. 10 (4) (1995) 2008–2013.

[2] P.H. Henault, R.B. Eastvedt, J. Peschon, L.P. Hajdu, Power system long term planning in the presence of uncertainty, IEEE Trans. Power Apparatus Syst. PAS-89 (1970) 156–164.

[3] G.S. Christensen, A. Rouhi, S.A. Soliman, A new technique for unconstrained and constrained LAV parameter estimation, Can. J. Elect. Comp. Eng. 14 (1) (1989) 24–30.

[4] Ministry of Electricity and Energy, Egyptian Electricity Authority, Load and energy forecast for the period 1996/1997 to 2009/2010, Report, February 1998.

[5] H.K. Temraz, K.M. El-Nagar, M.M.A. Salama, Applications of non-iterative least absolute value estimation for forecasting annual peak electric power demand, Can. J. Elect. Comp. Eng. 23 (4) (1998) 141–146.

[6] D. Srinivasan, T.S. Swee, C.S. Cheng, E.K. Chan, Parallel neural network-fuzzy expert system strategy for short-term load forecasting: system implementation and performance evaluation, IEEE Trans. Power Syst. 14 (3) (1999) 1100–1106.

[7] I. Drezga, S. Rahman, Short-term load forecasting with local ANN predictors, IEEE Trans. Power Syst. 14 (3) (1999) 844–850.

[8] A.A. Ding, Neural-network prediction with noisy predictors, IEEE Trans. Neural Netw. 10 (5) (1999) 1196–1203.

[9] H.C. Wu, C. Lu, Automatic fuzzy model identification for short term load forecast, IEE Proc. Gener. Transm. Distrib. C 146 (5) (1999) 477–482.
[10] W. Charytoniuk, M.S. Chen, Very short-term load forecasting using artificial neural networks, IEEE Trans. Power Syst. 15 (1) (2000) 263–268.
[11] K.H. Kim, H.S. Youn, Y.C. Kang, Short-term forecasting for special days in anomalous load conditions using neural networks and fuzzy inference method, IEEE Trans. Power Syst. 15 (2) (2000) 559–565.
[12] P.A. da Silva, L.S. Moulin, Confidence intervals for neural network based short-term load forecasting, IEEE Trans. Power Syst. 15 (4) (2000) 1191–1196.
[13] S.A. Villalba, C.A. Bel, Hybrid demand model for load estimation and short-term load forecasting in distribution electric systems, IEEE Trans. Power Syst. 15 (2) (2000) 764–769.
[14] R.H. Liang, C.C. Cheng, Combined regression-fuzzy approach for short-term load forecasting, IEE Proc. Gener. Transm. Distrib. 147 (4) (2000) 261–266.
[15] H.S. Hippert, C.E. Pedreira, R.C. Souza, Neural networks for short-term load forecasting: a review and evaluation, IEEE Trans. Power Syst. 16 (1) (2001) 44–55.
[16] M. Huang, H.T. Yang, Evolving wavelet-based networks for short-term load forecasting, IEE Proc. Gener. Transm. Distrib. 148 (3) (2001) 222–228.
[17] D. Srinivasan, M.A. Lee, Survey of hybrid fuzzy neural approach to electric load forecasting, IEEE Int. Conf. Syst. Man Cybern. 5 (1995) 4004–4008.
[18] F.K. Selker, W.R. Wroblewski, Medium-term load forecasting and wholesale transaction profitability, IEEE Trans. Power Eng. Soc. (summer meeting) (1990) 1268–1272.
[19] N.X. Jia, R. Yokoyama, Y.C. Zuou, Z.Y. Gao, A flexible long-term forecasting approach based on a new dynamic simulation theory—GSIM, Int. J. Electr Power Energ. Syst. 23 (2001) 549–556.
[20] V.M. Vlahovic, I.M. Vujosevic, Long-term forecasting: a critical review of direct-trend extrapolation methods, Int. J. Electr. Power Energ. Syst. 9 (1) (1987) 2–8.
[21] E. Handchin, Ch. Dornemann, Bus load modeling and forecasting, IEEE Trans. Power Syst. 3 (2) (1988) 627–633.
[22] A.G. Parlos, E. Oufi, J. Muthusami, A.D. Patton, Development of an intelligent long-term electric load forecasting system, in: Proceedings, ISAP '96, International Conference, 28 Jan.-2 Feb 1996. Intell. Syst. Appl. Power Syst. (1996) 288–292.
[23] E.H. Barakat, S.A. Al-Rashid, Long-term peak demand forecasting under conditions of high growth, IEEE Trans. Power Syst. 7 (4) (1992) 1483–1486.
[24] M.R. Gent, Electric supply and demand in the US: next 10 years, IEEE Power Eng. Rev. 12 (4) (1992) 8–13.
[25] S.C. Terpathy, Demand forecasting in a power system, Energ. Conv. Manage. 38 (14) (1997) 1475–1481.
[26] Y. Minato, Y. Yokoi, Development of a forecasting method of a region's electric power demand (1)-forecasting economic and social indices, Tans. IEE Jpn 116B (1996) 147–154.
[27] S. Tomonobu, S. Hirokazu, T. Yoshinori, U. Katsumi, Next-day load curve forecasting using neural network based on similarity, Electr. Power Compon. Syst. 29 (2001) 939–948.
[28] H.M. Al-Hamadi, S.A. Soliman, Short-term electric load forecasting based on Kalman filtering algorithm with moving windows weather and load model, Electr. Power Syst. Res. 68 (1) (2004) 47–59.
[29] R.G. Brown, Introduction to Random Signal Analysis and Kalman Filtering, John Wiley and Sons, New York, 1983.
[30] G.F. Franklin, J.D. Powel, M.L. Workman, Digital Control of Dynamic Systems, second ed. Addison-Wesley, Reading, MA, 1990.

[31] C.W. Fu, T.T. Nguyen, Models for long-term energy forecasting, IEEE Power Eng. Soc. Gen. Meeting 1 (2003) 235–239.
[32] K. Nagasaka, M. Al Mamun, Long-term peak demand prediction of 9 Japanese power utilities using radial basis function networks, IEEE Power Eng. Soc. Gen. Meeting 1 (2004) 315–322.
[33] S. Phimphachanh, K. Chamnongthai, P. Kumhom, A. Sangswang, Using neural network for long term peak load forecasting in Vientiane municipality, TENCON 2004, 2004 Region 10 Conference, C, (2004) 147–154.
[34] B. Kermanshahi, H. Iwamiya, Up to 2020 load forecasting using neural nets, Int. J. Electr. Power Energ. Syst. 24 (2002) 789–797.
[35] X. Da, Y. Jiangyan, Y. Jilai, The physical series algorithm of mid-long term load forecasting of power systems, Electr. Power Res. J. 53 (2000) 31–37.
[36] M.S. Kandil, S.M. El-Debeiky, N.E. Hasanien, Overview and comparison of long-term forecasting techniques for fast developing utilities: Part I, Electr. Power Res. J. 58 (2001) 11–17.
[37] M.S. Kandil, S.M. El-Debeiky, N.E. Hasanien, The implementation of long term forecasting strategies using a knowledge-based expert system: Part II, Electr. Power Res. J. 58 (2001) 19–25.
[38] D. Srinivasan, M.A. Lee, Survey of hybrid fuzzy neural approach to electric load forecasting, IEEE Trans. Power Syst. 5 (1995) 4004–4008.
[39] V.M. Vlahovic, I.M. Vujosevic, Long-term forecasting: a critical review of direct-trend extrapolation methods, Electr. Power Energ. Syst. 9 (1) (1987) 2–8.
[40] L. Chenhui, Theory and Methods of Load Forecasting of Power Systems, Haerbin Institute of Technology Press, China, 1987.
[41] Y. Tamura, Z. Deping, N. Umeda, K. Sakashita, Load forecasting using grey dynamic model, IEEE Trans. Power Syst. (1995) 361–365.
[42] V.M. Vlahovic, I.M. Vujosevic, Long-term forecasting: a critical review of direct-trend extrapolation methods, Electr. Power Energ. Syst. 9 (1) (1987) 2–8.
[43] D.W. Bunn, E.D. Farmer (Eds.), Comparative Models for Electric Load Forecasting, John Wiley and Sons, Hoboken, NJ, 1985.
[44] H.K. Temraz, V.H. Quintana, Analytic spatial electric load forecasting methods: a survey, Can. J. Elect. Comp. Eng. 17 (1) (1992) 34–41.
[45] J. Nazarko, W. Zalewski, The fuzzy regression approach to peak load estimation in power distribution systems, IEEE Trans. Power Syst. 14 (3) (1999) 809–813.
[46] J.R. Yokoyama, Y.C. Zhou, A novel approach to long term load forecasting where functional relations and impact relations coexist. Int. Conf. Elect. Power Tech., Budapest, August 29–September 2, 1999.
[47] S.K. Padmakumari, K.P. Mohandas, D. Thiruvengadam, Application of fuzzy system theory in land use based long-term distribution load forecasting, in: Proceedings of EMPD'98, IEEE Catalog Number 98EX137, Int. Conf. on Energy Management and Power Delivery, 311–316.
[48] H.K. Temraz, K.M. El-Naggar, M.M. Salama, Application of noniterative least absolute value estimation for forecasting annual peak electric load power demand, Can. J. Elect. Comp. Eng. 23 (4) (1998) 141–146.
[49] A.G. Parlos, E. Oufi, J. Muthusami, A.D. Patton, A.F. Atiya, Development of an intelligent long term electric load forecasting system, in: Proceedings of ISAP'96, IEEE Catalog Number 96TH8152, Int. Conf. on Intelligent Systems Applications to Power Systems (1996) 288–292.
[50] A.G. Parlos, A.D. Patton, Long-term electric load forecasting using a dynamic neural network architecture, Joint Int. Power Conference, Athens Power Tech'1993, APT'93, 816–820.

[51] P.Y. Wang, G.S. Wang, Power system load forecasting with ANN and fuzzy logic control, Int. Conf. on Computer Communication, Control and Power Engineering, TENCON'93, IEEE Catalog Number 93CH3286–2, 5, 359–362.
[52] E.H. Barakat, J.M. Al-Qassim, S.A. Al Rashed, New model for peak demand forecasting applied to highly complex load characteristics of fast developing area, IEE Proc. C 139 (2) (1992) 136–140.
[53] B.L. Bowerman, R.T. O'Connell, A.B. Koehler, Forecasting, Time Series, and Regression: An Applied Approach, fourth ed., Thomson Brooks/Cole, Duxbury, CA, 2005.
[54] D.C. Montgomery, L.A. Johnson, J.S. Gardiner, Forecasting & time series analysis, second ed., McGraw-Hill, New York, 1990.
[55] G.A.N. Mbamalu, M.E. El-Hawary, Load forecasting via suboptimal seasonal autoregressive models and iteratively reweighted least squares estimation, IEEE Trans. Power Syst. 8 (1) (1993) 343–348.
[56] J.H. Chow, F.F. Wu, J.A. Momoh, Applied Mathematics for Restructured Electric Power Systems, Springer, New York, 2005.
[57] G. Janacek, L. Swift, Time Series: Forecasting, Simulation, Applications, Ellis Horwood Limited, West Sussex, 1993.
[58] I. Moghram, S. Rahman, Analysis and evaluation of five short-term load forecasting techniques, IEEE Trans. Power Syst. 4 (4) (1989) 1484–1491.
[59] J.Y. Fan, J.D. McDonald, A real-time implementation of short-term load forecasting for distribution power systems, IEEE Trans. Power Syst. 9 (2) (1994) 988–994.
[60] G.E.P. Box, G.M. Jenkins, G.C. Reinsel, Time Series Analysis, Forecasting and Control, third ed., Prentice Hall, Englewood Clifs, NJ, 1994.

Index

Kolibri 360 GmbH
Ferdinand-Jühlke-Straße 7
99095 Erfurt, Deutschland
produktsicherheit@kolibri360.de